T0138567

כתבי האקדמיה הלאומית הישראלית למדעים

PUBLICATIONS OF THE ISRAEL ACADEMY
OF SCIENCES AND HUMANITIES

SECTION OF SCIENCES

—

Wood Anatomy
and Identification of Trees and Shrubs
from Israel and Adjacent Regions

Wood Anatomy and Identification of Trees and Shrubs from Israel and Adjacent Regions

by

ABRAHAM FAHN ELLA WERKER

The Hebrew University of Jerusalem, Israel

and

PIETER BAAS

Rijksherbarium, Leiden, The Netherlands

JERUSALEM 1986

THE ISRAEL ACADEMY OF SCIENCES AND HUMANITIES

ISBN 965-208-073-x

Printed in Israel
Set by Monoline, Benei Beraq

CONTENTS

PREFACE

For many years the Israeli authors of this book have been engaged in tackling various specific anatomical problems related to the cambium and secondary xylem of woody species growing in Israel. They were also frequently approached by archaeologists with requests for the identification of wood specimens discovered in the extensive excavations carried out in this so historically important area. For these and other reasons these authors have felt the need to undertake a comprehensive study of the wood anatomy of the trees and shrubs of Israel and adjacent areas and summarize it in a book.

The climatic and edaphic conditions vary considerably in this area, as does the woody vegetation. It was felt, therefore, that while undertaking the task of describing the wood anatomy of the various plants, it would be interesting also to analyze the wood structure in the light of the recent approach to the relation between secondary xylem characteristics and ecological conditions.

At the meeting of the International Association of Wood Anatomists in Amsterdam in August 1979, discussions with the third author led to his participation in this undertaking, and from then on the three authors worked jointly on all aspects of the project.

We are much indebted to Dr Avinoam Danin of the Department of Botany of The Hebrew University of Jerusalem for providing us with many identified wood samples and for assisting us in the ecological chartacterization of the species dealt with in this book. He and Dr David Heller of the same Department also extended help in solving problems relating to nomenclature. Dr A. Shmida and Mr Aaron Liston were helpful in providing wood samples. Thanks are due to Mr Paul van Veen from the Rijksherbarium, Leiden, for sectioning a number of particularly hard and breakable woods. Our thanks also go to Mr Ya'akov Gamburg of The Hebrew University for printing most of the photographs, and to Mrs Esther Kamer for helping in the typing of the manuscript. Mr Herbert Vial from the Rijksherbarium was also involved in part of the photographic work.

We wish to acknowledge the financial support provided under the Cultural Agreement between Israel and The Netherlands, which enabled the second and third authors to make exchange visits in order to work jointly on the project.

The Authors

1. INTRODUCTION

1.1 *Aims*

The literature dealing with the wood anatomy of plants growing naturally in Israel and its vicinity is very limited. The existing publications cover only a small number of species, and in some cases the wood anatomy is mentioned but briefly in connection with wood developmental problems (Fahn, 1953, 1955, 1958, 1959; Fahn & Sarnat, 1963; Grundwag & Werker, 1976). The only available source to date for the wood identification of a number of the most common tree species of Israel is the work published by Chudnoff (1956). Furthermore, several of the Mediterranean species from Europe have been described by Greguss (1959), Jacquiot et al. (1973), and Schweingruber (1978). A more comprehensive survey of all woody plants of the region is obviously needed for a variety of purposes.

Primarily such a study should facilitate identification. In Israel wood identification is of prime importance for archaeological research. Wooden artefacts, construction materials and firewood abound amongst excavated materials. A botanical identification based on the often unique combination of wood anatomical features in each species, genus or family, helps one to understand the preference of ancient cultures for certain kinds of wood and can provide circumstantial evidence on timber trade routes in the past. When remnants of wood used for burning are found during archaeological excavations, one may conclude that such firewood was probably originally collected in the vicinity of these sites. These remnants may show whether there have been any changes in the vegetation and thus in the ecological conditions at these sites.

Wood identification also serves contemporary needs in forensic studies and the timber trade. In criminal investigations the matching of wood fragments found on a suspected person or his tools with wooden objects at the site of the crime, may tip the balance of the evidence in favour of the defence council or the public prosecutor. In the timber trade, wood identification has been, and always will be, an important means of checking the proper name, and thus of guaranteeing a fair price for commercial species. Although very few native species in the region under study yield commercial timber, the value of anatomical identification in this field should not be underrated.

Another potential use of wood anatomical data is in the promising field of structure-property relations. It is not the purpose of this book to predict end-use properties of timber from its microscopical structure. However, future students engaged in applied wood research hopefully will find much useful information for such predictive analyses in our data.

A totally different, yet major goal of the present study has been the search for correlations between the ecology of the species and wood anatomical parameters. The region in and around Israel offers a diversity of habitats with great differences in temperature, annual precipitation and water retention capacity of the soil. The search for ecological trends in the xylem anatomy of a rich, woody flora is a legitimate complement to the analysis of such trends within closely related taxa (species, genera, families) with a wide ecological range, as has been demonstrated in the studies by Carlquist (1975, 1980) and Baas (1973, 1976, 1982). Apart from offering a comparison of rough ecological categories within the flora of the region, the woody flora in and around Israel as a whole lends itself well to comparison with the more temperate and mesic floras of Central and Northwestern Europe on the one hand, and the mesic tropical lowland floras on the other. Ultimately, ecological trends and — if feasible — their critical functional interpretation are crucial to the understanding of the major evolutionary trends in wood anatomy established by Bailey and his school (Bailey & Tupper, 1918; and many later publications), and supported by recent studies which approach phylogenetic wood anatomy from various angles (Carlquist, 1975, 1980; Page, 1981; Baas, 1982). Functional hypotheses based on well-established ecological trends should also serve as an inspiration in the field of functional anatomy in a strictly physiological sense (i.e., based on experimental work), as pursued, for example, by Zimmermann (1983, and many papers cited there). However, not all trends have to have a functional basis, as was pointed out by Van den Oever et al. (1981) and Baas (1982).

Last but not least, it is hoped that the descriptive information on numerous species (either previously undescribed or only incompletely described in the wood anatomical literature) gathered in this book will be a useful source of data for systematic wood anatomy and all the other, diverse applications of comparative wood anatomy.

1.2 *Origin and Limitations of the Material Studied*

As indicated in the title of this book, the species studied are from Israel and adjacent areas. The geographical delimitation has been dictated by two practical aims: 1. to cover the entire area of greatest interest for archaeological wood identification in the region; and 2. to include the greatest possible range of habitat types in order to make the results of the ecological analysis as meaningful as possible. Consequently, materials for our studies come from the entire area west of the River Jordan, extending up to Mt. Hermon in the north and to the Sinai Peninsula in the south.

Introduction

All woody species native to the region, as well as some of those introduced in ancient times (e.g., *Citrus, Vitis*) or imported as timber (*Cedrus*), are included. Recently introduced cultivated trees, such as *Eucalyptus* and other forest species, and the great number of ornamental trees and shrubs are excluded. Wood anatomical information on these species and their identification is often readily available in the wood anatomical literature.

The distinction between woody and herbaceous plants is a rather arbitrary one, and we do not claim to have followed an absolutely consistent approach. Generally, all species described are truly woody and develop into fair-sized shrubs or trees; however, some common small shrubs or semi-woody perennials are also included.

It should be emphasized that the number of specimens per species studied by us was usually limited to not more than one or two, except for the conifers, of which more samples were examined. We are fully aware of serious drawbacks of this limitation in assessing the value of the features described for wood identification. In constructing identification keys, we have deliberately refrained from using futile differences which might also be found within a single species. Comparison with data from the literature on the same or related species has in most cases confirmed the diagnostic value of the characters that were ultimately used. Quantitative characters in the descriptions and tables are only meant to give a rough idea of the values within the taxa studied. They should not be regarded as accurate means for the species.

In several cases, with regard to rare species, we were obliged to have recourse to branchwood instead of peripheral wood from adult stems. The disadvantages implied have been considered in weighing the diagnostic value of the characters. On the other hand, samples for identification, including archaeological material, are often from branchwood, and thus the inclusion of branches in our materials is perhaps justified. It should also be recalled that the maximum stem diameter of most desert shrubs is quite small.

The material studied has been collected over the years by the authors and other botanists; identification of floral or fruiting material was usually carried out by taxonomists from the Department of Botany of The Hebrew University of Jerusalem. In most cases nomenclature and taxonomic treatment follow *Flora Palaestina* (1966, 1972, 1978), which has also been our main source for ecological and phytogeographical data.

2. SURVEY OF THE WOOD ANATOMICAL CHARACTERS WITH COMMENTS ON THEIR DIAGNOSTIC VALUE

In order to make profitable use of this book, the reader should have a basic knowledge of the three-dimensional microscopic structure of wood. Explanations can be found in general textbooks on plant anatomy (Esau, 1965; Fahn, 1982) or on wood structure in particular (Jane, 1980; Core et al., 1979; Panshin & De Zeeuw, 1980; Desch, 1981). Here the different, often diagnostic, wood anatomical characters will be explained and discussed briefly in order to facilitate the use of the keys and the understanding of the descriptions. Generally, the definitions of the IAWA Committee on Nomenclature (1964) and the recommendations of the IAWA Committee for the Standardization of Characters Suitable for Computerized Hardwood Identification (1981) are followed.

2.1 *Vessel-less Woods*

In contrast to most angiosperms, the conifers or softwoods are characterized by the absence of vessels and paucity of axial parenchyma. Thus the characters most effective in separating hardwoods are missing in conifers, and a special jargon and emphasis on minute details have been developed in softwood anatomy and identification (Phillips, 1948; Greguss, 1955, 1972; Jacquiot, 1955). For the differentiation of conifers from Israel and adjacent regions the following characters are relevant.

Growth Rings and Transition from Early- to Latewood

Distinctness of growth ring boundaries and the gradual or abrupt transition from thin-walled earlywood tracheids to thick-walled, often flattened latewood tracheids are considered important characters for the separation of certain softwoods. In the species treated here, growth rings are virtually always distinct, and the early- to latewood transition is commonly gradual. *Pinus halepensis,* with mostly abrupt transitions from early- to latewood, is the sole exception. However, this character has no diagnostic value in this species either.

Tracheid Pits and Wall Ornamentation

The structure of the bordered pit in conifers is diagrammatically illustrated in Fig. 1A. Bordered pits are usually confined to the radial walls of the tracheids, but may occur in low to fairly high frequencies in the tangential walls of the latewood. The absence or relative frequency of such pits is of some diagnostic value in distinguishing the coniferous genera dealt with in this book.
The torus of the bordered pits usually has a more or less smooth, circular outline (Figs. 1B,C; 2A). The genus *Cedrus* can immediately be recognized by the scalloped or fringed margin of the torus (Figs. 1D; 2B).

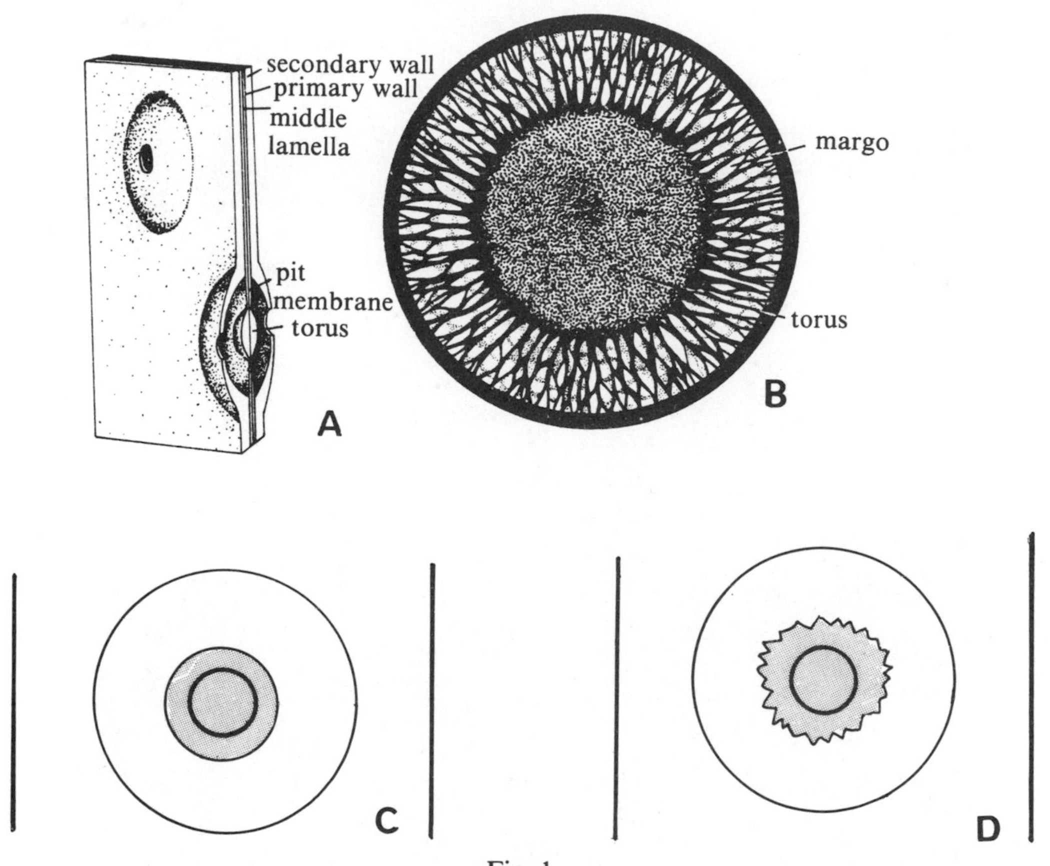

Fig. 1.
Bordered pits
A — three-dimensional diagram of a portion of the adjacent walls of two tracheids showing the structure of bordered pit-pairs
B — diagram of pit membrane and torus of *Pinus* showing the perforations in the membrane
C — diagram of a bordered conifer pit, the torus (stippled) of which has a smooth, circular outline
D — diagram of a bordered pit of *Cedrus*; the torus (stippled) has scalloped margins
(*A* and *B* from Fahn, 1982)

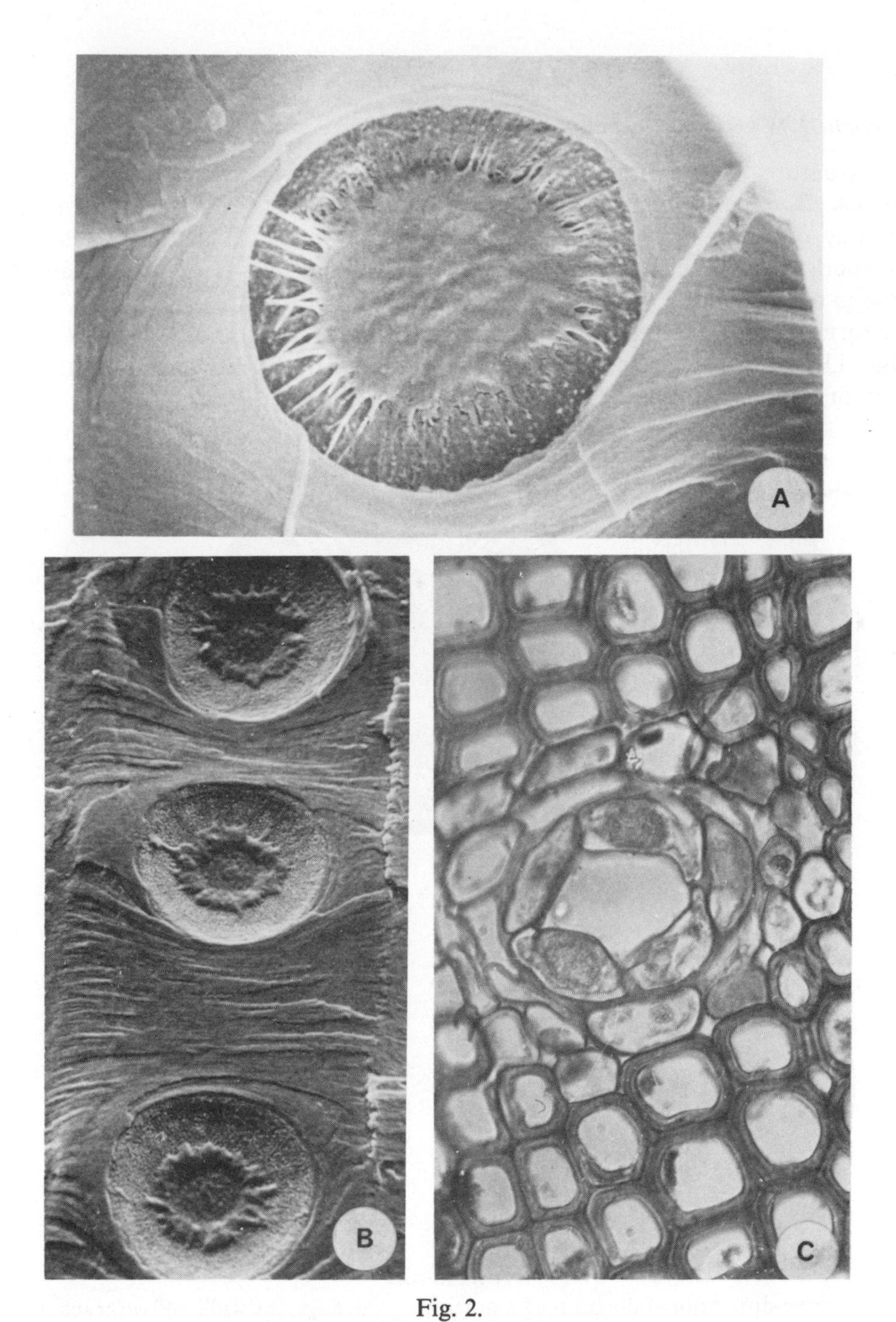

Fig. 2.

A and *B* — scanning electron micrographs of bordered pits of conifers. A radial longitudinal cut in the primary wall area has exposed the pit membranes

A — pit membrane of *Cupressus sempervirens*, showing a torus with a relatively smooth, circular outline from which fibrillar bundles extend across the margo, × 6,000

B — membranes of three pits of *Cedrus libani* showing tori with scalloped margins. The tori are depressed in the centre, ×4,800

C — transverse section of secondary xylem of *Pinus halepensis*, showing a vertical resin duct, ×650

(*C* from Werker & Fahn, 1969)

In many conifers horizontal thickenings of the primary wall and middle lamella may occur between the inter-tracheid pits: so-called crassulae (synonymous with the more confusing term 'bars of Sanio'). Since crassulae, if present, may vary in distinctness within a species (for instance in *Pinus halepensis*), their diagnostic value is very limited (cf. Panshin & De Zeeuw, 1980).

Spiral thickenings, a very useful character for the identification of certain conifers from other parts of the world, notably *Pseudotsuga* and *Taxus*, are absent from all coniferous species indigenous to Israel. Spiral markings due to slits in the wall, parallel to the microfibrils of the S_2 cell wall layer, are a feature of compression wood (i.e., reaction wood on the lower part of leaning stems or branches) and may be found in all species.

Trabeculae or rod- to spool-shaped radial projections of the wall through the lumina of a row of tracheids, occur sporadically in many conifers, and have no diagnostic value. Their presence has only been included in the descriptions to provide a record of the haphazard occurrence of these structures, which are probably of traumatic origin.

Axial Parenchyma

In conifers the axial or longitudinal parenchyma is usually much less abundant than in angiosperm woods. Axial parenchyma cells are distinguished from tracheids by their simple pitting, the presence of horizontal end walls, their mostly thinner walls, and the presence of dark-stainable cell contents. The presence and conspicuousness of pitting in the horizontal end walls, often resulting in a beaded appearance, can be a useful character for distinguishing between genera or species (cf. rays). In the limited number of conifers from the Israel region, the axial parenchyma is of common occurrence only in the genera *Juniperus* and *Cupressus.* In *Cedrus* it is virtually limited to the growth ring boundary, and in *Pinus* it is absent.

Parenchyma associated with resin ducts will be discussed separately.

Rays and Cross-Field Pitting

In conifers rays are either exclusively composed of procumbent parenchyma cells or of these in combination with ray tracheids. Ray tracheids mostly occur in single or multiple rows along the margins of the rays; sometimes they are also intercalated between the ray parenchyma cells (probably as a result of ray fusion). They are provided with bordered pits, which are invariably smaller than the radial pit pairs between the axial tracheids. The outer horizontal walls of ray tracheids are often irregularly curved or undulate. Near bordered pit pairs the walls may be smooth or have dentate thickenings.

The tangential end walls of the ray parenchyma cells may be smooth or conspicuously pitted, resulting in a nodular or beaded appearance (Plates 1D and 2D). More or less nodular end walls are characteristic for all conifers indigenous to

the region of study, except *Cupressus*. In conifers the rays are usually almost exclusively uniseriate, but rays with biseriate portions occur in low frequency in most species of the region. Here, rays containing normal or traumatic resin ducts — so-called fusiform rays — are not considered: these are always multiseriate.
Ray height, measured as the number of cells, varies within wide limits in each tangential section. If one considers average values for different specimens of the same species, the intraspecific variation is also considerable and thus of very limited diagnostic value. For reasons of convenience, painstaking counts to arrive at a statistically meaningful average value are often avoided, and if at all used, the absence or extreme scarcity of rays exceeding a certain height category is a good enough criterion for distinguishing a taxon with low average ray height from one with higher rays. It should be emphasized that in branchwood of conifers the rays are usually much lower than those in adult stemwood.
Ray frequency for conifers, expressed as the number of rays per square tangential millimetre, which is often inversely proportional to ray height, shows the same type of variability as ray height. Nevertheless, ray height and frequency offer the only reliable means for generic wood anatomical distinction between *Juniperus* and *Cupressus* in the region.
Diversity in cross-field pitting, i.e., in the type and number of pit contacts between ray parenchyma and individual axial tracheids, is one of the most important diagnostic variables in softwood identification. In accordance with Phillips (1948), five types are distinguished as seen in radial section (Fig. 3):

a. Fenestriform or window-like pits: large, virtually simple pits, often entirely occupying a cross-field.
b. Pinoid pits: smaller, simple or weakly bordered pits; if bordered, the borders are often of unequal size or 'lopsided'.
c. Piceoid pits: fairly small pits with wide borders and narrow, slit-like, often slightly extended apertures.
d. Cupressoid pits: fairly small pits with wide, included apertures (apertures at most touching the outline of the pit border at the opposite ends of their long axes).
e. Taxodioid pits: fairly small simple pits, of which the pit aperture is wider than the pit membrane or pit floor. Careful focussing is needed to distinguish taxodioid pits from cupressoid and weakly bordered pinoid pits.

For identification purposes only the cross-field pits of the earlywood are traditionally considered. In our descriptions we have also included the deviations in the latewood cross-field pits in order to emphasize variation which can be crucial for the identification of pulps or tiny wood fragments lacking earlywood.

Resin Ducts (Fig. 2C)

Schizogenous resin ducts in the axial and radial direction occur in certain conifers. If of regular occurrence in each growth ring, they are of great diagnostic

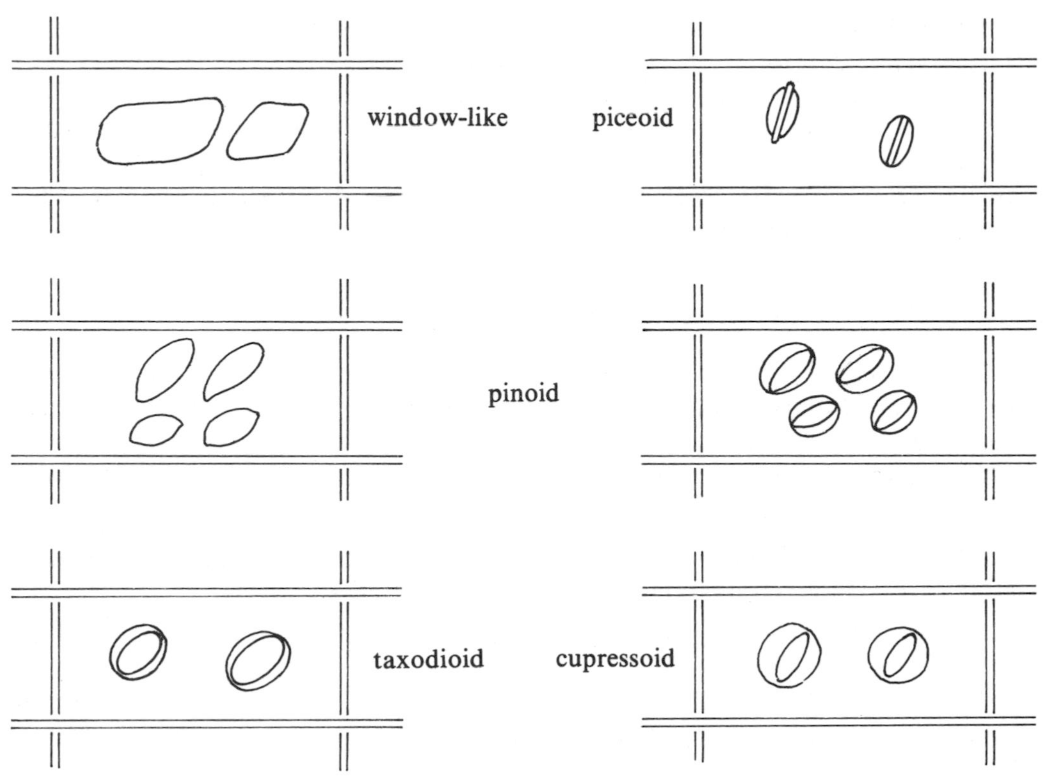

Fig. 3.
Diagrams of pit-pairs occurring in cross-fields of coniferous woods

value — in and around Israel these normal resin ducts are confined to *Pinus*. Traumatic resin ducts which are the result of some type of injury to the cambium are of more haphazard occurrence. They occur frequently in *Cedrus* and are characterized in this genus by their closely spaced arrangement. The parenchyma surrounding the ducts may be either thin-walled and unlignified or relatively thick-walled and lignified (Werker & Fahn,1969; Fahn et al., 1979).

2.2 *Vessel-Bearing Woods*

Except for two palm species that belong to the monocotyledons and the genus *Ephedra* of the gymnosperms, all woods dealt with in the following are of dicotyledonous plants.

The wood of the dicotyledons is much more complex than that of the conifers. It comprises elements that vary in type, size, shape and pattern of arrangement. In the wood of *Quercus,* for instance, vessel members, tracheids, fibre-tracheids, libriform fibres (sometimes gelatinous), wood-parenchyma and rays of different sizes are present. All these types of element, with the exception of the ray cells, are of a common origin — the fusiform initials of the cambium. In a relatively small number of dicotyledonous species vessels are absent, and the function of conduction is performed by tracheids only. Such species do not grow in Israel and its adjacent areas. Tracheids or the entire range of fibres are not present in all vessel-bearing species.

Included Phloem (Interxylary Phloem)

In certain plants, e.g., all species of the Chenopodiaceae, there are strands of secondary phloem within the wood (Plates 20–25). Such plant species are said to have wood with included phloem (Chalk & Chattaway, 1937; Metcalfe & Chalk, 1950). When viewed in transverse sections, the woods with included phloem can be divided into three types:

1. *Concentric type;* the phloem groups are embedded in continuous bands of unlignified or lignified conjunctive parenchyma (Plate 21 A).
2. *Foraminate type;* the phloem groups are not or only rarely linked by conjunctive parenchyma (Plate 23 C).
3. *Foraminate to concentric type* (intermediate type); the phloem groups are quite often linked by conjunctive parenchyma, but do not form long tangential bands.

Growth Rings

The presence of growth rings is determined by several characters: a. differences between late- and earlywood fibres (wall thickness, radial diameter; Plate 9 C); b. occurrence of marginal parenchyma (Plate 52 D); c. progressive changes in the number or width of parenchyma bands (Plate 48 A).

Vessels

When vessels are fairly uniform or change only gradually in size and distribution throughout a growth ring, the wood is termed *diffuse-porous* (Plate 6 A). When the wood contains vessels of different diameters and those produced at the beginning of the growing season are distinctly larger than those produced later, the wood is termed *ring-porous* (Plate 7 A and D). When the earlywood is marked by a zone of occasional large vessels or numerous small vessels, the wood is termed *semi-ring-porous.*

Vessel frequency (i.e., the number of vessels per square millimetre) varies greatly within and between species. In many cases this character can be used as a supportive diagnostic feature, if the differences are sufficiently great.

As seen in transverse section, the vessels may be distinctly circular, oval or angular. The first two types are described here as rounded. Both radial and

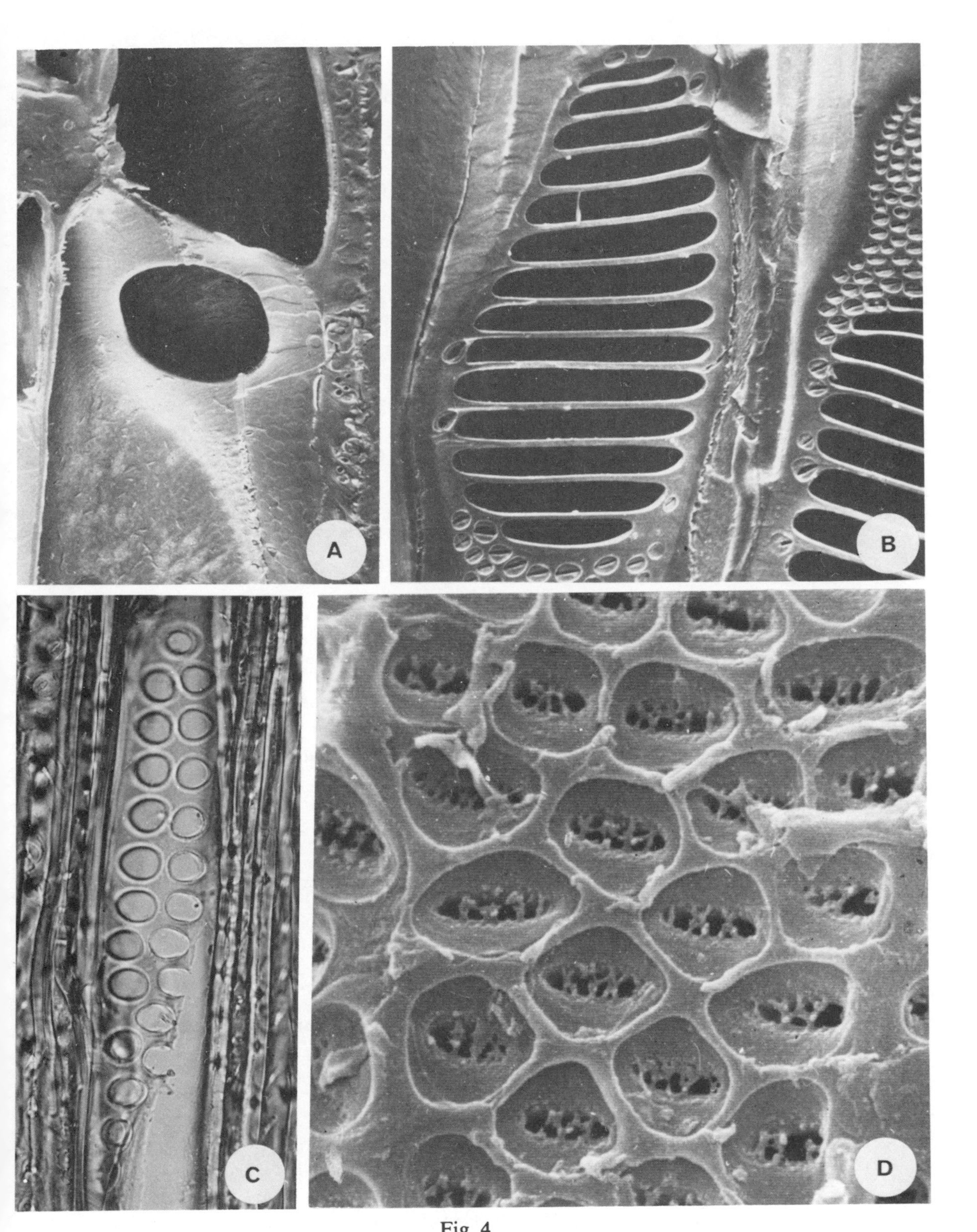

Fig. 4.
A–C — perforation plates
A and *B* — scanning electron micrographs; *A* — a simple perforation plate, ×600
B — a scalariform perforation plate, ×1,000
C — a foraminate perforation plate of *Ephedra alata*, ×480
D — scanning electron micrograph of vestured pits of *Acacia raddiana*, ×4,500

tangential diameters are given when pores are oval, elongated predominantly in one direction. When the larger diameter of the pore is oriented at random or the pores are circular, measurements of one diameter are given. In measurements of vessel-member length, tails, when present, are included.

Vessels may occur solitary or in groups of different size and shape. The groups may consist of radial, oblique or tangential rows of two to many vessels, the walls of which are in contact with one another. Such groups are termed *multiples* (Plates 68C, 69A). When the group consists of varying numbers of vessels in both radial and tangential directions, we refer to it as a *cluster* (Plates 13A, 76D).

Even when the vessels are actually solitary they may have a close topographic relationship and form groups that run radially or oblique. Such groups may form various patterns, e.g., a dendritic pattern (Plate 52D). The end walls of the vessel members may bear simple (Fig. 4A), or scalariform (Fig. 4B), or foraminate (Fig. 4C) perforation plates. Vessels with scalariform perforation plates are rare in the woods of Israel and its adjacent areas. Those with foraminate perforation plates occur here only in the genus *Ephedra.*

Pits

Pits may be simple, bordered or half-bordered. Inter-vessel pits are bordered. Pits between vessels and axial parenchyma or ray cells are usually half-bordered. In some plants sculpturings or outgrowths occur on the wall of the pit chamber or around the pit aperture. Such pits are termed *vestured pits* (Fig. 4D).

The bordered pits present on vessels vary in their shape and arrangement. When the pits are elongated and form ladder-like tiers, the arrangement is termed *scalariform pitting* (Fig. 5B). When the pits are circular or elliptic they may be arranged in horizontal lines, i.e., *opposite pitting,* or in diagonal lines, i.e., *alternate pitting.* In the latter type of pitting the pits may sometimes be polygonal in outline (Fig. 5A). The apertures of bordered vessel pits are usually slit-like and included within the circumference of the pit chamber; sometimes they extend over several pit pairs, and are then termed coalescent.

Pits between vessels (so-called inter-vessel or intervascular pits) may be similar in size and arrangement to the pits between vessels and axial parenchyma or ray

---

Fig. 5.

A — alternate, polygonal inter-vessel pits of *Salix*, ×480
B — scalariform inter-vessel pits of *Vitis vinifera*, ×480
C — a vessel of *Pistacia atlantica* with spiral thickenings, × 480
D — a vessel and vasicentric tracheids of *Quercus boissieri* (Tr), ×300
E — vessels of two size classes of *Nitraria retusa,* the narrow ones intergrading with vascular tracheids, ×300
F — septate fibres of *Hedera helix*, ×480

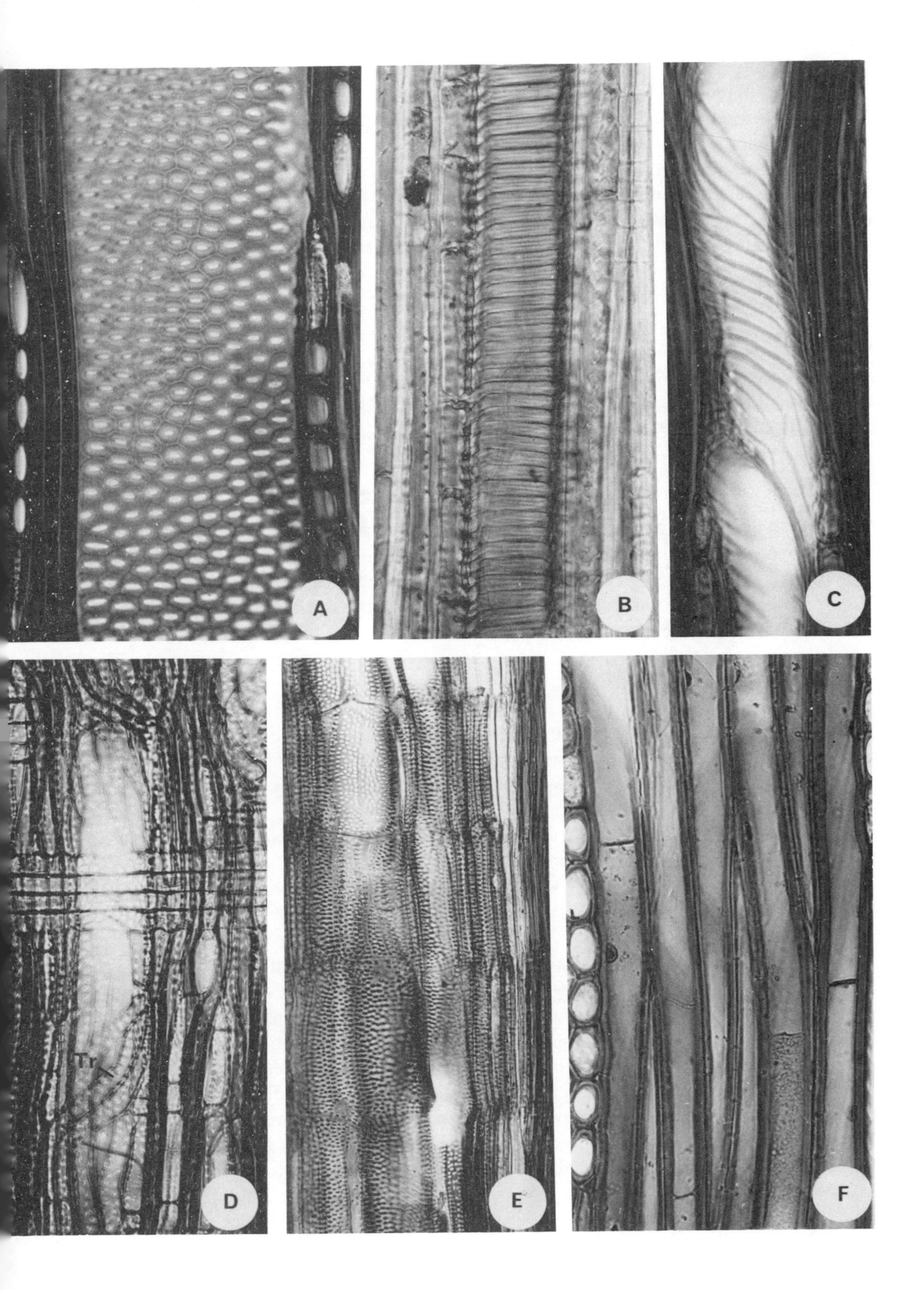
A
B
C
Tr
D
E
F

parenchyma cells. In some genera the latter pits are large and simple and show an arrangement different from that of the inter-vessel pits. If the vessel-ray and vessel-parenchyma pits are simple and elongate they are sometimes referred to as gash-like.
In some species a single pit on the parenchyma side subtends two or more pits on the vessel side; such a pit pair is termed *unilaterally compound.* Very rarely the contact area between a ray cell and a vessel may contain a perforation. Such perforated ray cells are not diagnostic for the few species from Israel in which they were recorded.

Wall Sculpturing

On the inner side of the secondary cell wall of the vessels of some woods spiral thickenings of various distinctness are present (Fig. 5C). Such thickenings may sometimes also occur in fibres.

Tyloses and Gummy Substances

The occurrence of tyloses, outgrowths of ray or axial parenchyma cells into the lumen of vessels, is characteristic of certain plant species (Plate 9A and B). Gummy substances filling the lumen of vessels may also be found in some woods. Tyloses and gummy substances occur in vessels which stopped functioning as conductors.

Vascular Tracheids and Vasicentric Tracheids

For practical reasons, in the following descriptions the term *vascular tracheids* will be used for very narrow tracheary elements, usually occurring in groups, in which it is difficult to distinguish between elements with perforations and those without them (Fig. 5E).
Tracheids (always imperforate) associated with vessels are termed *vasicentric tracheids* (Fig. 5D).

Fibres

In the descriptions no distinction was made between libriform fibres and fibre-tracheids. However, the type of pits, either simple or bordered, is mentioned. Within woody plants as a whole there is a gradual variation from fibres with distinctly bordered pits (Diameter 3.0 μm or more) in both tangential and radial walls, i.e., fibre-tracheids, to fibres with minutely bordered (less than 3 μm) to simple pits confined to the radial walls, i.e., libriform fibres. In view of the intermediates which occur between these two extremes, we have refrained from using the terms fibre-tracheids and libriform fibres in the descriptions, and have described the type and distribution of fibre wall pits in each species.

Fibre wall thickness is determined here in relation to the fibre diameter. Fibres with a wall thickness less than one-quarter of their diameter are regarded as thin-walled; when the wall equals one-quarter of the fibre diameter, the fibre is described as medium thick-walled; when the wall exceeds one-quarter but is less than half the diameter of the fibre, it is termed thick; when the wall is so thick that only a very narrow cell lumen is left, the fibre is said to be very thick-walled.

Gelatinous fibres (the inner unlignified wall layer of which has the capacity to absorb water and to swell more strongly than the outer lignified wall layers) are characteristic of reaction wood but may occur in some species in randomly distributed patches. Fibres which become subdivided by thin transverse walls, after secondary wall layers have been formed, are called *septate fibres* (Fig. 5F).

Axial Parenchyma

The axial parenchyma in dicotyledons is generally much more abundant than that of the conifers, and its distribution pattern is sometimes of great diagnostic value. There are two main types of distribution of axial parenchyma: *apotracheal,* in which the parenchyma is typically independent of the vessels though it may come into contact with them here and there; and *paratracheal,* in which the parenchyma is distinctly associated with the vessels. Both these types are subdivided into the following variations. When the apotracheal parenchyma is in the form of single cells or small bands scattered irregularly among the fibres, it is said to be *diffuse* (Fig. 6) or *diffuse-in-aggregates* (Plate 48C), respectively. When the axial parenchyma in transverse section of the wood is seen to form concentric bands, it is said to be *banded* (Fig. 6). Single apotracheal parenchyma cells or those arranged in more or less continuous layers which may be of variable width, at the border between two growth rings, are termed here *marginal parenchyma* (Plate 52D).

The paratracheal parenchyma, when not forming a continuous sheath around the vessels, is said to be *scanty paratracheal* (Fig. 6). When the paratracheal parenchyma occurs on one side, either abaxial or adaxial, of the vessels, it is described as *unilaterally paratracheal.* When the paratracheal parenchyma forms entire sheaths, which may be of different width, around the vessel, it is termed *vasicentric parenchyma* (Fig. 6). When the sheaths, as seen in transverse section, have lateral extensions, the parenchyma is termed *aliform* (Fig. 6). The aliform parenchyma may form tangential or diagonal bands; this type is termed *confluent parenchyma* (Fig. 6).

The fusiform initials of the cambium, as seen in tangential section of some species, are arranged in horizontal rows and their ends are approximately at the same level. This type of cambium is referred to as *storied* (Plate 80B). The vessel elements and the axial parenchyma developing from such a cambium retain a storied structure. Cells of storied parenchyma may sometimes undergo transverse subdivisions, but the parenchyma maintains the basic storied structure.

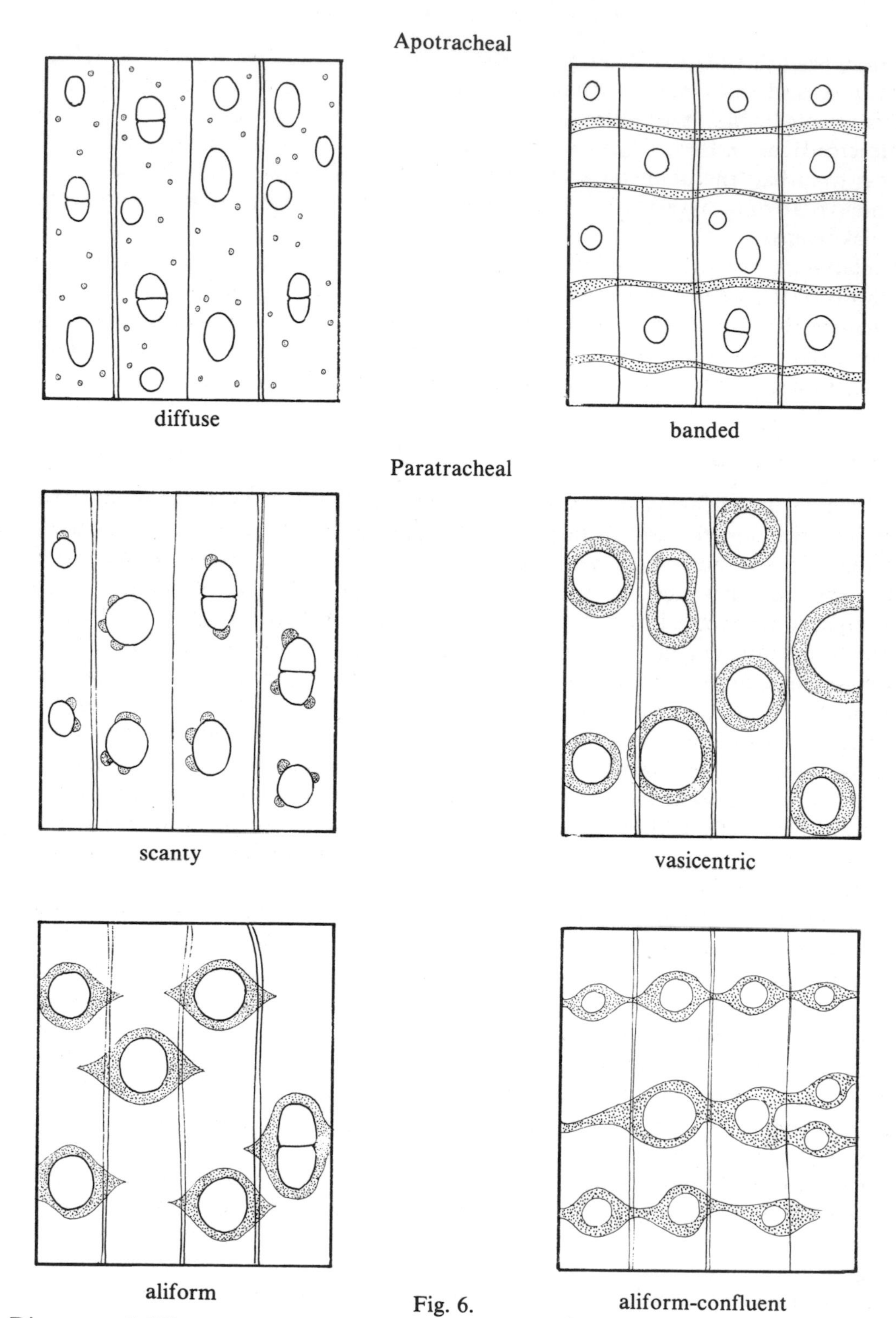

Fig. 6.
Diagrams of different types of parenchyma (dotted areas) distribution in vessel-bearing woods, as seen in transverse sections

Axial parenchyma in a vertical row, derived from one fusiform initial which has transversely divided into a number of cells, is termed a *parenchyma strand.* Each strand is terminated at either end by a wedge-shaped cell. When no transverse divisions have taken place in the fusiform initial, the parenchyma cell developed from it is termed a *fusiform parenchyma cell.*

Rays

In the dicotyledons, with the exception of a few species, the rays consist of parenchyma cells only. Ray frequency was established here by counting the rays along a horizontal line in tangential longitudinal sections.
Rays of the dicotyledons show a very wide range of variation. They may be uniseriate and sometimes only a few cells high, or multiseriate and sometimes very large (several millimetres high and many cells wide). In some species the apparently large rays are, in fact, bundles of several smaller rays separated by axial fibres. Such rays are termed *aggregate rays.*
Although the rays of the dicotyledonous woods consist of parenchyma cells only, great variations can be noted in the shape of these cells. If all the cells are similar in shape and orientation, the rays are said to be *homocellular* or *homogeneous* (Fig. 7). When the ray consists of two or more types of cells, it is termed *heterocellular* or *heterogeneous* (Fig. 7). Kribs (1935, 1950, 1968) proposed a classification of rays based on their appearance in tangential and radial sections. In this book we use the classification of heterogeneous rays suggested by Kribs (1968) only when our descriptions precisely fit his definitions. In other cases we have employed the terms hetero- and homocellular.

1. *Heterogeneous type I* (Fig. 7)
   a. Uniseriate rays composed of vertically elongate (upright) cells which differ from the cells of the multiseriate part of the multiseriate rays.
   b. Multiseriate rays with uniseriate wings as high as or higher than the multiseriate portion of the ray, and composed of vertically elongate (upright) cells similar to those of the uniseriate rays. The cells of the multiseriate portion are rounded to oval in tangential section, and radially elongate (procumbent) in radial section.
2. *Heterogeneous type II* (Fig. 7)
   a. Uniseriate rays composed of vertically elongate (upright) cells which differ from those of the multiseriate portion of the multiseriate rays.
   b. Multiseriate rays with one large, vertically elongate cell, or with uniseriate wings shorter than the multiseriate portion of the ray, and composed of vertically elongate cells similar to those of the uniseriate rays. The cells of the multiseriate portion are round to oval in tangential section, and radially elongate (procumbent) in radial section.
3. *Heterogeneous type III* (Fig. 7)
   a. Uniseriate rays usually of two types; some of them consist of vertically

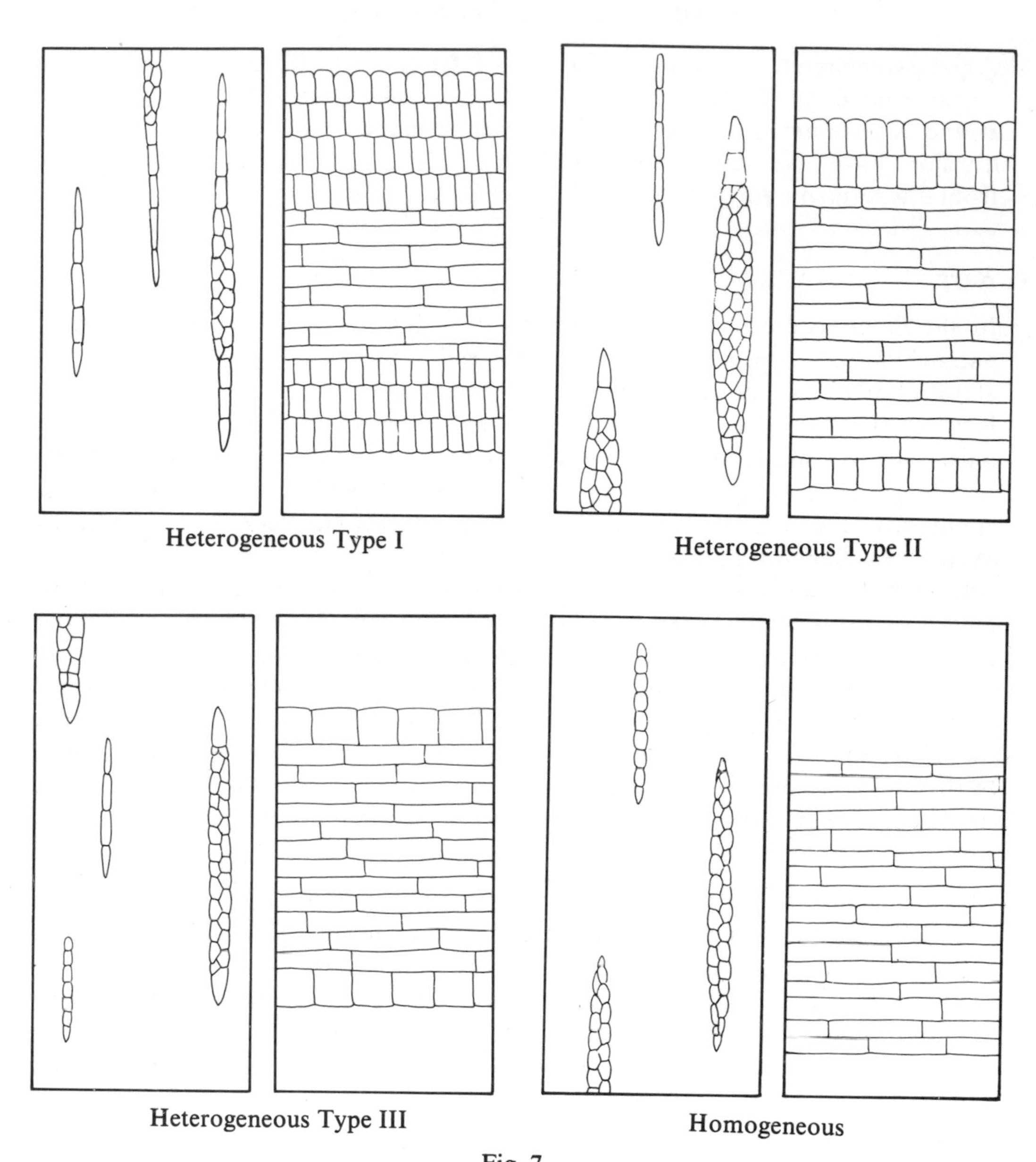

Fig. 7.
Diagrams of different types of rays in vessel-bearing woods, as seen in tangential and radial longitudinal sections

elongate cells, and some of cells which are nearly identical to those of the multiseriate portion of the multiseriate rays.

b. Multiseriate rays with square marginal cells, usually in a single row; if wings are present, marginal cells are square, not vertically elongate. The cells of the multiseriate portion are rounded to oval in tangential section, and radially elongate (procumbent) in radial section.

Within the large category of homocellular rays as defined here, there is a fundamental difference between homocellular rays composed of procumbent cells only (uniseriate and/or multiseriate rays of the homogeneous type in the classical terminology) and those entirely composed of upright or square cells. The latter are often an early ontogenetic phase of heterocellular ray types, and are sometimes referred to as juvenile.

Trabeculae

The woods of certain dicotyledons, like those of conifers, sometimes feature rod-shaped extensions of cell wall material, which traverse the cell lumen from one tangential wall to another. Such structures are termed *trabeculae* (Fig. 8A) and may occur in tracheids, vessels, fibres and parenchyma cells.

Crystals

Crystals consisting largely of calcium oxalate are often found in axial and ray parenchyma cells. The cells in which they occur are termed *crystalliferous cells.* In axial parenchyma cells they may occupy each cell of a strand or occur in some cells only. Some of the parenchyma cells sometimes undergo subdivisions, and the new compartments become containers of crystals. Such parenchyma is referred to as *chambered* (Fig. 8B). Sometimes crystals may also be found in fibres.

Rhomboidal or prismatic crystals may occur singly (Fig. 8B) or in aggregates in a cell. Regular star-shaped aggregates of crystals are referred to as *druses* (Fig. 8C). Less regular aggregates are termed *clusters.* Druses and clusters are very uncommon in woody tissues, but are highly diagnostic in those few species in which they do occur. Very small crystals occurring in masses are called *crystal sand* (Fig. 8D). Very thin needle-like crystals (*raphides,* Fig. 8E) occurring in bunches and long prismatic tapered crystals (*styloids,* Fig. 8F) are rarely found in wood parenchyma of dicotyledons.

Another diagnostic type of mineral inclusion is exemplified by silica bodies of various shapes and distribution. They have not been encountered in any of the species from Israel and the adjacent regions, although their occurrence in tropical tree families, for example, is not uncommon.

Secretory Structures in Wood

Various types of structures secreting a variety of chemical substances occur in plants: secretory cells, cavities, ducts or canals and laticifers. Secretory structures are more commonly found in the primary plant tissues and in the secondary phloem (Fahn, 1979) than in the secondary xylem. However, when they occur in the wood they are of diagnostic importance.

The secretory ducts and cavities in the wood may occur normally or as a result

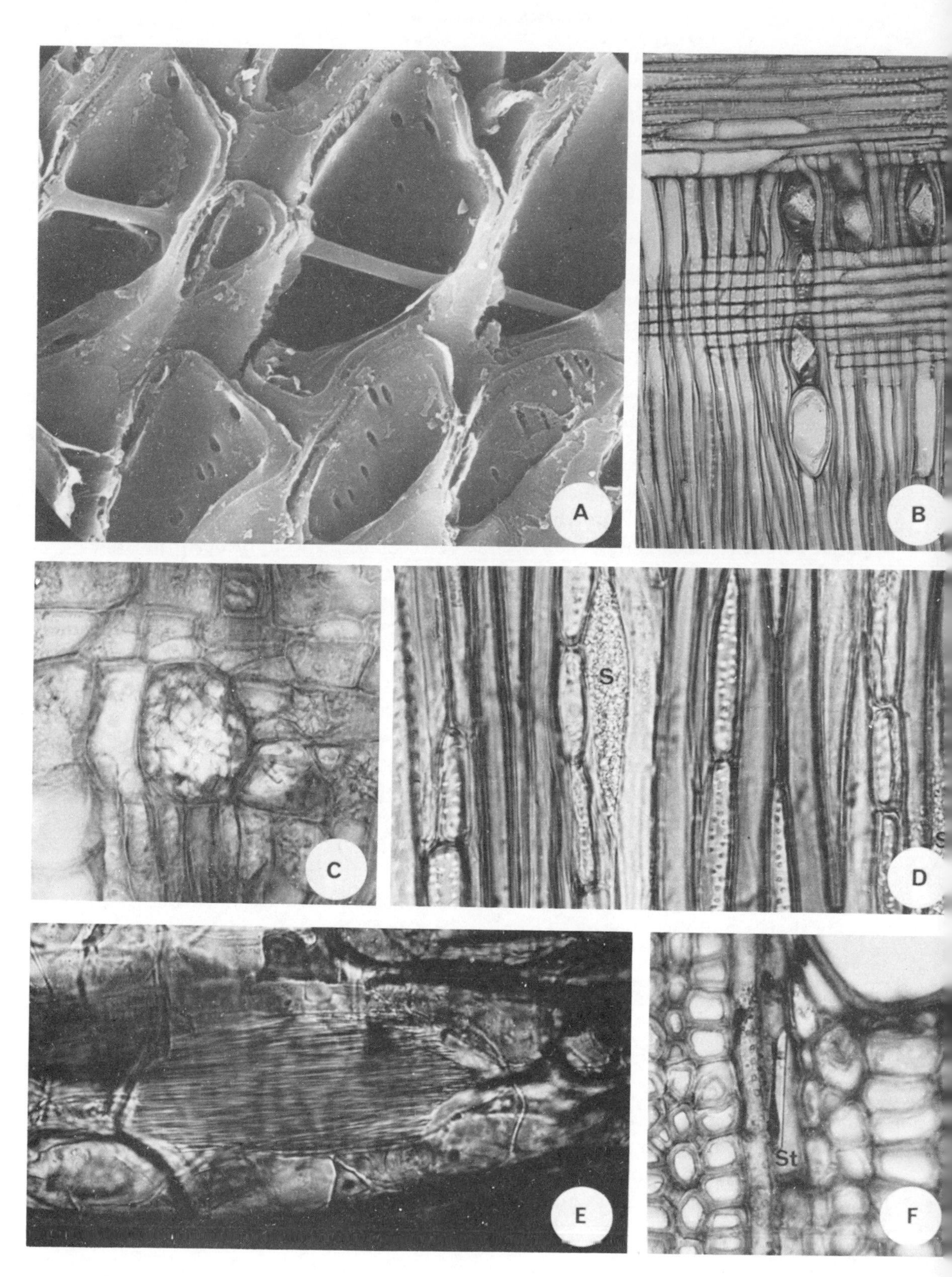
A
B
C
S
D
E
St
F

of injury (traumatic ducts or cavities). The ducts are either axially aligned in the stem or radially situated in the rays, but the two types occur very rarely together (Metcalfe & Chalk, 1983). The ducts (elongate structures) and cavities (spherical or only slightly elongate structures) consist of intercellular spaces usually surrounded by epithelial cells. They contain resin consisting mainly of a variety of terpenes, or gums consisting of polysaccharides, or gum-resins consisting of both types of compounds.

In the wood of the Israeli plants radial resin ducts occur, for instance, in *Pistacia* species (Fig. 9A), and axial traumatic gum ducts are found in *Citrus* species and in *Armeniaca* (Fig. 9C).

Laticifers are specialized cells or rows of cells containing latex, a suspension or emulsion of many small particles in a liquid with a different refractive index. The coarse dispersed phase of the latex consists of various lipophylic substances, and in laticifers of various plants proteins, starch grains, tannins, alkaloids or certain other substances may be present. The colour of the latex varies in the different plant species; it may be white and milky, yellow-brown, orange or colourless.

The laticifers are classified into two main types: *non-articulated laticifers* — single, long, multinucleate cells; and *articulated laticifers* — developing from rows of cells, the end walls of which remain entire or become porous or disappear completely. Both types of laticifers may be simple unbranched or branched.

Laticifers are to be found in the wood of relatively few plant species. In the wood of Israeli plants they were observed in rays of *Euphorbia hierosolymitana, Calotropis procera* and *Ficus spp.* (Fig. 9B and D).

Secretory cells, which may or may not differ in size and shape from the neighbouring cells but contain secreted material, occur in the wood of only a limited number of plants. Secretory oil cells, for instance, occur in *Laurus nobilis.*

Fig. 8.

A — scanning electron micrograph of trabeculae in a row of fibres of *Inula viscosa*, ×1,100

B — solitary prismatic crystals in chambered parenchyma cells in *Citrus aurantium*; some cells enlarged, ×300

C — an enlarged ray cell of *Rhamnus palaestinus,* containing a druse, ×480

D — tangential longitudinal section of the wood of *Lycium shawii* showing parenchyma cells with crystal sand (S), ×480

E — an enlarged conjunctive parenchyma cell of *Commicarpus africanus*, containing a bundle of raphides, ×480

F — a styloid (St) in a ray cell of *Laurus nobilis,* × 480

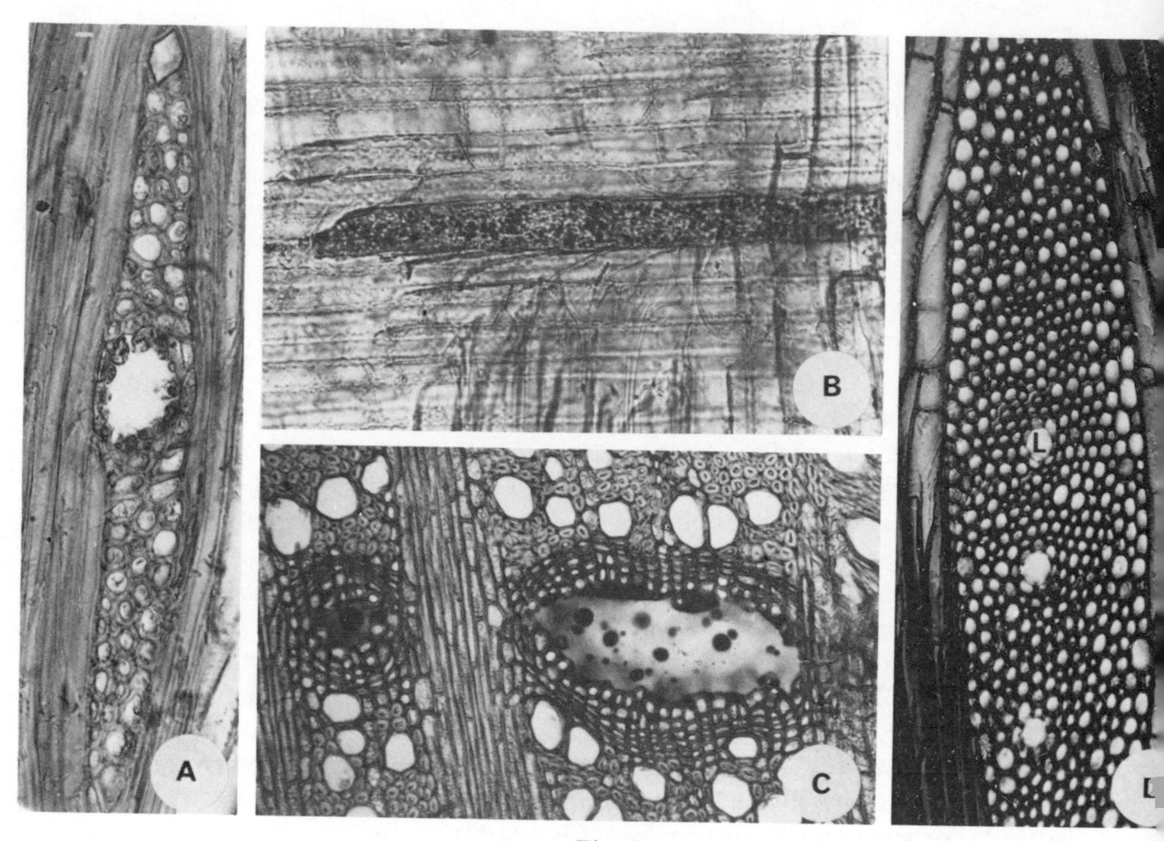

Fig. 9.
Secretory structures
A — resin duct in a ray of *Pistacia palaestina*, tangential longitudinal section, ×300
B — a laticifer in a ray of *Ficus pseudo-sycomorus*, radial longitudinal section, ×126
C — traumatic gum ducts of *Armeniaca vulgaris*, transverse section, ×126
D — laticifers (L) in a ray of *Ficus sycomorus*, tangential longitudinal section, ×300
(*A* from Grundwag & Werker, 1976)

3. THE IDENTIFICATION OF SHRUBS AND TREES BY THEIR WOOD STRUCTURE

3.1 *Introduction and Explanatory Notes*

In this chapter the descriptive information on the wood anatomy of shrubs and trees treated in this book is presented in such a form as to facilitate identification of wood fragments. Two means are provided: 1. the classical form of the dichotomous key; and 2. the listing of species with certain diagnostic features. The latter has been restricted to the hardwood species and *Ephedra,* because one can readily separate the small number of conifers and monocotyledons using the first part of the dichotomous key.

Lists of species with diagnostic characters can serve as a so-called synoptical key, and have the advantage that an unknown wood with one or more striking (i.e., uncommon) characters can be identified very rapidly, when these characters are considered in combination with only a few other, more common characters also represented in the lists. This approach has shown its potential in the time-honoured use of lists in Metcalfe & Chalk's *Anatomy of the Dicotyledons* (1950, 1979, 1983) for identification down to the family level. Not all characters for which lists of species are given in this chapter are of equal diagnostic weight. Species in which a certain character is either variable or poorly developed are marked with superscript numerals, and from the large number of species so marked in some lists it is evident that the character concerned is only of limited diagnostic value. The same applies to those characters for which species show overlap. For quantitative characters only extreme categories have been used: the species with intermediate values are so numerous and would often overlap to such an extent that there is little value in listing them.

The dichotomous key should be used with great caution. In view of the limited number of samples examined for this study, the possibility exists that the total range of variation of some species was not recorded by us, and that some leads in the key may head the user in the wrong direction for samples exceeding the range observed by us. To minimize this risk, the variational range of especially quantitative features has been artificially extended beyond the observed range for the construction of the key. Nevertheless, it is recommended that both alternative leads be followed in borderline cases or in case of doubt. Due to infraspecific variation observed by ourselves, or deduced from data in the literature

(especially in Greguss, 1959; Jacquiot et al., 1973; Schweingruber, 1978), many species appear twice or even several times in the key.
A number of leads end with a group of species, sometimes even belonging to different genera. In these cases the differences in our material were considered insufficient to separate these taxa on the basis of wood anatomy. However, the user may find it helpful to consult the descriptions and plates in order to narrow down the possible identity of the unknown sample.
Some characters used in the key may cause frequently recurring problems. These are briefly discussed below.

1. *Ring-porosity, semi-ring-porosity and diffuse-porosity.* In most instances there will be no problem in distinguishing between the latter option and the two former ones. However, in some specimens with very narrow annual increments, ring-porous woods may give the impression of diffuse vessel distribution. In these growth rings only the wide-vesselled earlywood zones are represented, thus obscuring the ring-porous nature (Plate 46A and B). In such cases typical distribution patterns of latewood vessels (oblique, tangential or dendritic in a number of woods) are also not developed. Moreover, some basically diffuse-porous woods may show some gradient in vessel diameter and frequency from early- to latewood in part of the growth rings (Plate 66D), thus complicating the distinction from semi-ring-porosity.
2. *Storying.* The storied arrangement of vessel members and parenchyma strands is a highly diagnostic feature of many species from the studied region. However, in young stems or thin branches this character may not be clearly developed, or its conspicuousness may vary with position in the growth ring, being dependent on the local abundance of parenchyma strands and vessel members.
3. *Ray composition and size.* The proportion of upright or procumbent cells usually varies markedly with the age of the stem. Upright cells are generally much more common near the pith than at the periphery of stems exceeding a certain diameter. Therefore the leads in the key relating to ray composition should be used with extra caution, especially if stems thinner than 1 cm are involved as unknowns. By the same token, it is possible that species which we have recorded in the descriptions and key as possessing exclusively or predominantly upright cells in their rays, may show a higher percentage of square to procumbent cells in wood more mature than that at our disposal. The same caution should be exercized with regard to ray width and height, because in hardwoods the rays are usually narrower and taller near the pith than at the periphery of thick stems.
4. *Crystal complement.* The type and distribution of crystals are highly diagnostic. However, crystals have been used only sporadically in the key, because crystals, even when generally present in the wood, may occasionally or locally be missing. Therefore the user is advised to consult the lists of taxa with various crystal types in those cases where the unknown sample contains crystals. Crystals can only be used as a positive character, never as a negative one, because of the

above-mentioned fluctuations in their frequency. By inference it follows that some species for which no crystals were observed in our material may exhibit crystals in other specimens. We have attempted to reduce this possibility by supplementing our observations with data from the literature.

Finally, it should be stressed that both the key and the lists of species with certain diagnostic features are only imperfect aids in identification. The final step should always be a comparison with the descriptions and illustrations, and preferably also with sections of reference specimens, reliably identified on the basis of macromorphological characters of the entire plant, including its reproductive parts.

3.2 *Key*

1. Vessels absent — Coniferae 2
- Vessels present — Gnetales, Angiospermae 5

2. Resin ducts present in all growth rings; cross-field pits pinoid. — **Pinus halepensis**, **Pinus pinea**
- Resin ducts absent or of the traumatic type; cross-field pits different — 3

3. Inter-tracheid pits with fringed torus margins; ray tracheids present; traumatic ducts sometimes present. — **Cedrus libani**
- Inter-tracheid pits with more or less smooth torus margins; ray tracheids absent; resin ducts always missing — 4

4A. Adult stemwood: rays 1–10(14) cells high, rarely exceeding 10 cells in height, 55–100/mm² in tangential section; end walls of ray cells mostly nodular. — **Juniperus drupacea**, **Juniperus excelsa**, **Juniperus oxycedrus**, **Juniperus phoenicea**
- Rays 1–40 cells high, often exceeding 10 cells in height, 18–40/mm² in tangential section; end walls of ray cells smooth to very faintly nodular. — **Cupressus sempervirens**

4B. Branchwood or juvenile stemwood: rays more than 70/mm² in tangential section; end walls of ray cells mostly nodular. — **Juniperus drupacea**, **Juniperus excelsa**, **Juniperus oxycedrus**, **Juniperus phoenicea**
- Rays less than 65/mm² in tangential section; end walls of ray cells smooth to very faintly nodular. — **Cupressus sempervirens**

5. All perforation plates foraminate. — **Ephedra alata**, **Ephedra aphylla**, **Ephedra campylopoda**, **Ephedra foliata**
- Perforation plates simple or scalariform, rarely reticulate — 6

6. Discrete vascular bundles accompanied by fibre caps scattered at random in parenchymatous ground tissue — Monocotyledoneae 7
- Wood not as above — Dicotyledoneae 8
7. Perforation plates scalariform or partly reticulate. — **Phoenix dactylifera**
- Perforation plates simple. — **Hyphaene thebaica**
8. Wood with included phloem — 9
- Wood without included phloem — 35
9. Tangential bands of sclereids or sclerified parenchyma cells present in unsclerified tissue — 10
- Not as above — 12
10. Sclereid bands intruding centripetally between phloem strands; rays over 15 cells wide present. — **Cocculus pendulus**
- Sclereid arrangement not as above; rays less than 10 cells wide — 11
11. Sclereids mostly rectangular in cross-section, with very thick walls leaving a very narrow cell lumen. — **Avicennia marina**
- Sclerified cells irregularly shaped with a relatively large lumen. — **Cleome droserifolia**
12. Fibrous tissue, including vessels, in isolated radial strips embedded in abundant unlignified parenchyma. — **Chenolea arabica**
- Wood differing from the above — 13
13. Rays clearly discernible in cross-section — 14
- Rays absent or indistinct or some of the conjunctive parenchyma forming ray-like extensions — 21
14. Included phloem of the foraminate to concentric type — 15
- Included phloem distinctly of the concentric type — 16
15. Rays very infrequent (1–3/mm). — **Salsola tetrandra**
- Rays 7–13/mm. — **Salvadora persica**
16. All rays narrow, 1–3(4) cells wide — 17
- Some rays 5 or more cells wide — 19
17. Rays usually uniseriate; crystals abundant. — **Anabasis articulata**
- Rays 1–3(4) cells wide; crystals absent or scarce — 18
18. Conjunctive parenchyma largely unlignified; many rays only few cells high. — **Anabasis setifera**
- Conjunctive parenchyma largely lignified; most rays more than 8 cells high. — **Salvadora persica**

19 (16). At least some ray cells show marked radial elongation (in cross-section) — 20
- Ray cells not as above. — **Suaeda fruticosa**

20. Tangential diameter of wide vessels 50 μm or wider. — **Haloxylon persicum**
- Tangential diameter of wide vessels 45 μm or narrower. — **Aellenia lancifolia**

21 (13). Raphides present in conjunctive parenchyma. — **Commicarpus africanus**
- Raphides absent — 22

22. Rays indistinct or conjunctive parenchyma forming ray-like extensions — 23
- Rays and radial parenchyma extensions absent — 28
23. Maximal radial vessel diameter 120 μm or more. — **Atriplex halimus**
- Maximal radial vessel diameter less than 120 μm — 24
24. Included phloem partly of the foraminate type and conjunctive parenchyma partly forming short tangential bands — 25
- Conjunctive parenchyma forming concentric rings or long tangential bands — 26

25. Conjunctive parenchyma linking phloem groups thin-walled, partly unlignified or lignified. **Seidlitzia rosmarinus**
– Conjunctive parenchyma linking phloem groups lignified, partly thick-walled. **Salsola tetrandra**
26 (24). Conjunctive parenchyma unlignified. **Anabasis syriaca**
– Conjunctive parenchyma mostly lignified 27
27. Prismatic crystals mostly over 10 μm in diameter; smaller crystals present or absent. **Suaeda asphaltica**
Suaeda fruticosa
Suaeda monoica
– Prismatic crystals mostly less than 10 μm in diameter. **Salsola baryosma**
Salsola vermiculata
28 (22). Included phloem of the foraminate type 29
– Included phloem of the concentric or foraminate to concentric type 30
29. Tangential diameter of phloem groups small, 35–80 μm in cross-section; vessels solitary and in multiples of 2–3 or in small clusters. **Arthrocnemum perenne**
– Tangential diameter of phloem groups not smaller than 65 μm and up to 130 μm or larger; vessels in clusters or radial multiples of up to 6. **Hammada negevensis**
Hammada salicornica
Hammada scoparia
30 (28). Conjunctive parenchyma forming short bands 31
– Conjunctive parenchyma forming very long bands or concentric rings 33
31. Tangential vessel diameter up to 35 μm. **Noaea mucronata**
– Tangential vessel diameter up to 45 μm or larger 32
32. Vessels solitary or in multiples of 2–3 or in small clusters; tangential diameter of phloem groups 35–80 μm in cross-section. **Arthrocnemum perenne**
– Vessels in relatively large clusters; phloem groups with a tangential diameter of more than 100 μm present. **Halogeton alopecuroides**
33 (30). Conjunctive parenchyma varies considerably in width along each band; fibre bands 3 or more times the width of parenchyma bands. **Suaeda palaestina**
– Parenchyma not as above; fibre bands less than 3 times the width of parenchyma bands 34
34. Most conjunctive parenchyma lignified. **Arthrocnemum macrostachium**
– Conjunctive parenchyma unlignified. **Noaea mucronata**
35 (8). Rays exclusively uniseriate or almost so (some biseriate, rarely triseriate) 36
– Many rays 2 or more cells wide 47
36. Vessel arrangement dendritic 37
– Vessel arrangement not dendritic 39
37. Many ray cells upright, some square or very weakly procumbent. **Lycium europaeum**
Lycium schweinfurthii
Lycium shawii
– Ray cells procumbent, some square 38
38. Vessels over 120 μm in diameter present. **Lycium europaeum**
– All vessels less than 100 μm in diameter. **Thymelaea hirsuta**
39 (36). Parenchyma in wide confluent bands. **Acacia albida**
– Parenchyma not as above 40

40. Fibre pits distinctly bordered, numerous in radial and tangential walls 41
– Fibre pits simple to minutely bordered 42
41. Many rays only 1 cell high. **Gymnocarpos decandrum**
– Rays mostly over 5 cells high. **Jasminum fruticans**
42 (40). Fibres septate, often crystalliferous. **Punica granatum**
– Fibres not septate; crystals, if present, occur in parenchyma or ray cells 43
43. Fibre walls thin to medium-thick; paratracheal parenchyma scanty 44
– Fibre walls medium-thick to thick; paratracheal parenchyma mostly vasicentric. **Ziziphus lotus** **Ziziphus spina-christi** **Paliurus spina-christi**
44. Procumbent ray cells rare or absent. **Withania somnifera** (very young stems)
– Procumbent ray cells fairly to very common 45
45. Vessel-ray pits half-bordered and small. **Nerium oleander**
– Vessel-ray pits large and mostly simple, mainly restricted to ray margins 46
46. Rays typically homocellular or slightly heterocellular with marginal cells slightly less procumbent than central cells, or square. **Populus euphratica**
– Rays heterocellular with 1–3 rows of upright, sometimes irregular, marginal cells and procumbent, sometimes square or even upright central cells. **Salix acmophylla** **Salix alba** **Salix babylonica**
47 (35). Perforation plates exclusively or partly scalariform 48
– Virtually all perforation plates simple 53
48. All perforation plates scalariform 49
– Only part of perforation plates scalariform 50
49. Vessels with spiral thickenings, mostly solitary, some in radial multiples of 2–3. **Viburnum tinus**
– Vessels without spiral thickenings, often in radial multiples. **Styrax officinalis**
50 (48). Rays not more than 5 cells wide 52
– Rays over 8 cells wide present 51
51. Inter-vessel pits round to elongate; rays homocellular up to ca. 3 mm in height. **Platanus orientalis**
– Inter-vessel pits scalariform, in large vessels in several rows; rays heterocellular up to 20 mm or more in height. **Vitis vinifera**
52(50). Rays with uniseriate wings often taller than the multiseriate part. **Lonicera etrusca**
– Rays usually with a single row of marginal cells or uniseriate wings usually much shorter than the multiseriate part. **Arbutus andrachne**
53 (47). Vessels hardly distinguishable from parenchyma cells in cross-section. **Viscum cruciatum**
– Vessels not as above 54
54. Vessels up to 350 μm in diameter, many of them with dense spiral thickenings; clusters of narrow vessels occupying only a minor part of latewood. **Melia azedarach**
– Not as above in at least one feature 55
55. Parenchyma and/or vessel members storied in longitudinal tangential view 56
Parenchyma and vessel members non-storied in longitudinal tangential view 81

56. Rays over 17 cells wide present 57
- Rays not more than 16 cells wide 60

57. Groups of stone cells present in rays. **Loranthus acaciae**
- Rays not as above 58

58. Rays not well defined in longitudinal tangential view, composed of mostly upright and square cells, some weakly procumbent. **Clematis cirrhosa**
- Rays well defined in longitudinal tangential view, composed of strongly procumbent central cells and square, rarely upright marginal or sheath cells 59

59. Apotracheal parenchyma diffuse and diffuse-in-aggregates. **Balanites aegyptiaca**
- Apotracheal parenchyma absent or rare. **Tamarix aphylla**
Tamarix passerinoides

60 (56). Vessels of two size classes, together forming a dendritic pattern or partly arranged in oblique rows 61
- Vessels not as above 62

61. Tangential bands of apotracheal parenchyma 1–2 cells wide, common.
Retama raetam
- Tangential parenchyma bands absent or rare. **Anagyris foetida**
Calycotome villosa
Genista fasselata
Gonocytisus pterocladus
Spartium junceum
(**Colutea cilicica** and **C. istria** — dendritic or oblique pattern poorly developed)

62 (60). Marginal parenchyma present 63
- Marginal parenchyma absent or indistinct 72

63. Rays not more than 13 cells high 64
- Rays more than 20 cells high present 66

64. Confluent parenchyma in conspicuous bands 2–5 cells wide. **Cordia sinensis**
- Parenchyma not as above 65

65. Wood semi-ring-porous; fibres with distinctly bordered pits.
Zygophyllum dumosum
- Wood diffuse-porous; fibres with simple pits. **Reaumuria hirtella**

66 (63). Vessel diameter up to 250 μm or larger 67
- Vessel diameter not exceeding 200 μm 68

67. Rays up to 8 cells wide. **Melia azedarach**
- Rays not more than 4 cells wide. **Calligonum comosum**

68 (66). Rays not taller than 40 cells (less than 750 μm), usually much smaller 69
- Rays up to 1 mm or more in height 70

69. Fibres with distinctly bordered pits in radial and tangential walls. **Fagonia mollis**
- Fibres with simple to minutely bordered pits mainly in radial walls.
Ochradenus baccatus

70 (68). Rays up to 10 cells or more in width. **Ononis natrix**
- Rays up to 6 cells wide 71

71. Vasicentric to confluent parenchyma present. **Colutea cilicica**
Colutea istria
- Paratracheal parenchyma scanty or seemingly absent. **Zilla spinosa**

72 (62). Parenchyma aliform to confluent in 2–5-seriate, more or less tangential bands
73
- Parenchyma not as above 74

73. Vessels exceeding 150 μm in diameter present; very narrow vessels infrequent. **Cordia sinensis**
– Maximum vessel diameter less than 150 μm; numerous very narrow vessels present in addition to wider ones. **Nitraria retusa**
74 (72). Rays up to 6 cells wide 75
– Rays up to 8 cells or more in width 79
75. Vessels with spiral thickenings. **Cercis siliquastrum**
– Spiral thickenings absent or very faint 76
76. Vessel frequency less than 50/mm^2 77
– Vessel frequency exceeding 80/mm^2 78
77. Maximal radial vessel diameter less than 150 μm; rays weakly heterocellular.
Capparis cartilaginea
Capparis decidua
Capparis ovata
Capparis spinosa
Maerua crassifolia
– Maximal radial vessel diameter exceeding 200 μm; rays homocellular.
Calligonum comosum
78 (76). Tangential fibre walls with numerous pits. **Zilla spinosa**
– Tangential fibre walls without or with infrequent pits. **Artemisia arborescens**
79 (74). Multiseriate rays more than 4 cells wide present, many cells high 80
– Rays usually not more than 3 cells wide, less than 10 cells high. **Moringa peregrina**
80. Apotracheal parenchyma abundant, diffuse-in-aggregates. **Balanites aegyptiaca**
– Apotracheal parenchyma not as above. **Tamarix amplexicaulis**
Tamarix chinensis
Tamarix gennessarensis
Tamarix nilotica
Tamarix palaestina
Tamarix parviflora
Tamarix tetragyna
81 (55). Parenchyma distinctly banded or abundant aliform to confluent 82
– Parenchyma not as above 87
82. Rays up to 14 cells wide. **Ficus sycomorus**
– Rays not more than 7 cells wide 83
83. Rays distinctly heterocellular, composed of procumbent, square and upright cells 84
– Rays homocellular or weakly heterocellular, composed of procumbent, sometimes also square cells 85
84. Vessel-ray pits at least partly large and simple. **Ficus carica**
Ficus pseudo-sycomorus
– Vessel-ray pits half-bordered. **Abutilon pannosum**
(**Abutilon fruticosum** — wide parenchyma bands only locally developed)
85 (83). All parenchyma bands less than 12 cells wide. **Citrus aurantium**
– Parenchyma bands up to 15 cells or more in width present 86
86. Vessel frequency less than 14/mm^2; vessel clusters rather infrequent.
Acacia gerrardii
Acacia raddiana
Acacia tortilis

– Vessel frequency exceeding 15/mm^2; vessels often in clusters, including very narrow vessels. **Prosopis farcta**

87 (81). Rays of 2 distinct sizes: narrow rays 1(–6) cells wide and wide rays up to 16 cells or more in width 88

– Rays vary gradually in width, or if there are two sizes, the wide rays are not more than 12 cells wide 91

88. Vessels mostly in radially arranged clusters, some in radial multiples; vasicentric tracheids absent. **Euphorbia hierosolymitana**

– Vessels almost exclusively solitary; vasicentric tracheids present 89

89. Wood diffuse-porous; vasicentric tracheids distinct in cross-section. **Quercus calliprinos**

– Wood ring-porous, sometimes semi-ring-porous, and if diffuse-porous, vasicentric tracheids difficult to distinguish from fibres in cross-section 90

90. Latewood vessels numerous and angular to slightly rounded in cross-section. **Quercus boissieri**

– Latewood vessels infrequent, mostly rounded in cross-section. **Quercus ithaburensis**
Quercus libani

91 (87). Wood ring-porous or semi-ring-porous 92

– Wood diffuse-porous or somewhat semi-ring-porous; if semi-ring-porous, vessels in a dendritic or oblique pattern 120

92. Rays up to 1.7 mm and more in height 93

– Ray height less than 1.6 mm 96

93. Rays 1–3 cells wide. **Cerasus prostrata**

– Rays up to 5 or more cells wide present 94

94. Fibres with distinctly bordered pits. **Rosa arabica**
Rosa canina
Rosa phoenicea
Rosa pulverulenta

– Most fibres with simple to very minutely bordered pits 95

95. Crystals absent; fibres non-septate. **Inula viscosa**

– Crystals present in some of the ray cells; fibres partly septate. **Sarcopoterium spinosum**

96 (92). Rays composed mainly of upright, sometimes also square or weakly procumbent cells 97

– Rays composed mainly of procumbent or procumbent and square cells, sometimes also of upright cells 99

97. Vessels forming a dendritic or flame-like pattern. **Rosmarinus officinalis**

– Vessels not as above 98

98. Fibres thin- to medium thick-walled; vessels mostly in multiples and clusters. **Prasium majus**

– Fibres very thick-walled; vessels mostly solitary. **Cerasus prostrata**

99 (96). At least narrower vessels with distinct spiral thickenings 100

– Vessels without or with very faint spiral thickenings 115

100. Radial resin ducts present. **Pistacia atlantica**
Pistacia khinjuk
Pistacia lentiscus
Pistacia palaestina

100. *(continued)* **Pistacia saportae**
Pistacia vera
- Radial resin ducts absent 101

101. Latewood vessels clustered, tending to form conspicuous tangential to oblique bands 102
- Not as above 104

102. Rays mainly homocellular, composed of procumbent cells. **Ulmus canescens**
- Rays heterocellular 103

103. Ray sheath cells present (often only weakly differentiated). **Celtis australis**
- Sheath cells absent. **Morus alba**

104 (101). Rays not more than 3 cells wide 105
- Rays up to 4 cells or more in width 108

105. Fibre pits distinctly bordered; maximum vessel diameter less than 60 μm. **Cotoneaster nummularia**
- Fibre pits simple; vessels 70 μm in diameter or wider present 106

106. Vessels forming a dendritic pattern. **Lycium europaeum**
- Vessels not as above 107

107. Tangential vessel multiples common in earlywood; vessel diameter less than 110 μm; latewood vessels solitary or in short multiples. **Salvia fruticosa**
- Tangential vessel multiples rare; vessels wider than 120 μm present; very narrow latewood vessels in large clusters. **Rhus coriaria**

108 (104). Maximum vessel diameter exceeding 200 μm 109
- Maximum vessel diameter less than 180 μm 111

109. Inter-vessel pits mostly less than 7 μm in diameter; vessel-ray pits half-bordered. **Melia azedarach**
- Inter-vessel pits mostly exceeding 7 μm in diameter; vessel-ray pits at least partly simple or with strongly reduced borders 110

110. Sheath cells present (often only weakly differentiated). **Celtis australis**
- Sheath cells absent. **Morus alba**

111 (108). Earlywood vessels embedded in vasicentric tracheids 112
- Earlywood not as above 113

112. Parenchyma in 2–6-celled strands. **Armeniaca vulgaris**
- Parenchyma fusiform and in 2-celled strands, with a tendency to be storied. **Elaeagnus angustifolia**

113 (111). Rays more than 7 cells wide present. **Amygdalus communis**
Amygdalus korschinskii
- Rays less than 6 cells wide 114

114. Inter-vessel pits vestured; fibres without spiral thickenings. **Cercis siliquastrum**
- Inter-vessel pits non-vestured; some fibres with spiral thickenings. **Prunus ursina**

115 (99). Rays 5 or fewer cells wide 116
- Rays up to 7 cells or more in width. **Amygdalus communis**
Amygdalus korschinskii

116. Apotracheal parenchyma absent or very infrequent 118
- Apotracheal parenchyma (diffuse and diffuse-in-aggregates) common 117

117. Most vessels solitary, less than 80 μm in diameter. **Eriolobus trilobatus**
- Vessel multiples common; vessels exceeding 100 μm in diameter present. **Juglans regia**

118 (116). Rays homocellular, up to 20 cells high. **Fraxinus syriaca**

– Rays heterocellular, partly more than 30 cells high 119

119. Fibres with distinctly bordered pits in radial and tangential walls. **Cistus creticus**
Cistus salvifolius

– Fibres not as above. **Vitex agnus-castus**
Vitex pseudo-negundo

120 (91). Rays up to 5 mm or more in height present 121

– Rays not higher than 4 mm 122

121. Ray cells mostly procumbent. **Hedera helix**

– Ray cells mostly upright or square. **Pluchea dioscoridis**

122 (120). Rays up to 10 cells or more in width 123

– Rays not more than 8 cells wide 125

123. Vessel-ray pits simple and fairly large; most ray cells strongly procumbent. **Hedera helix**

– Vessel-ray pits half-bordered and small; many or most ray cells square to upright 124

124. Vessel frequency exceeding 100/mm^2; rays, including uniseriates, 5–7/mm. **Sarcopoterium spinosum**

– Vessel frequency up to 70/mm^2; rays 2–3/mm. **Inula crithmoides**
Inula viscosa

125 (122). Vessels forming a dendritic pattern 126

– Vessels not forming a dendritic pattern 131

126. Marginal parenchyma in conspicuous initial bands, often more than 1 cell wide. **Phillyrea latifolia**

– Marginal parenchyma bands usually not more than 1 cell wide or absent 127

127. Most rays uniseriate 128

– Most rays more than 1 cell wide 129

128. Vessels exceeding 120 μm in diameter present. **Lycium europaeum**

– All vessels less than 100 μm in diameter. **Thymelaea hirsuta**

129 (127). Vessels in earlywood much wider than those in latewood **Lycium depressum**

– Vessel diameter more or less uniform throughout growth ring 130

130. Prismatic crystals present in ray cells; vessels with spiral thickenings. **Rhamnus alaternus**
Rhamnus dispermus
Rhamnus palaestinus
Rhamnus punctatus

– Crystals and spiral thickenings absent. **Ruta chalepensis**

131 (125). Many vessels with distinct spiral thickenings 132

– Spiral thickenings in vessels absent or very faint 140

132. Vessels almost exclusively solitary 133

– Vessels not as above 134

133. Rays heterocellular (Kribs' heterogeneous type I), with many upright cells. **Myrtus communis**

– Rays predominantly homocellular; upright cells very rare. **Cydonia oblonga**

134 (132). Most ray cells upright to square, some slightly procumbent. **Teucrium creticum**

– Ray cells not as above 135

135. Vessels in radial multiples and clusters, arranged in more or less tangential rows. **Cercis siliquastrum**

- – Vessels not as above 136
- 136. Resin ducts present in some of the rays; vessels of two distinct size classes. **Pistacia lentiscus** **Pistacia saportae**
- – Ducts absent; vessels not as above 137
- 137. Rays homocellular or only weakly heterocellular; most ray cells strongly procumbent. **Acer hermoneum** **Acer obtusifolium**
- – Rays distinctly heterocellular 138
- 138. Fibres thin- to medium thick-walled; some narrow vessel members with scalariform perforation plates containing 1–6 bars 139
- – Fibres thick- to very thick-walled; all perforation plates simple. **Prunus ursina**
- 139. Rays with uniseriate wings often taller than multiseriate parts. **Lonicera etrusca**
- – Rays usually with a single row of marginal cells or uniseriate wings usually much shorter than multiseriate part. **Arbutus andrachne**
- 140 (131). Fibre walls very thin and lumina wide 141
- – Fibre walls thin, medium thick or thick 144
- 141. Rays storied, less than 10 cells high. **Moringa peregrina**
- – Rays non-storied, partly more than 10 cells high 142
- 142. Maximum vessel diameter less than 120 μm. **Nicotiana glauca**
- – Maximum vessel diameter exceeding 130 μm 143
- 143. Vessel members up to 6 times as long as wide; procumbent ray cells infrequent; vessel-ray pits often large and simple. **Ricinus communis**
- – Vessel members less than 3 times as long as wide; procumbent ray cells fairly common; vessel-ray pits typically half-bordered. **Calotropis procera**
- 144 (140). All or most fibres septate 145
- – All or most fibres non-septate 148
- 145. Vessel frequency less than 100/mm^2 146
- – Vessel frequency exceeding 150/mm^2. **Hedera helix**
- 146. Crystals solitary, prismatic 147
- – Crystals present as small styloids. **Laurus nobilis**
- 147. Paratracheal parenchyma scanty; vessel-ray pits large and simple; inter-vessel pits non-vestured. **Rhus pentaphylla** **Rhus tripartita**
- – Paratracheal parenchyma vasicentric to aliform, rarely scanty; vessel-ray pits half-bordered; inter-vessel pits vestured. **Ceratonia siliqua**
- 148 (144). All ray cells upright or upright and square, rarely procumbent 149
- – Ray cells not as above 157
- 149. All ray cells upright and very elongate; rays hardly distinguishable in tangential section, mostly 1 cell wide and only very few cells high. **Coridothymus capitatus**
- – Rays not as above 150
- 150. Many rays uniseriate 151
- – Most rays 2 or more cells wide 153
- 151. Most vessels in long radial multiples. **Nicotiana glauca**
- – Vessels not as above 152
- 152. Vessel frequency up to 40/mm^2; prismatic crystals present. **Periploca aphylla**
- – Vessel frequency 100–125/mm^2; crystals absent. **Myrtus communis**
- 153 (150). Many rays up to 2 mm or more in height. **Pluchea dioscoridis**

– Rays less than 1.6 mm high 154
154. Vessel diameter up to 170 μm or larger. **Ricinus communis**
– Vessel diameter less than 140 μm 155
155. Fibres thin-walled 156
– Fibres medium thick-walled to thick-walled. **Teucrium creticum**
156. Fibres with distinctly bordered pits. **Nicotiana glauca**
– Fibres with minutely bordered pits. **Withania somnifera**
157 (148). Fibres storied as seen in tangential view. **Artemisia arborescens**
– Fibres non-storied 158
158. Marginal parenchyma present 159
– Marginal parenchyma absent or very scanty 166
159. Vessels exceeding 140 μm in diameter present 160
– Vessels less than 100 μm in diameter 161
160. Parenchyma fusiform or in 2-celled strands; inter-vessel pits vestured, 5–8 μm in diameter. **Prosopis farcta**
– Parenchyma in 4–8-celled strands; inter-vessel pits non-vestured, 7–12 μm in diameter. **Juglans regia**
161 (159). Fibre pits distinctly bordered, numerous in radial and tangential walls. **Fagonia mollis**
– Fibre pits not as above 162
162. Rays 4 or more cells wide present 163
– All rays less than 4 cells wide. **Reaumuria hirtella**
163. Multiples and clusters of up to 15 vessels; fibres thin-walled to medium thick-walled. **Artemisia monosperma**
– Multiples of 2–6 vessels, clusters small; fibres medium thick- to very thick-walled 164
164. Rays tending to be of two size classes. **Abutilon fruticosum**
– Rays not as above 165
165. Solitary prismatic crystals in chambered cells; rays less than 5 cells wide, well delimited. **Grewia villosa**
– Crystals absent; rays more than 6 cells wide present, irregular or compound as seen in tangential section. **Hibiscus micranthus**
166 (158). Numerous conspicuous vasicentric tracheids present. **Hedera helix**
– Vasicentric tracheids absent, or at most vascular tracheids intergrading with narrow vessels in a vasicentric position present 167
167. Vessels often forming a pattern of tangential or oblique bands, usually 1 vessel wide. **Zygophyllum coccineum**
– Vessel arrangement not as above 168
168. Rays composed of procumbent, square and distinctly elongate upright cells 169
– Rays composed of procumbent cells or also of square to weakly upright cells 172
169. Vessels almost exclusively solitary 171
– Many vessels in radial multiples 170
170. Fibres thin-walled; rays hardly distinguishable in cross-section; vessel diameter not exceeding twice the fibre diameter. **Nerium oleander**
– Fibres medium thick- to very thick-walled; rays conspicuous in cross-section, many of the ray cells rounded; diameter of large vessels more than three times that of fibres. **Olea europaea**
171 (169). Parenchyma strands more than 4 cells long common. **Myrtus communis**

\- Parenchyma strands at most 4 cells long. **Cistus creticus**
Cistus salvifolius

172 (168). Apotracheal parenchyma common 173
\- Apotracheal parenchyma sparse or absent 177

173. Maximum vessel diameter exceeding 130 μm. **Juglans regia**
\- Vessel diameter up to 100 μm 174

174. Crystals absent. **Cydonia oblonga**
Pyrus syriaca
\- Solitary crystals present in chambered parenchyma cells 175

175. Crystals present in enlarged parenchyma cells. **Crataegus aronia**
Crataegus azarolus
Crataegus monogyna
Crataegus sinaica
\- Crystalliferous parenchyma cells not enlarged 176

176. Rays mainly heterocellular. **Eriolobus trilobatus**
\- Rays mainly homocellular. **Pyrus syriaca**

177 (172). Vessel frequency not exceeding 80/mm^2; vessel diameter up to 100 μm or larger 178
\- Vessel frequency 100/mm^2 or more; vessel diameter not exceeding 70 μm 179

178. Growth rings very distinct; vessel size more or less uniform; vessel-ray pits simple; styloids present in ray cells. **Laurus nobilis**
\- Growth rings faint to fairly distinct; vessels tending to be of two size classes intermingled; vessel-ray pits half-bordered; crystals, if present, prismatic.
Capparis cartilaginea
Capparis decidua
Capparis ovata
Capparis spinosa
Maerua crassifolia

179 (177). Vessels mainly arranged in multiples of 2–6. **Ruta chalepensis**
\- Vessels mainly solitary. **Pyrus syriaca**

3.3 *Lists of Species of Gnetales and Dicotyledons with Certain Diagnostic Features*

Vessels

([1]=character of doubtful diagnostic value for the taxon listed or vaguely pronounced)

Wood ring-porous to semi-ring-porous

Gnetales
Ephedra alata
Anacardiaceae
Pistacia spp. (except P. lentiscus)
Rhus coriaria

Elaeagnaceae
Elaeagnus angustifolia
Fagaceae
Quercus boissieri
Q. ithaburensis

Q. libani
Labiatae
Prasium majus
Rosmarinus officinalis
Salvia fruticosa[1]
Leguminosae
Calycotome villosa[1]
Colutea spp.[1]
Gonocytisus pterocladus[1]
Spartium junceum
Meliaceae
Melia azedarach
Moraceae
Morus alba
Oleaceae
Fraxinus syriaca
Jasminum fruticans
Polygonaceae
Calligonum comosum
Ranunculaceae
Clematis cirrhosa
Rosaceae
Amygdalus spp.
Armeniaca vulgaris
Cerasus prostrata
Cotoneaster nummularia[1]
Prunus ursina[1]
Rosa arabica
R. phoenicea
R. pulverulenta
Solanaceae
Lycium europaeum
L. schweinfurthii[1]
Styracaceae
Styrax officinalis
Tamaricaceae
Tamarix chinensis[1]
T. gennessarensis[1]
T. nilotica
T. palaestina
T. passerinoides
T. tetragyna
Ulmaceae
Celtis australis
Ulmus canescens
Verbenaceae
Vitex spp.
Vitaceae
Vitis vinifera
Zygophyllaceae
Zygophyllum dumosum[1]

Dendritic pattern

Anacardiaceae
Pistacia khinjuk[1]
P. vera[1]
Rhus coriaria[1]
Fagaceae
Quercus spp.
Labiatae
Rosmarinus officinalis[1]
Leguminosae
Anagyris foetida
Calycotome villosa
Colutea spp.[1]
Genista fasselata
Gonocytisus pterocladus[1]
Retama raetam
Spartium junceum
Meliaceae
Melia azedarach[1]
Oleaceae
Olea europaea[1]
Phillyrea latifolia
Rhamnaceae
Rhamnus spp.
Rutaceae
Ruta chalepensis
Solanaceae
Lycium spp.
Thymelaeaceae
Thymelaea hirsuta

Oblique pattern[1]

Asclepiadaceae
Periploca aphylla
Fagaceae
Quercus boissieri
Q. calliprinos
Q. ithaburensis
Leguminosae
Anagyris foetida
Calycotome villosa
Colutea spp.
Genista fasselata
Gonocytisus pterocladus
Retama raetam
Spartium junceum
Moraceae
Ficus sycomorus
Ulmaceae
Celtis australis
Ulmus canescens
Zygophyllaceae
Balanites aegyptiaca[1]

Tangential bands

Araliaceae
Hedera helix[1]
Labiatae
Teucrium creticum[1]
Moraceae
Morus alba[1]
Ulmaceae
Celtis australis
Ulmus canescens
Zygophyllaceae
Zygophyllum coccineum[1]

*Of two distinct size classes**

Anacardiaceae
Pistacia spp.[1]
Rhus spp.[1]
Capparidaceae
Capparis spp.[1]
Maerua crassifolia[1]
Chenopodiaceae
Aellenia lancifolia
Anabasis spp.
Atriplex halimus
Halogeton alopecuroides
Haloxylon persicum
Hammada spp.
Noaea mucronata
Salsola spp.
Seidlitzia rosmarinus
Suaeda spp.
Compositae
Inula spp.
Labiatae
all spp.[1]
Leguminosae
Anagyris foetida
Calycotome villosa[1]
Colutea spp.
Genista fasselata
Gonocytisus pterocladus
Ononis natrix
Retama raetam
Spartium junceum
Meliaceae
Melia azedarach
Moraceae
Ficus pseudo-sycomorus[1]
Nyctaginaceae
Commicarpus africans
Polygonaceae
Commicarpus africanus
Resedaceae
Ochradenus baccatus[1]
Rosaceae
Sarcopoterium spinosum[1]

* This list does not include species in which the presence of two vessel size classes is due only to ring-porosity.

Salvadoraceae
 Salvadora persica
Solanaceae
 Lycium spp.
Tamaricaceae
 Tamarix passerinoides[1]
Ulmaceae
 Celtis australis[1]
 Ulmus canescens[1]
Vitaceae
 Vitis vinifera
Zygophyllaceae
 Nitraria retusa

Vessels mostly solitary (over 80%)

Gnetales
 Ephedra spp.
Caprifoliaceae
 Lonicera etrusca[1]
 Viburnum tinus
Caryophyllaceae
 Gymnocarpos decandrum
Cistaceae
 Cistus spp.
Fagaceae
 Quercus spp.
Menispermaceae
 Cocculus pendulus
Myrtaceae
 Myrtus communis
Oleaceae
 Jasminum fruticans
Rosaceae
 Amygdalus korschinskii
 Armeniaca vulgaris
 Cerasus prostrata
 Cotoneaster nummularia
 Crataegus spp.
 Cydonia oblonga
 Eriolobus trilobatus
 Pyrus syriaca
 Rosa spp.
Tamaricaceae
 Tamarix spp.
Verbenaceae
 Vitex pseudo-negundo[1]
Zygophyllaceae
 Fagonia mollis
 Zygophyllum spp.[1]

Radial or tangential multiples of 6 or more present

Anacardiaceae
 Pistacea spp.
 Rhus spp.
Asclepiadaceae
 Calotropis procera
Avicenniaceae
 Avicennia marina
Capparidaceae
 Capparis cartilaginea
 C. ovata
 C. spinosa
 Cleome droserifolia
 Maerua crassifolia
Chenopodiaceae
 Aellenia lancifolia
 Atriplex halimus
 Chenolea arabica
 Hammada negevensis
 Salsola baryosma
Compositae
 Artemisia spp.
 Inula viscosa
Cruciferae
 Zilla spinosa
Ericaceae
 Arbutus andrachne
Euphorbiaceae
 Ricinus communis
Lauraceae
 Laurus nobilis
Leguminosae
 Cercis siliquastrum
 Prosopis farcta
Loranthaceae
 Loranthus acaciae
 Viscum cruciatum

Malvaceae
  Abutilon pannosum[1]
Moraceae
  Ficus pseudo-sycomorus
Polygonaceae
  Calligonum comosum
Rutaceae
  Ruta chalepensis
Salvadoraceae
  Salvadora persica
Solanaceae
  Lycium depressum
  Nicotiana glauca
Tamaricaceae
  Reaumuria hirtella
Tiliaceae
  Grewia villosa
Vitaceae
  Vitis vinifera

Perforations exclusively scalariform

Caprifoliaceae
  Viburnum tinus
Styracaceae
  Styrax officinalis

Perforations partly scalariform

Caprifoliaceae
  Lonicera etrusca
Ericaceae
  Arbutus andrachne
Platanaceae
  Platanus orientalis
Vitaceae
  Vitis vinifera

Perforations foraminate (=ephedroid)

Gnetales
  Ephedra spp.

Spiral thickenings on vessel walls

([2]=faint or only in a small part of the vessels)

Aceraceae
  Acer spp.
Anacardiaceae
  Pistacia spp.
  Rhus coriaria
Araliaceae
  Hedera helix[2]
Caprifoliaceae
  Lonicera etrusca
  Viburnum tinus[2]
Chenopodiaceae
  Noaea mucronata[2]
  Salsola tetrandra[2]
Cistaceae
  Cistus spp.[2]
Compositae
  Inula crithmoides[2]
Elaeagnaceae
  Elaeagnus angustifolia
Ericaceae
  Arbutus andrachne
Labiatae
  all spp. (some[2])
Leguminosae
  Acacia gerrardii[2]
  A. raddiana[2]
  Anagyris foetida[2]
  Calycotome villosa
  Cercis siliquastrum
  Colutea spp. ([2] in C. istria)
  Genista fasselata
  Gonocytisus pterocladus

Retama raetam
Spartium junceum
Meliaceae
Melia azedarach
Moraceae
Morus alba
Myrtaceae
Myrtus communis
Oleaceae
Jasminum fruticans
Phillyrea latifolia
Rhamnaceae
Rhamnus spp.[2]
Rosaceae
Amygdalus spp.
Armeniaca vulgaris[2]
Cerasus prostrata
Cotoneaster nummularia
Cydonia oblonga
Eriolobus trilobatus[2]
Prunus ursina
Rosa arabica[2]
R. canina[2]
R. phoenicea
R. pulverulenta
Solanaceae
Lycium spp.
Thymelaeaceae
Thymelaea hirsuta[2]
Ulmaceae
Celtis australis[2]
Ulmus canescens
Vitaceae
Vitis vinifera[2]
Zygophyllaceae
Nitraria retusa[2]

Maximum vessel diameter less than 50 μm

Aceraceae
Acer hermoneum
Chenopodiaceae
Aellenia lancifolia
Anabasis articulata
Arthrocnemum spp.
Chenolea arabica
Hammada scoparia
Noaea mucronata
Salsola baryosma
S. vermiculata
Seidlitzia rosmarinus
Suaeda palaestina
Loranthaceae
Viscum cruciatum
Myrtaceae
Myrtus communis
Oleaceae
Phillyrea latifolia
Rhamnaceae
Rhamnus punctatus
Rosaceae
Cerasus prostrata
Cotoneaster nummularia
Crataegus spp.
Cydonia oblonga
Pyrus syriaca

Maximum vessel diameter more than 200 μm

([3]=rarely so)

Asclepiadaceae
Calotropis procera
Fagaceae
Quercus boissieri
Q. ithaburensis
Juglandaceae
Juglans regia[3]
Leguminosae
Acacia albida
A. gerrardii
A. tortilis[3]
Meliaceae
Melia azedarach
Menispermaceae
Cocculus pendulus
Moraceae
Ficus carica
F. sycomorus

Morus alba
Moringaceae
Moringa peregrina
Oleaceae
Fraxinus syriaca
Tamaricaceae
Tamarix aphylla
Ulmaceae
Celtis australis
Ulmus canescens
Vitaceae
Vitis vinifera

Inter-vessel pits scalariform

([4]=not exclusively so)

Caprifoliaceae
Viburnum tinus[4]
Compositae
Artemisia monosperma[4]
Euphorbiaceae
Euphorbia hierosolymitana[4]
Leguminosae
Colutea istria[4]
Loranthaceae
Loranthus acaciae[4]
Vitaceae
Vitis vinifera

Pits vestured

([5]=weakly so or difficult to distinguish)

Apocynaceae
Nerium oleander
Asclepiadaceae
Calotropis procera
Periploca aphylla
Capparidaceae
Capparis spp.[5]
Cleome droserifolia[5]
Maerua crassifolia
Caprifoliaceae
Lonicera etrusca (?)[5]
Viburnum tinus (?)[5]
Cistaceae
Cistus spp.[5]
Cruciferae
Zilla spinosa
Leguminosae
all spp.
Myrtaceae
Myrtus communis
Punicaceae
Punica granatum
Thymelaeaceae
Thymelaea hirsuta
Zygophyllaceae
Balanites aegyptiaca

Inter-vessel pits not larger than 3 μm

Avicenniaceae
Avicennia marina
Chenopodiaceae
Hammada salicornica
Rutaceae
Citrus aurantium
Tamaricaceae
Tamarix parviflora
Zygophyllaceae
Balanites aegyptiaca
Nitraria retusa
Zygophyllum dumosum

Inter-vessel pits larger than 8 μm

([6]=rarely so)

Aceraceae
Acer hermoneum
Anacardiaceae
Pistacia atlantica

P. khinjuk
Rhus coriaria
Araliaceae
Hedera helix
Caprifoliaceae
Lonicera etrusca
Chenopodiaceae
Atriplex halimus
Compositae
Inula viscosa[6]
Ericaceae
Arbutus andrachne[6]
Euphorbiaceae
Euphorbia hierosolymitana
Ricinus communis
Fagaceae
Quercus boissieri[6]
Juglandaceae
Juglans regia
Leguminosae
Acacia albida
Anagyris foetida
Colutea istria
Ononis natrix
Spartium junceum[6]
Moraceae
Ficus sycomorus
Morus alba
Moringaceae
Moringa peregrina
Platanaceae
Platanus orientalis
Rhamnaceae
Ziziphus spina-christi
Rosaceae
Crataegus azarolus[6]
Salicaceae
Populus euphratica
Salix acmophylla
Solanaceae
Nicotiana glauca
Ulmaceae
Celtis australis
Ulmus canescens

Vessel-ray pits simple (usually large)
([7]=of doubtful diagnostic value)

Anacardiaceae
Pistacia spp. ([7] in P. vera)
Rhus spp. ([7] in R. coriaria)
Elaeagnaceae
Elaeagnus angustifolia[7]
Fagaceae
Quercus spp.
Lauraceae
Laurus nobilis[7]
Loranthaceae
Loranthus acaciae[7]
Viscum cruciatum
Menispermaceae
Cocculus pendulus[7]
Moraceae
Ficus spp.[7]
Morus alba[7]
Moringaceae
Moringa peregrina[7]
Myrtaceae
Myrtus communis[7]
Salicaceae
all spp.
Solanaceae
Nicotiana glauca[7]
Ulmaceae
Celtis australis[7]
Ulmus canescens[7]

Tracheids and Fibres

Vasicentric tracheids (sensu stricto)
([8]=of doubtful diagnostic value)

Asclepiadaceae
Periploca aphylla
Elaeagnaceae
Elaeagnus angustifolia

Fagaceae
 Quercus spp.
Rosaceae
 Armeniaca vulgaris[8]
Zygophyllaceae
 Balanites aegyptiaca
 Fagonia mollis[8]

Fibres with distinctly bordered pits in the radial and tangential walls

([9]=intermediate, or libriform fibres also occur)

Gnetales
 Ephedra spp.
Asclepiadaceae
 Periploca aphylla[9]
Caprifoliaceae
 Lonicera etrusca
 Viburnum tinus
Caryophyllaceae
 Gymnocarpos decandrum
Chenopodiaceae
 Suaeda monoica[9]
Cistaceae
 Cistus spp.
Elaeagnaceae
 Elaeagnus angustifolia[9]
Ericaceae
 Arbutus andrachne[9]
Loranthaceae
 Loranthus acaciae
Menispermaceae
 Cocculus pendulus
Myrtaceae
 Myrtus communis
Oleaceae
 Jasminum fruticans
Platanaceae
 Platanus orientalis
Rosaceae
 Amygdalus spp.
 Armeniaca vulgaris[9]
 Cerasus prostrata
 Cotoneaster nummularia
 Crataegus spp.
 Cydonia oblonga
 Eriolobus trilobatus
 Prunus ursina
 Pyrus syriaca
 Rosa spp.
Thymelaeaceae
 Thymelaea hirsuta[9]
Zygophyllaceae
 Balanites aegyptiaca[9]
 Fagonia mollis
 Zygophyllum spp.

Spiral thickenings on fibre walls

([10]=of doubtful diagnostic value)

Caprifoliaceae
 Lonicera etrusca
 Viburnum tinus
Cistaceae
 Cistus spp.[10]
Elaeagnaceae
 Elaeagnus angustifolia[10]
Ericaceae
 Arbutus andrachne
Oleaceae
 Jasminum fruticans
Rosaceae
 Cerasus prostrata[10]
 Cotoneaster nummularia
 Crataegus spp.[10]
 Cydonia oblonga[10]
 Eriolobus trilobatus[10]
 Prunus ursina
 Rosa spp. (some[10])

Fibres septate

Anacardiaceae
 Rhus spp. (some[10])
Araliaceae
 Hedera helix[10]

Ericaceae
 Arbutus andrachne
Lauraceae
 Laurus nobilis[10]
Leguminosae
 Acacia raddiana[10]
 Ceratonia siliqua
 Prosopis farcta[10]
Oleaceae
 Phillyrea latifolia[10]
Punicaceae
 Punica granatum
Rosaceae
 Sarcopoterium spinosum
Vitaceae
 Vitis vinifera

Parenchyma

Apotracheal diffuse or diffuse-in-aggregates
(i.e., as at least one of the main distribution types)
([11]=vaguely pronounced)

Gnetales
 Ephedra spp.
Aceraceae
 Acer spp.
Anacardiaceae
 Pistacia spp.[11]
Apocynaceae
 Nerium oleander
Asclepiadaceae
 Calotropis procera
 Periploca aphylla
Caprifoliaceae
 Lonicera etrusca[11]
 Viburnum tinus[11]
Caryophyllaceae
 Gymnocarpos decandrum
Cistaceae
 Cistus spp.
Elaeagnaceae
 Elaeagnus angustifolia
Ericaceae
 Arbutus andrachne[11]
Euphorbiaceae
 Euphorbia hierosolymitana
 Ricinus communis
Fagaceae
 Quercus spp.
Juglandaceae
 Juglans regia
Leguminosae
 Gonocytisus pterocladus[11]
 Retama raetam
Loranthaceae
 Loranthus acaciae
 Viscum cruciatum
Menispermaceae
 Cocculus pendulus
Myrtaceae
 Myrtus communis
Oleaceae
 Jasminum fruticans
Platanaceae
 Platanus orientalis
Rosaceae
 Amygdalus spp.[11]
 Armeniaca vulgaris
 Cerasus prostrata[11]
 Cotoneaster nummularia
 Crataegus spp.
 Cydonia oblonga
 Eriolobus trilobatus
 Prunus ursina
 Pyrus syriaca
 Rosa spp. (some[11])
Salicaceae
 Salix alba[11]
Solanaceae
 Lycium spp. (some[11])
 Nicotiana glauca
Styracaceae
 Styrax officinalis
Thymelaeaceae
 Thymelaea hirsuta
Tiliaceae

Grewia villosa[11]
Zygophyllaceae
Balanites aegyptiaca
Fagonia mollis
Zygophyllum spp.

In broad bands (apart from conjunctive parenchyma)
([12]=bands irregular)

Leguminosae
Acacia spp. (some[12])
Prosopis farcta[12]
Malvaceae
Abutilon spp.[12]
Moraceae
Ficus spp.
Rutaceae
Citrus aurantium

Marginal parenchyma present
([13]=of doubtful diagnostic value)

Caryophyllaceae
Gymnocarpos decandrum
Cistaceae
Cistus spp.[13]
Compositae
Artemisia monosperma
Juglandaceae
Juglans regia
Labiatae
Prasium majus
Salvia fruticosa
Leguminosae
Anagyris foetida
Calycotome villosa
Ceratonia siliqua
Cercis siliquastrum
Colutea spp.
Genista fasselata
Gonocytisus pterocladus
Ononis natrix
Prosopis farcta
Retama raetam
Spartium junceum
Malvaceae
Abutilon spp.[13]
Hibiscus micranthus
Meliaceae
Melia azedarach
Oleaceae
Fraxinus syriaca
Jasminum fruticans[13]
Phillyrea latifolia
Polygonaceae
Calligonum comosum
Resedaceae
Ochradenus baccatus
Rhamnaceae
Paliurus spina-christi
Ziziphus spp.
Salicaceae
Populus euphratica[13]
Salix spp. (some[13])
Solanaceae
Lycium schweinfurthii
L. shawii
Tamaricaceae
Reaumuria hirtella
Tiliaceae
Grewia villosa
Zygophyllaceae
Fagonia mollis
Zygophyllum dumosum

Vasicentric to confluent
([14]=also scanty paratracheal)

Anacardiaceae
Rhus coriaria[14]
Boraginaceae
Cordia sinensis

Capparidaceae
- Capparis spp. (some[14])
- Cleome droserifolia[14]

Compositae
- Inula crithmoides[14]
- Pluchea dioscoridis

Juglandaceae
- Juglans regia

Leguminosae
- Acacia gerrardii
- A. raddiana
- A. tortilis
- Anagyris foetida[14]
- Calycotome villosa
- Ceratonia siliqua[14]
- Cercis siliquastrum[14]
- Colutea spp. ([14] in C. istria)
- Genista fasselata
- Gonocytisus pterocladus
- Ononis natrix[14]
- Prosopis farcta
- Retama raetam[14]
- Spartium junceum

Loranthaceae
- Loranthus acaciae

Malvaceae
- Abutilon spp.
- Hibiscus micranthus[14]

Meliaceae
- Melia azedarach

Moraceae
- Morus alba

Moringaceae
- Moringa peregrina

Oleaceae
- Fraxinus syriaca
- Olea europaea

Polygonaceae
- Calligonum comosum[14]

Resedaceae
- Ochradenus baccatus[14]

Rhamnaceae
- Paliurus spina-christi
- Ziziphus spp.

Rutaceae
- Citrus aurantium

Solanaceae
- Lycium depressum[14]
- L. europaeum

Tamaricaceae
- Tamarix spp.

Tiliaceae
- Grewia villosa[14]

Ulmaceae
- Celtis australis
- Ulmus canescens

Verbenaceae
- Vitex spp.[14]

Zygophyllaceae
- Nitraria retusa

Fusiform or in 2-celled strands

([15]=occasionally in longer strands)

Anacardiaceae
- Pistacia atlantica
- P. khinjuk

Capparidaceae
- Cleome droserifolia

Chenopodiaceae
- all spp. (rarely[15])

Cistaceae
- Cistus spp. ([15] in C. creticus)

Compositae
- Artemisia spp. ([15] in A. arborescens)
- Inula viscosa

Cruciferae
- Zilla spinosa

Elaeagnaceae
- Elaeagnus angustifolia

Labiatae
- Prasium majus
- Rosmarinus officinalis
- Salvia fruticosa

Leguminosae
- Acacia spp. ([15] in A. albida)
- Anagyris foetida
- Calycotome villosa
- Colutea spp.
- Genista fasselata

Gonocytisus pterocladus
Ononis natrix[15]
Prosopis farcta[15]
Retama raetam
Spartium junceum[15]
Loranthaceae
Loranthus acaciae
Viscum cruciatum
Malvaceae
Hibiscus micranthus[15]
Menispermaceae
Cocculus pendulus[15]
Ranunculaceae
Clematis cirrhosa
Resedaceae
Ochradenus baccatus
Rosaceae
Prunus ursina[15]
Rosa pulverulenta
Sarcopoterium spinosum
Rutaceae
Ruta chalepensis
Salvadoraceae
Salvadora persica
Tamaricaceae
Reaumuria hirtella
Tamarix spp.
Thymelaeaceae
Thymelaea hirsuta
Zygophyllaceae
Balanites aegyptiaca
Fagonia mollis
Nitraria retusa
Zygophyllum spp.

Included Phloem

Included phloem present

Avicenniaceae
Avicennia marina
Capparidaceae
Cleome droserifolia
Chenopodiaceae
all spp.
Menispermaceae
Cocculus pendulus
Nyctaginaceae
Commicarpus africanus
Salvadoraceae
Salvadora persica

Conjunctive parenchyma including sclerified layers

Avicenniaceae
Avicennia marina
Capparidaceae
Cleome droserifolia
Menispermaceae
Cocculus pendulus

Conjunctive parenchyma in continuous bands
([16]=vaguely pronounced)

Avicenniaceae
Avicennia marina
Capparidaceae
Cleome droserifolia
Chenopodiaceae
Aellenia lancifolia
Anabasis articulata
A. syriaca[16]
Arthrocnemum macrostachyum
Halogeton alopecuroides[16]
Noaea mucronata[16]
Salsola spp. ([16] in S. tetrandra)
Seidlitzia rosmarinus
Suaeda spp. ([16] in S. palaestina)
Menispermaceae
Cocculus pendulus

Nyctaginaceae
Commicarpus africanus[16]

Salvadoraceae
Salavadora persica

Rays

Rays absent
([17]=rays very poorly delimited)

Capparidaceae
Cleome droserifolia[17]
Chenopodiaceae
Anabasis syriaca[17]
Arthrocnemum spp.
Atriplex halimus[17]
Chenolea arabica[17]
Halogeton alopecuroides
Hammada spp.
Noaea mucronata
Salsola baryosma[17]
S. vermiculata[17]
Seidlitzia rosmarinus[17]
Suaeda spp. ([17] in some)
Labiatae
Coridothymus capitatus[17] (in tangential section)
Nyctaginaceae
Commicarpus africanus[17] (in cross-section)

Exclusively uniseriate (or almost so)
([18]=occasionally wider)

Caryophyllaceae
Gymnocarpos decandrum
Chenopodiaceae
Anabasis articulata
Labiatae
Coridothymus capitatus[18]
Leguminosae
Acacia albida
Oleaceae
Jasminum fruticans[18]
Punicaceae
Punica granatum
Rhamnaceae
Paliurus spina-christi
Ziziphus spp.
Salicaceae
all spp.
Solanaceae
Lycium europaeum[18]
L. schweinfurthii
L. shawii
Thymelaeaceae
Thymelaea hirsuta[18]

Maximum ray width 10 or more cells
([19]=rarely exceeding 10)

Gnetales
Ephedra campylopoda
Araliaceae
Hedera helix
Chenopodiaceae
Aellenia lancifolia
Atriplex halimus[19]
Haloxylon persicum
Seidlitzia rosmarinus
Suaeda fruticosa[19]
Compositae
Inula spp.
Euphorbiaceae
Euphorbia hierosolymitana
Fagaceae
Quercus spp.
Leguminosae
Ononis natrix

Loranthaceae
Loranthus acaciae
Menispermaceae
Cocculus pendulus
Moraceae
Ficus sycomorus
Morus alba
Platanaceae
Platanus orientalis
Ranunculaceae
Clematis cirrhosa
Rosaceae
Armeniaca vulgaris
Rosa phoenicea[19]
Sarcopoterium spinosum
Tamaricaceae
Tamarix amplexicaulis
T. aphylla
T. chinensis
T. nilotica
T. parviflora
T. passerinoides
T. tetragyna
Ulmaceae
Celtis australis[19]
Vitaceae
Vitis vinifera
Zygophyllaceae
Balanites aegyptiaca

Rays not more than 10 cells high

Caryophyllaceae
Gymnocarpos decandrum
Chenopodiaceae
Anabasis articulata
A. setifera
Salsola tetrandra
S. vermiculata
Labiatae
Coridothymus capitatus
Moringaceae
Moringa peregrina

Rays homocellular (procumbent cells only)

([20]=transitions to Kribs' heterogeneous type III)

Aceraceae
Acer spp. ([20] in A. hermoneum)
Araliaceae
Hedera helix[20]
Elaeagnaceae
Elaeagnus angustifolia
Fagaceae
Quercus spp.
Juglandaceae
Juglans regia[20]
Leguminosae
Acacia spp.
Cercis siliquastrum
Genista fasselata[20]
Gonocytisus pterocladus[20]
Prosopis farcta[20]
Spartium junceum[20]
Meliaceae
Melia azedarach[20]
Moringaceae
Moringa peregrina[20]
Oleaceae
Fraxinus syriaca
Platanaceae
Platanus orientalis[20]
Polygonaceae
Calligonum comosum
Rhamnaceae
Rhamnus palaestinus[20]
R. punctatus[20]
Rosaceae
Amygdalus spp.[20]
Crataegus aronia[20]
C. azarolus[20]
C. sinaica[20]
Cydonia oblonga[20]
Eriolobus trilobatus[20]
Pyrus syriaca[20]

Rutaceae
Citrus aurantium[20]
Salicaceae
Populus euphratica
Ulmaceae
Ulmus canescens
Zygophyllaceae
Zygophyllum dumosum

Storied Structure

Present in: f=fibres; p=parenchyma; ph=included phloem elements; r=rays; t=tracheids; v=vessel members
([21]=weakly storied)

Capparidaceae
Capparis cartilaginea[21] p
C. decidua p
Chenopodiaceae
Aellenia lancifolia p
Anabasis spp. p, v, ph ([21] in A. setifera)
Arthrocnemum spp. p, v, ph
Atriplex halimus p, v, ph
Chenolea arabica[21] v
Hammada spp. p, v, ph
Noaea mucronata p, v, ph
Salsola spp. p, v, ph
Seidlitzia rosmarinus p, v, ph
Suaeda spp. p, v, ph
Compositae
Artemisia arborescens[21] p, v, f
Cruciferae
Zilla spinosa p, v
Leguminosae
Acacia albida p, r[21]
Anagyris foetida p, v
Calycotome villosa p, v
Cercis siliquastrum[21] p, r
Colutea spp. p, v
Genista fasselata p, v
Gonocytisus pterocladus p, v
Ononis natrix[21] p, v
Retama raetam p, v
Spartium junceum p, v
Loranthaceae
Loranthus acaciae p, v
Viscum cruciatum p, v
Meliaceae
Melia azedarach[21] p
Moringaceae
Moringa peregrina p, f, r
Nyctaginaceae
Commicarpus africanus p, ph, v
Polygonaceae
Calligonum comosum p, v
Ranunculaceae
Clematis cirrhosa p, v, f
Resedaceae
Ochradenus baccatus p, v, f
Salvadoraceae
Salvadora persica p, v, ph
Tamaricaceae
Reaumuria hirtella[21] p, v
Tamarix spp. p, v
Zygophyllaceae
Balanites aegyptiaca p, v
Nitraria retusa p, v
Zygophyllum dumosum p, v, f

Secretory Structures

Laticifers

Asclepiadaceae
Calotropis procera
Euphorbiaceae
Euphorbia hierosolymitana
Moraceae
Ficus pseudo-sycomorus
F. sycomorus

Resin or gum ducts: r=radial; v=vertical

([22]=occasionally present, probably traumatic)

Anacardiaceae
- Pistacia spp. r
- Rhus pentaphylla[22] r

Rosaceae
- all spp.[22] (probably)

Rutaceae
- Citrus aurantium[22]

Zygophyllaceae
- Balanites aegyptiaca[22]

Crystals

c=chambered; f=in fibres; p=in parenchyma; r=in ray cells;
ty=in vessel members in tyloses

([23]=infrequent or sometimes absent)

Prismatic

Aceraceae
- Acer hermoneum cp, f
- A. obtusifolium cp, r

Anacardiaceae
- Pistacia atlantica r, ty
- P. khinjuk r
- P. lentiscus[23] r
- P. palaestina r
- P. × saportae[23] r
- P. vera[23] r
- Rhus coriaria r
- R. pentaphylla r
- R. tripartita r, p

Asclepiadaceae
- Periploca aphylla cr, p

Boraginaceae
- Cordia sinensis r, p[23]

Capparidaceae
- Capparis decidua r (water soluble)
- C. spinosa r (water soluble)
- Maerua crassifolia r (water soluble)

Chenopodiaceae
- Aellenia lancifolia r, p
- Anabasis articulata p, f, r
- A. syriaca p, r
- Atriplex halimus p
- Chenolea arabica p
- Halogeton alopecuroides cp
- Haloxylon persicum r, f, p
- Hammada negevensis[23] cp, f
- H. salicornica cp, f
- H. scoparia[23] p
- Salsola baryosma cp
- S. tetrandra p, r, cf
- S. vermiculata cf, cp
- Seidlitzia rosmarinus p, r, f
- Suaeda asphaltica p
- S. fruticosa c[23]p
- S. monoica cp, f[23]
- S. palaestina cp

Cistaceae
- Cistus salvifolius[23] r

Compositae
- Pluchea dioscoridis[23] r

Ericaceae
- Arbutus andrachne cp, cr, cf[23]

Euphorbiaceae
- Euphorbia hierosolymitana cr
- Ricinus communis c[23] r

Fagaceae
- Quercus spp. c ([23] in some) p, (c)r

Juglandaceae
- Juglans regia[23] r

Leguminosae
- Acacia spp. cp (also cr[23] in A. albida)
- Ceratonia siliqua cp, r[23]
- Cercis siliquastrum cp
- Ononis natrix c[23]p
- Prosopis farcta cp

Loranthaceae
- Loranthus acaciae r
- Viscum cruciatum r

Malvaceae
- Abutilon fruticosum p, r

A. pannosum[23] r
Meliaceae
Melia azedarach cp
Moraceae
Ficus carica cp, r
F. pseudo-sycomorus c[23]p, r
F. sycomorus r, p
Morus alba[23] r
Moringaceae
Moringa peregrina r
Oleaceae
Phillyrea latifolia r, p
Platanaceae
Platanus orientalis cr
Punicaceae
Punica granatum cf
Rhamnaceae
Paliurus spina-christi r
Rhamnus spp. r (also cf in R. palaestinus)
Ziziphus lotus r
Z. spina-christi r, p[23]
Rosaceae
Cotoneaster nummularia cp
Crataegus spp. cp ([23] in C. monogyna)
Eriolobus trilobatus cp
Pyrus syriaca[23] cp
Rosa arabica r
R. canina r
R. phoenicea r
Sarcopoterium spinosum c[23]r
Rutaceae
Citrus aurantium cp
Salicaceae
Salix alba[23] cf, cp
Salvadoraceae
Salvadora persica r
Styracaceae
Styrax officinalis cp
Tamaricaceae
Tamarix amplexicaulis r
T. aphylla r
T. nilotica[23] r
T. passerinoides r
Tiliaceae
Grewia villosa cp, cr
Ulmaceae
Celtis australis r
Ulmus canescens p
Zygophyllaceae
Balanites aegyptiaca cr, cp
Nitraria retusa cp, r, cf
Zygophyllum dumosum p

Crystal sand

Boraginaceae
Cordia sinensis[23] r, p
Chenopodiaceae
Aellenia lancifolia[23] r, p
Anabasis syriaca[23] r, p
Chenolea arabica[23] p
Solanaceae
Lycium europaeum p
L. schweinfurthii p
L. shawii p

Druses or clusters

Chenopodiaceae
Hammada scoparia[23] p (sphaerocrystals)
Euphorbiaceae
Ricinus communis[23] r
Rhamnaceae
Rhamnus spp. r
Rosaceae
Amygdalus spp.[23] r
Armeniaca vulgaris[23] r
Prunus ursina r, cp

Raphides

Nyctaginaceae
Commicarpus africanus p
Vitaceae
Vitis vinifera r

Minute crystals (as main type)

Avicenniaceae
Avicennia marina r
Chenopodiaceae
Salsola tetrandra[23] r, p, cf
Labiatae
Prasium majus r, p (in masses)
Menispermaceae
Cocculus pendulus[23] r
Oleaceae
Olea europaea r
Phillyrea latifolia r, p

Acicular crystals

Oleaceae
Olea europaea r

Styloids

Lauraceae
Laurus nobilis r

Crystals of various shapes and sizes, often in aggregates

Avicenniaceae
Avicennia marina r
Chenopodiaceae
Aellenia lancifolia r, p
Anabasis syriaca p, r
Arthrocnemum macrostachyum p
Chenolea arabica p
Haloxylon persicum r, p, f
Hammada negevensis p, f
H. salicornica cp, f
H. scoparia p
Salsola baryosma p, cf
S. vermiculata cp, cf
Seidlitzia rosmarinus p, r, f
Suaeda asphaltica cp
S. fruticosa cp, p, r
S. palaestina cp
Ericaceae
Arbutus andrachne cp, cr, cf
Euphorbiaceae
Euphorbia hierosolymitana cr
Malvaceae
Abutilon fruticosum r
Menispermaceae
Cocculus pendulus r
Rosaceae
Rosa pulverulenta r
Zygophyllaceae
Nitraria retusa cp, r, f

4. WOOD ANATOMICAL DESCRIPTIONS

4.1 *Vessel-less Woods*

CUPRESSACEAE

Cupressus sempervirens L.
Plate 1; Table 4. 1

Var. *horizontalis* (Mill.) Gordon. Tree; Upper Galilee, Lebanon, Gilead, Edom, and in cultivation in the Mediterranean region of Israel.
Var. *pyramidalis* (Targ.-Tozz.) Nym. Tree; in cultivation in the Mediterranean region of Israel.

Growth rings distinct. Transition from early- to latewood gradual. Tracheids 1,700 (700–2,800) μm long, with uniseriate (rarely biseriate) bordered pits confined to the radial walls, but also common in the tangential walls in the latewood near the growth ring boundary. Pit apertures in earlywood mainly circular in var. *horizontalis,* elliptic to spindle-shaped in var. *pyramidalis.* Latewood pits with spindle-shaped apertures in both varieties. Faint crassulae somctimes present.
Axial parenchyma diffuse or in tangential bands, with smooth to faintly nodular horizontal end walls.
Rays 18–40/mm^2 in mature stemwood, 40–55/mm^2 in branchwood, mostly uniseriate, rarely biseriate, (1)3–20(40) cells high in stemwood, 1–8(16) cells high in branchwood, entirely composed of fairly thick-walled parenchyma cells with smooth to very faintly nodular end walls. Cross-field pits 2–4 per field, cupressoid with elliptic to slit-like, included apertures in both early- and latewood.
Resin ducts absent.

Archaeological records: Identified in 14 locations throughout Israel and Sinai; periods ranging from the late Bronze Age to 16th century A. D. (A. Fahn; E. Werker; N. Liphschitz & Y. Waisel).

Juniperus L.
Plate 2; Table 4. 1

J. drupacea Labill. — Shrub; Mt. Hermon.
J. excelsa Bieb. — Tree; Mt. Hermon.

J. oxycedrus L. — Shrub or tree; Upper Galilee (rare).
J. phoenicea L. — Shrub or tree; Edom and Sinai.

Growth rings distinct. Transition from early- to latewood gradual. Tracheids 1,760 (760–2,700) μm long in *J. phoenicea*; 1,150 (710–1,440) μm long in *J. drupacea* (stemwood); 810 (490–1,140) μm long in *J. excelsa* (in thick branchwood); 1,450 (700–1,700) μm long in *J. oxycedrus* (branchwood), with uniseriate bordered pits confined to the radial walls in the earlywood, but also common in the tangential walls of the latewood. Pit apertures round to elliptic, rarely slit-like in the latewood.
Axial parenchyma diffuse or in short tangential bands (especially in the first three species) throughout the growth ring, with pitted to smooth horizontal end walls.
Rays of *J. oxycedrus* and *J. phoenicea* 55–85/mm² and of *J. drupacea* 90–100/mm² in mature stemwood, 90–115/mm² in branchwood, mostly uniseriate, rarely biseriate, 1–8 (14) cells high in stemwood, only slightly lower in branchwood, entirely composed of fairly thick-walled parenchyma cells with nodular end walls. Cross-field pits 2–4 per field in *J. oxycedrus* and *J. phoenicea* and 1–3 (4) in *J. drupacea* and *J. excelsa,* cupressoid with elliptic to slit-like included apertures in both earlywood and latewood.
Resin ducts absent.

Notes: Within the genus *Juniperus* the species studied are all very similar. The small quantitative differences recorded in Table 4.1 are based on the study of very few samples and are probably not of diagnostic value. *Juniperus* is wood anatomically very similar to *Cupressus,* especially if branchwood or juvenile stemwood is concerned. Ray height and frequency give a very good separation of mature stemwood; for branchwood only ray frequency appears to offer a reliable means for separation. Other characters traditionally used to separate the two genera, such as the occurrence of intercellular spaces between the tracheids (Jacquiot, 1955), curvature or degree of beading of the ray parenchyma end walls, vary somewhat within both genera.

Archaeological records: Identified in Timna — 4th millennium B. C. (E. Werker); Masada — Herod's palace (A. Fahn). *Juniperus* sp., Jerusalem — Second Temple (A. Fahn).

Table 4.1
Ray height and frequency in *Cupressus* and *Juniperus.*

	Stemwood		Branchwood	
	number/mm²	height	number/mm²	height
Cupressus sempervirens				
var. *horizontalis*	22–40	1–over 20	40–55	1–8(16)
var. *pyramidalis*	18–35	1–over 20	45	1–8
Juniperus				
J. drupacea	90–100	1–8(13)		
J. excelsa	100	1–7(12)		
J. oxycedrus	55–85	1–8(14)		
J. phoenicea	70–85	1–8(10)	100	1–6

Wood Anatomical Descriptions

PINACEAE

Cedrus libani Loud.

Figs. 1D and 2B; Plate 3

Tree; mountains of Lebanon, Turkey, Cyprus and N. W. Africa and in cultivation in the Mediterranean region of Israel.

Growth rings distinct. Transition from early- to latewood gradual, sometimes abrupt. Tracheids 1,600 (940–2,230) μm long, with uniseriate (very rarely biseriate, alternate) bordered pits with fringed torus margins confined to the radial walls, except for some tangential pits on and near the growth ring boundary in the latewood. Pit apertures circular to elliptic in the earlywood, slit-like (included to extended) in the latewood. Faint crassulae sometimes present.
Axial parenchyma with pitted horizontal walls confined to the growth ring boundary.
Rays ca. 25/mm^2, mostly uniseriate, rarely biseriate, (1)3–20(35) cells high, composed of marginal ray tracheids in single rows, and fairly thick-walled ray parenchyma cells with nodular end walls. Ray tracheids with smooth walls. Cross-field pits with 2–4 taxodioid pits per field in the earlywood (piceoid in the latewood).
Resin ducts absent or of the traumatic type and close to each other, with mostly thick-walled, lignified epithelial cells. Axial ducts more common than horizontal ducts.

Archaeological records: Wood remnants identified in 12 locations throughout Israel and Sinai; periods ranging from the late Bronze Age to 9th century A.D. (N. Liphschitz & Y. Waisel). Coffin, Ein Gedi, 1st century B.C. (A. Fahn).

Pinus L.

Plate 4

P. halepensis Mill. — Tree; hills and mountains; coast of Galilee, Upper Galilee, Mt. Carmel, Samaria, Judean Mts., Upper Jordan Valley, Gilead, Ammon.
P. pinea L. — Tree; Mediterranean; cultivated in Israel.

Growth rings distinct, very rarely vague. Transition from early- to latewood usually abrupt, sometimes gradual. Tracheids 2,450 (1,270–3,870) μm long, with uniseriate bordered pits with round apertures in the earlywood and elliptic to slit-like apertures in the latewood, confined to the radial walls, except for some very rare tangential pits in the latest formed latewood. Crassulae often present. Trabeculae occasionally observed.
Axial parenchyma restricted to resin ducts.
Rays 30–40/mm^2, except for fusiform ones, uniseriate, very rarely biseriate, (1)3–16(22) cells high, composed of marginal (rarely also intercalated) tracheids in one or several rows and fairly thick-walled parenchyma cells with nodular end walls. Walls of ray tracheids sometimes weakly dentate. Cross-field pits 1–3(4)

per field, pinoid, usually with fairly large, lop-sided borders, ca. 10 μm in diameter in *P. halepensis,* and smaller, up to ca. 7 μm in *P. pinea.*
Axial resin ducts with thin-walled epithelial cells common in transition zone or latewood. Horizontal ducts much narrower, common in fusiform rays.

Archaeological records: *P. halepensis* identified in 10 locations throughout Israel: periods from the late Bronze Age to 8th century A. D. (N. Liphschitz & Y. Waisel; E. Werker; A. Fahn & E. Zamski).

4.2 *Vessel-bearing Woods*

Explanatory Note

The descriptions of vessel-bearing woods follow a consistent sequence in which growth rings, vessels, fibres, axial parenchyma, rays and crystals are always treated. Unusual features such as included phloem and secretory structures are only described if present. Also for other features like vasicentric tracheids and vascular tracheids, vestured pits, presence of spiral thickenings, septate and gelatinous fibres, and storied structure only positive character states are included in the descriptions, and if these features are not mentioned at all, this implies that they are absent from the wood involved.
For each species a full description of the wood is given. If a genus is represented in Israel and adjacent regions by more than one species, a summary of common characters is usually given (not for the Chenopodiaceae) as well as an indication of the wood anatomical separation of species (as far as possible). For those families of which more than one genus occurs in the flora of Israel, and for some other families, there are short introductory remarks on their shared wood anatomical characters, often comparing our results with data summarized by Metcalfe & Chalk (1950). Where appropriate, these comparisons are complemented with taxonomic notes.
No attempt has been made to give detailed comparisons with wood anatomical accounts of the same or related species in the literature. At the end of the book there is a separate list of selected references concerning the species treated.
The archaeological materials of the region abound in wooden artefacts and charcoals (*inter alia*, firewood), and when available, records of wood anatomical identifications are cited under the species, genus, or family concerned. Only few of the records are accessible in the published literature; many of them are from internal reports or private communications, and in this case the source of information is acknowledged by listing the names of the investigators only.

GYMNOSPERMAE

GNETALES, EPHEDRACEAE

Ephedra alata Decne.

Plate 5A; Fig. 4C

Shrub; stony desert; S. Negev; Saharo-Arabian; very rare.

Growth rings distinct. Wood ring-porous. Vessels virtually confined to narrow earlywood zone, mostly solitary, sometimes in very short tangential multiples; angular to rounded in cross-section, diameter 30–70 μm, walls 2–4 μm thick. Vessel member length 790(400–1,130) μm. Perforations foraminate in very oblique end walls. Vessel-tracheid pits diffuse, circular, 10–14 μm in diameter, with rounded to slit-like apertures. Vessel-parenchyma and vessel-ray pits similar but half-bordered.
Tracheids 710(420–1,240) μm long, with bordered pits on both radial and tangential walls, much smaller than vessel-tracheid pits and with slit-like apertures.
Parenchyma very sparse, scanty paratracheal and diffuse apotracheal, in few-celled strands.
Rays 2–4/mm; 1–5(7)-seriate, of various heights up to ca. 1.5 mm, sometimes compound; heterocellular, consisting of procumbent, square, irregular and upright cells, the latter often as sheath cells.
Crystals not observed.

Ephedra aphylla Forssk. (*E. alte* C. A. Mey.)*

Shrub; steppes and deserts, also in maquis and on hedges of the coastal sandy soil belt; Sharon Plain, Philistean Plain, Judean Desert, Negev, Upper and Lower Jordan Valley, Dead Sea area, Arava Valley, Moav, Edom; Saharo-Arabian element.

Growth rings distinct. Vessels diffuse, 80–120/mm^2, diminishing in number in latewood, typically solitary and seemingly in tangential or diagonal pairs due to overlapping of end walls; angular in cross-section, diameter 25–55 μm, walls ca. 3 μm thick. Vessel member length 940(650–1,200) μm. Perforations foraminate in very oblique end walls. Vessel-tracheid pits diffuse, circular, 12 μm in diameter, with rounded to slit-like apertures. Vessel-ray pits smaller and half-bordered.
Tracheids 880(430–1,130) μm long, with bordered pits in both radial and tangential walls, much smaller than vessel-tracheid pits, with oblique slit-like apertures.
Parenchyma sparse, very scanty paratracheal and diffuse apotracheal, in few-celled strands.

*Nomenclature follows Danin & Hedge, 1973.

Rays 3–6/mm; 1–6(9)-seriate, up to 3.5 mm or more in height, often compound; heterocellular, consisting of procumbent, square and upright cells, the latter often as sheath cells.
Crystals not observed.

Ephedra campylopoda C. A. Mey.
Plate 5B and C

Shrub; hills, stony ground; mostly in batha and garigue or on bare rocks; Acco Plain, Upper and Lower Galilee, Mt. Carmel, Esdraelon Plain, Samaria, Shefela, Judean Mts., Moav, Edom; East Mediterranean element.

Growth rings distinct. Vessels diffuse, ca. 200/mm², diminishing in number in latewood, typically solitary and seemingly in tangential or diagonal pairs due to overlapping of end walls; angular in cross-section, diameter 30–70 μm, walls ca. 3 μm thick. Vessel member length 710(410–950) μm. Perforations foraminate in very oblique end walls. Vessel-tracheid pits diffuse, circular, ca. 12 μm in diameter, with rounded to slit-like apertures. Vessel-ray pits smaller and half-bordered.
Tracheids 710(300–1,000) μm long, with bordered pits in both radial and tangential walls, much smaller than vessel-tracheid pits, with oblique slit-like apertures.
Parenchyma sparse, very scanty paratracheal and diffuse apotracheal in few-celled strands.
Rays 2–5/mm; 1–10-seriate, up to about 1 cm high, some compound; heterocellular, consisting of procumbent, square and upright cells, the latter often as sheath cells.
Crystals not observed.

Ephedra foliata Boiss. et Ky.
Plate 5D

Shrub; deserts or semi-deserts, rocks; Lower Arava Valley, S. Negev, Sinai; Sudanian element.

Growth rings distinct. Vessels diffuse, 70–100/mm², solitary, sometimes in tangential, rarely diagonal, multiples of 2(4); angular in cross-section, diameter 20–65 μm, walls ca. 4 μm thick. Vessel member length 980 (750–1,240) μm. Perforations foraminate in very oblique end walls. Vessel-tracheid pits diffuse, circular, ca. 12 μm in diameter, with rounded to slit-like apertures. Vessel-ray pits smaller and half-bordered.
Tracheids 970(580–1,430) μm long, with bordered pits in both radial and tangential walls, much smaller than vessel-tracheid pits, with oblique slit-like apertures.
Parenchyma sparse, very scanty paratracheal and diffuse apotracheal, in few-celled strands.
Rays 4–6/mm, 1–5-seriate, up to 2 mm in height, but usually much lower, some

compound; heterocellular, consisting of upright, square and slightly procumbent cells mixed together.
Crystals not observed.

Characteristics of the Genus Ephedra

Growth rings distinct. Vessels mostly solitary. Perforations foraminate in very oblique end walls. Vessel-tracheid pits diffuse, 10 μm or more in diameter. Tracheids with bordered pits much smaller than vessel-tracheid pits in both radial and tangential walls, with slit-like apertures. Rays uni- to multiseriate, including tall ones over 1.5 mm; heterocellular, consisting of procumbent, square and upright cells, the latter often as sheath cells.

Separation of Ephedra Species

Table 4.2 summarizes some of the mainly quantitative differences found between the species, based on the limited material studied. Apart from the ring-porosity of *E. alata,* there are no other characters to separate species reliably using wood anatomical characters only, without further tests of the variability within the species.

Table 4.2
Differences between *Ephedra* species.

Species	Wood porosity	Max. ray height in mm	Max. vessel diameter in μm	Vessel frequency per mm^2
alata	ring	1.5	70	largely confined to early-wood
aphylla	diffuse	3.5	55	80–120
campylopoda	diffuse-ring	10	70	200
foliata	diffuse	2	65	70–100

ANGIOSPERMAE

ACERACEAE

Acer hermoneum (Bornm.) Schwer.
Plate 6A–C

Shrub; deciduous; Mt. Hermon, 700–1,400 m; East Mediterranean element.

Growth rings distinct. Vessels diffuse, ca. 60/mm², solitary (ca. 25–30%) and in radial multiples of up to 4(5) and small clusters; rounded to somewhat angular in cross-section, diameter 15–50 μm, walls 2–3 μm thick. Vessel member length 230(170–330) μm. Perforations simple in oblique end walls. Inter-vessel pits alternate, rounded to polygonal, 4–8 μm in diameter, with slit-like apertures. Vessel-parenchyma and vessel-ray pits similar but half-bordered. Vessel walls with prominent spiral thickenings.
Fibres 530(310–670) μm long, medium thick- to thick-walled, with simple pits more numerous in radial than in tangential walls; some chambered crystalliferous.
Parenchyma sparse, scanty paratracheal and diffuse apotracheal, in 2–3-celled strands; often chambered crystalliferous.
Rays 8–13/mm, 1–4-seriate, up to ca. 30 cells high; weakly heterocellular to homocellular, largely composed of procumbent cells with weakly procumbent, sometimes square cells in the margins.
Crystals solitary, prismatic, in chambered parenchyma cells and fibres.

Acer obtusifolium ssp. **syriacum** (Boiss. et Gaill.) Holmboe
Plate 6D

Tree or shrub; evergreen; in more or less humid maquis, usually on rocky ground; Upper Galilee, Lebanon and Syria; East Mediterranean element.

Growth rings distinct. Vessels diffuse, ca. 100/mm², solitary (40–50%) and in radial multiples of 2–5, up to 16 in latewood, very rarely in small clusters, rounded to somewhat angular in cross-section, tangential diameter 25–60 μm, radial diameter up to 80 μm, walls 2–4 μm thick. Vessel member length 280(230–340) μm. Perforations simple in oblique end walls. Inter-vessel pits alternate, polygonal, ca. 9 μm in diameter, with slit-like apertures. Vessel-parenchyma and vessel-ray pits more rounded and half-bordered. Vessel walls with prominent spiral thickenings.
Fibres 730(510–920) μm long, medium thick-walled, with numerous simple pits mainly in radial walls; some fibres gelatinous.
Parenchyma sparse, scanty paratracheal and diffuse apotracheal, in 2–4-celled strands; often chambered crystalliferous.
Rays ca. 11/mm, 1–4-seriate, up to ca. 50 cells high, almost always homocellular, composed of procumbent cells, rarely with some square marginal cells; infrequently crystalliferous.
Crystals solitary, prismatic, in chambered parenchyma and ray cells.

Characteristics of the Genus Acer

Growth rings distinct. Vessels diffuse; solitary and in radial multiples or small clusters; rounded to somewhat angular in cross-section. Perforations simple in oblique end walls. Vessel walls with prominent spiral thickenings. Fibres with simple pits. Parenchyma sparse, scanty paratracheal and diffuse apotracheal; chambered crystalliferous. Rays 1–4-seriate. Crystals solitary, prismatic.

Separation of Acer Species

The wood anatomical descriptions of the species show some quantitative differences only. It would, however, be unjustified to use the differences in maximum ray height, vessel diameter and vessel frequency in the material studied as reliable characters for separation of the two species without additional evidence.

ANACARDIACEAE

This family of mainly tropical trees and shrubs is represented in Israel by the genera *Pistacia* and *Rhus*. Both genera show the characteristic features of the family, i. e., vessels with coarse inter-vessel pits and large and simple vessel-ray and vessel-parenchyma pits, libriform fibres, scanty paratracheal parenchyma and heterocellular rays (cf. Metcalfe & Chalk, 1950). The two genera can be separated on the occurrence of radial resin ducts (always present in the species of *Pistacia* in Israel, and usually absent from *Rhus*) and septate fibres (absent from *Pistacia,* present in varying frequency in *Rhus*).

Pistacia atlantica Desf.

Plate 7A–C; Fig. 5C

Tree; deciduous. Var. *atlantica* mostly in semi-steppe areas and steppes; C. Negev, Golan, E. Gilead, Ammon, Edom, Sinai. Var. *latifolia* in more humid parts of the area, in maquis or forest; Acco Plain, Upper and Lower Galilee, Mt. Carmel, Esdraelon Plain, Samaria, Judean Mts., Dan Valley, Hula Plain, Upper Jordan Valley, Sinai. Irano-Turanian element.

Growth rings distinct. Wood ring- to semi-ring-porous. Vessels ca. 100/mm^2, solitary (ca. 10%), especially at the beginning of the growth ring, mainly in radial multiples of 2–10(13) or in clusters including vascular tracheids; rounded in cross-section, tangential diameter 15–150 μm, radial diameter up to 200 μm, walls ca. 2 μm thick. Vessel member length 230(110–290) μm. Perforations simple in horizontal end walls in earlywood and in oblique end walls in latewood. Inter-vessel pits alternate, round or polygonal, 8–10 μm in diameter, with small slit-like to round apertures. Vessel-parenchyma and vessel-ray pits simple, large, round to gash-like. Vessels, except for the widest, and vascular tracheids with prominent spiral thickenings. Tyloses, containing crystals of various sizes, very common.
Fibres 530(340–730) μm long, medium thick- to very thick-walled, with infrequent simple pits, mainly in radial walls; partly gelatinous.

Parenchyma sparse, scanty paratracheal and diffuse apotracheal, fusiform or in 2-celled strands.
Rays 8–13/mm, 1–5(6)-seriate, uniseriates 1–9 cells high, multiseriates 3–33; heterocellular, with square, upright or weakly procumbent, mostly crystalliferous marginal cells and strongly procumbent central cells.
Resin ducts often present in multiseriate rays, 1 sometimes 2 per ray.
Crystals prismatic, in marginal ray cells and in tyloses.

Pistacia khinjuk Stocks
Plate 7D

Tree; deciduous; among shrubs; Sinai, Edom; Irano-Turanian element.

Growth rings distinct. Wood ring-porous, with a single row of earlywood pores. Vessels 200–250/mm²; some solitary, mainly at the beginning of the growth ring, mainly in radial multiples of 2–10, or in clusters, often forming a dendritic pattern together with vascular tracheids; rounded in cross-section, diameter 15–140 μm, walls 1.5–4 μm thick. Vessel member length 250(160–350) μm. Perforations simple in horizontal end walls in earlywood and in oblique end walls in latewood. Inter-vessel pits alternate, polygonal or round, 7–11 μm in diameter, with small slit-like to round apertures. Vessel-parenchyma and vessel-ray pits simple, large, round to gash-like. Vessels, except for the widest, and vascular tracheids with prominent spiral thickenings. Tyloses present.
Fibres 610(350–800) μm long, medium thick- to very thick-walled, with infrequent simple pits mainly in radial walls; partly gelatinous.
Parenchyma scanty paratracheal, in 2-celled strands.
Rays ca. 5–7/mm, 1–3(4)-seriate, uniseriates 1–8 cells high, multiseriates 4–30; heterocellular, with square to upright or weakly procumbent marginal cells and procumbent central cells; many cells, especially marginal ones, crystalliferous.
Resin ducts often present in multiseriate rays.
Crystals solitary, prismatic, in many ray cells especially marginal ones.

Pistacia lentiscus L.
Plate 8A

Shrub; evergreen; maquis and garigue; coast and lower hills, usually up to 300(–500) m above sea level; coastal Galilee, Acco Plain, Sharon Plain, Philistean Plain, Upper and Lower Galilee, Mt. Carmel, Mt. Gilboa, Samaria, Judean Mts.; Mediterranean element.

Growth rings faint. Wood diffuse- to semi-ring-porous. Vessels ca. 120–160/mm², rarely solitary, especially at the beginning of the growth ring, mainly in radial multiples of 2–10(12), or in clusters, including vascular tracheids; rounded in cross-section; tangential diameter 15–100 μm, radial diameter up to 170 μm, walls 2–4 μm thick. Vessel member length 260(140–340) μm. Perforations simple in horizontal end walls in earlywood and in oblique end walls in

latewood. Inter-vessel pits usually alternate, rounded, ca. 7-8 μm in diameter, with small slit-like to round apertures. Vessel-parenchyma and vessel-ray pits simple, large, round to gash-like, elongated horizontally or vertically. Vessel walls, except for the widest, and vascular tracheids with prominent spiral thickenings.
Fibres 530(340–730) μm long, thick- to very thick-walled, with infrequent simple pits in radial walls; many gelatinous.
Parenchyma sparse, scanty paratracheal and diffuse apotracheal, in 2–3-celled strands.
Rays ca. 7–12/mm, 1–2(3)-seriate, uniseriates 1–10 cells high, multiseriates 6–28; heterocellular, with square to weakly procumbent marginal cells and strongly procumbent central cells; crystalliferous cells scarce.
Resin ducts often present in multiseriate rays, 1 sometimes 2 per ray.
Crystals absent or scarce, prismatic, in marginal ray cells.

Pistacia palaestina Boiss.
Plate 8B–D; Fig. 9A

Tree or shrub; deciduous; maquis and garigue; mainly on hills and mountains; Sharon Plain, Upper and Lower Galilee, Mt. Carmel, Samaria, Judean Mts., Dan Valley, Hula Plain, Golan, Gilead, Ammon, Moav, Edom; East Mediterranean element.

Growth rings distinct. Wood ring-porous. Vessels ca. 170/mm^2, some solitary, mainly in radial or oblique multiples of up to 7, or in clusters, including vascular tracheids; rounded in cross-section, diameter 15–160 μm, walls 1–2 μm thick. Vessel member length 290(150–360) μm. Perforations simple in horizontal end walls in earlywood and in oblique end walls in latewood. Inter-vessel pits alternate, rounded, ca. 8 μm in diameter, with slit-like apertures. Vessel-parenchyma and vessel-ray pits simple, large, round to gash-like. Vessels, except for the widest, and vascular tracheids with prominent spiral thickenings.
Fibres 600(300–790) μm long, medium thick- to very thick-walled, with infrequent simple pits in radial walls; many gelatinous.
Parenchyma sparse, scanty paratracheal and diffuse apotracheal, in 2–3-celled strands.
Rays 5–8/mm, 1–4-seriate, uniseriates 1–11 cells high, multiseriates 4–31; heterocellular, composed of square, weakly procumbent or upright marginal cells and procumbent central cells; many cells, especially the marginal ones, crystalliferous.
Resin ducts often present in multiseriate rays.
Crystals solitary, prismatic, in many ray cells especially marginal cells.

Pistacia × saportae Burnat.

Tree; evergreen; maquis; Upper and Lower Galilee, Mt. Carmel, Judean Mts.; Mediterranean element.

Growth rings distinct. Wood ring- to semi-ring-porous, sometimes diffuse-porous. Vessels ca. 170/mm², some solitary, mainly in radial to oblique multiples of 2–9, sometimes in clusters, including vascular tracheids; rounded in cross-section, diameter 20–160 μm, walls 1.5–3 μm thick. Vessel member length 330(260–540) μm. Perforations simple in horizontal end walls in earlywood and oblique end walls in latewood. Inter-vessel pits mostly alternate, round to polygonal, ca. 8 μm in diameter, with slit-like apertures. Vessel-parenchyma and vessel-ray pits simple, large, round to gash-like. Vessel walls, except for the widest, and vascular tracheids with prominent spiral thickenings.
Fibres 580(360–790) μm long, medium thick- to very thick-walled, with infrequent simple pits in radial walls; many gelatinous.
Parenchyma scanty paratracheal in 3–4-celled strands.
Rays 4–8/mm, 1–3-seriate, uniseriates 1–12 cells high, multiseriates 4–33; heterocellular, with square or upright marginal cells and strongly procumbent central cells; crystalliferous cells sometimes present.
Resin ducts often present in multiseriate rays, 1 sometimes 2 per ray.
Crystals solitary, prismatic, abundant in ray cells or absent.

Pistacia vera L.
Plate 9A and B

Cultivated fruit tree; deciduous; Irano-Turanian element.

Growth rings distinct. Wood ring- to semi-ring-porous, with earlywood pores in usually more than one row. Vessels 40–70/mm², solitary (ca. 25%) mainly at the beginning of the growth ring, in radial or oblique multiples of 2–10 or in small clusters, sometimes forming a dendritic pattern together with vascular tracheids; rounded in cross-section (diameter 20–190 μm), often grouped together with small ones (minimum diameter 20 μm); walls ca. 4–7 μm thick. Vessel member length 220(160–300) μm. Perforations simple in horizontal to somewhat oblique end walls. Inter-vessel pits alternate, rounded, 7–8 μm in diameter, with small elliptic to slit-like apertures. Vessel-parenchyma and vessel-ray pits simple to half-bordered, large, round to gash-like. Vessels, except for the widest, and vascular tracheids with prominent spiral thickenings. Tyloses present.
Fibres 580(300–880) μm long, medium thick- to very thick-walled, with simple pits more numerous in radial than tangential walls; partly gelatinous.
Parenchyma scanty paratracheal, in up to 7-celled strands.
Rays 3–5/mm, 1–5-seriate, uniseriates up to 10 cells high, multiseriates 4–30(55) cells high; heterocellular, with square to upright, sometimes slightly procumbent, rarely crystalliferous marginal cells, and procumbent central cells.
Resin ducts occasionally present in multiseriate rays.
Crystals rare, solitary, prismatic, in marginal ray cells.

Table 4.3
Differences between *Pistacia* species.

Species	Porosity*	Vessel frequency per mm²	Max. vessel diameter in μm	Vessel wall thickness in μm	Ray frequency per mm	Ray width (no. of cells) full range	Ray width (no. of cells) predominant width	Max. ray height (no. of cells)	Growth rings
atlantica	R–S	100	150	2	8–13	1–5(6)	4–5	33	Distinct; at the beginning of the growth ring usually one row of wide pores, the remainder of the vessels narrow with few of intermediate size
khinjuk	R	200–250	140	1.5–4	5–7	1–3(4)	3	30	Distinct; at the beginning of the growth ring one row of wide pores, the remainder of the vessels narrow
lentiscus	D–S	120–160	100	2–4	7–12	1–2(3)	2	28	Faint
palaestina	R	170	160	1–2	5–8	1–4	3	31	Distinct; gradual decrease in pore diameter
× *saportae*	R–S(D)	170	160	1.5–3	4–8	1–3	2–3	33	Distinct; gradual decrease in pore diameter
vera	R–S	40–70	190	4–7	3–5	1–5	3–4	30(55)	Distinct; at the beginning of growth ring usually several rows of wide pores

* R — ring-porous; D — diffuse-porous; S — semi-ring-porous.

Wood Anatomy

Characteristics of the Genus Pistacia

Vessels mostly in multiples and clusters, often including vascular tracheids. Some vessels, especially the wide ones, solitary. Perforations simple, in horizontal end walls in the largest vessels. Vessel-ray pits mostly large and simple. Vessel walls, except for the widest, and vascular tracheids with prominent spiral thickenings. Tyloses often present. Fibres with infrequent simple pits mainly or solely in radial walls; partly gelatinous. Parenchyma sparse. Rays uni- and multiseriate, heterocellular, with short margins of square, upright, sometimes slightly procumbent cells and procumbent central cells. Resin ducts present in some of the multiseriate rays.

Archaeological records: In many locations throughout Israel and Sinai; since very early periods (N. Liphschitz & Y. Waisel; A. Fahn; E. Werker).

Key to Pistacia Species

1. Wood semi-ring-porous to diffuse-porous; growth rings faint; rays mainly 1–2 cells wide. **P. lentiscus**
- Not as above 2
2. Rays more than 4 cells wide present 4
- Rays 1–4 cells wide 3
3. At the beginning of the growth ring one row of wide pores present, the other vessels narrow. **P. khinjuk**
- Pores decrease gradually in diameter. **P. palaestina**
  P. × saportae
4. At the beginning of the growth ring usually one row of wide pores; rays 8–13/mm, their marginal cells usually crystalliferous. **P. atlantica**
- At the beginning of the growth ring usually several rows of wide pores; rays 3–5/mm; crystals rare in marginal ray cells. **P. vera**

Rhus coriaria L.

Plate 9C and D

Tree or shrub; deciduous; neglected places near villages, rarely in maquis; Sharon Plain, Philistean Plain, Upper Galilee, Mt. Carmel, Samaria, Judean Mts., Golan, Gilead, Edom; distributed in the Mediterranean region.

Growth rings distinct. Wood ring-porous. Vessels 30–150/mm^2. Earlywood vessels solitary (ca. 40%) and in radial multiples of 2–4 or in clusters; latewood vessels mainly in clusters or in radial multiples of up to 12 cells, sometimes forming a dendritic pattern together with intergrading vascular tracheids and axial parenchyma. Vessels rounded in cross-section, tangential diameter 15–135 μm, radial diameter up to 145 μm, walls 2–4 μm thick. Vessel member length 230(150–310) μm. Perforations simple in oblique end walls. Inter-vessel pits usually alternate, polygonal, sometimes rounded, ca. 10 μm in diameter, with slit-like apertures. Vessel-parenchyma and vessel-ray pits mostly simple, rounded, or

polygonal to gash-like. Vessel walls, except for the widest, with prominent spiral thickenings. Thin-walled tyloses present.
Fibres 470(290–690) μm long, thin- to medium thick-walled, very infrequently septate, with simple pits mainly in radial walls.
Parenchyma scanty paratracheal in earlywood; in latewood more abundant sometimes forming a dendritic pattern together with the vessels and vascular tracheids, in 3–8(12)-celled strands.
Rays ca. 12/mm, 1–2(3)-seriate, up to 45 cells high; heterocellular, with usually 1–2 rows of square to upright marginal cells and procumbent to upright central cells.
Resin ducts not found.
Crystals prismatic, mostly solitary, in ray cells, sometimes 2 or 3 per cell.

Rhus pentaphylla (Jacq.) Desf.
Plate 10A and B

Shrub or shrublet; deciduous; coastal rocks; Coastal Galilee, Acco Plain; South Mediterranean element.

Growth rings faint to fairly distinct. Vessels diffuse, ca. 65/mm², solitary (ca. 25%), mostly in radial (up to tangential) multiples of 2–5 or in infrequent clusters, sometimes including very narrow vessels or vascular tracheids. Vessels rounded in cross-section, tangential diameter 25–110 μm, radial diameter up to 120 μm, walls 2–5 μm thick. Vessel member length 300(220–350) μm. Perforations simple in oblique to horizontal end walls. Inter-vessel pits alternate, round or polygonal, 5–7 μm in diameter, with slit-like, sometimes coalescent apertures. Vessel-parenchyma and vessel-ray pits large and simple, round to gash-like, horizontal to vertical. Thin-walled tyloses occasionally present. Spiral thickenings absent.
Fibres 520(250–670) μm long, medium thick-walled, with simple pits in radial walls, septate, partly also gelatinous.
Parenchyma scanty paratracheal, in 2–4-celled strands.
Rays ca. 7/mm, 1–2(3)-seriate, up to 35 cells high, heterocellular, composed of weakly procumbent central cells and several rows of square and upright marginal cells.
Wide resin ducts occasionally present in rays.
Crystals solitary or in pairs or groups of 3, prismatic, in ordinary or slightly enlarged ray cells.

Rhus tripartita (Bernard. da Ucria) Grande
Plate 10C and D

Shrub; deciduous; rocks in deserts and among shrubs; Sharon Plain, Samaria, Judean Desert, N., C. and S. Negev, Lower Jordan Valley, Sinai, Gilead, Ammon, Moav; Irano-Turanian element.

Growth rings faint. Vessels diffuse, ca. 60/mm², solitary (ca. 20%), usually in

radial (rarely oblique) multiples of 2–5(10) or in infrequent clusters; multiples and clusters often including very narrow vessels and/or vascular tracheids. Vessels rounded in cross-section, tangential diameter 20–100 μm, radial diameter up to 110 μm, walls 3–7 μm thick. Vessel member length 250(150–330) μm. Perforations simple in oblique end walls. Inter-vessel pits alternate, polygonal, or more widely spaced and rounded, with slit-like, sometimes coalescent apertures, ca. 7 μm in diameter. Vessel-parenchyma and vessel-ray pits large and simple, round to gash-like (mostly horizontal, sometimes oblique to vertical). Spiral thickenings absent. Thin-walled tyloses present in some vessels.
Fibres 560(360–860) μm long, medium thick-walled, with simple pits mainly in radial walls, mostly septate, partly also gelatinous.
Parenchyma scanty paratracheal, in 2–4-celled strands; occasionally crystalliferous.
Rays ca. 10/mm, mostly uniseriate, partly 2(4)-seriate, up to 25 cells high, heterocellular, with one to several rows of upright marginal cells and procumbent to upright central cells; sheath cells occasionally present.
Resin ducts not found.
Crystals solitary and large, prismatic, in ordinary to slightly enlarged ray cells, or smaller and solitary or in pairs in axial parenchyma.

Archaeological records: Dead Sea area and Sinai; Chalcolithic (A. Fahn & E. Werker; E. Werker).

Characteristics of the Genus Rhus

Vessels mainly in radial multiples; rounded in cross-section. Perforations simple. Vessel-ray and vessel-parenchyma pits mostly simple, round or polygonal to gash-like. Fibres with simple pits, mostly or infrequently septate. Parenchyma scanty paratracheal. Rays heterocellular. Crystals prismatic, in ray cells.

Key to Rhus Species

Wood ring-porous; vessel walls, except for those of the largest vessels, with prominent spiral thickenings. **R. coriaria**
Vessels diffuse; spiral thickenings absent. **R. pentaphylla**
R. tripartita

APOCYNACEAE

Nerium oleander L.

Plate 11

Shrub; evergreen; banks of lakes and streams and on stony wadi beds; Acco Plain, coast of Carmel, Upper and Lower Galilee, Mt. Carmel, Samaria, Judean Mts., Judean Desert, Hula Plain, Upper and Lower Jordan Valley, Golan,

Gilead, Ammon, Edom; Mediterranean element extending into the West Irano-Turanian and Saharo-Arabian regions.

Growth rings faint. Vessels diffuse, 40–50/mm², solitary (ca. 20–40%) and in radial multiples of 2–4(6), sometimes seemingly longer including fibres or parenchyma cells, rounded or angular in cross-section, tangential diameter (30)40–50(80) μm, radial diameter up to 70–90 μm, walls 4–7 μm thick. Vessel member length 430(270–630) μm. Perforations simple, mostly in oblique end walls. Inter-vessel pits vestured, alternate, rounded, 4–5 μm in diameter, with slit-like, sometimes coalescent, apertures. Vessel-parenchyma and vessel-ray pits similar but half-bordered, sometimes unilaterally compound.
Fibres 810(380–1,030) μm long, thin-walled, with minutely bordered pits more numerous in radial than in tangential walls.
Parenchyma scanty paratracheal and diffuse apotracheal, in 4(8)-celled strands.
Rays ca. 14–20/mm, 1–2(3)-seriate, the bi- and triseriate portions of the same width or hardly wider than the uniseriate margins, up to 17 cells high, heterocellular, the 2- and 3-seriates with low portions of strongly procumbent cells and tall margins of square to upright cells (Kribs' heterogeneous type I).
Crystals not observed.

Archaeological records: Dead Sea area; third to seventh centuries A. D. (N. Liphschitz & Y. Waisel).

ARALIACEAE

Hedera helix L.

Plate 12; Fig. 5F

Perennial climbing shrub; evergreen; rocks and maquis; Upper Galilee, Edom; Euro-Siberian and Mediterranean element.

Growth rings distinct to faint. Vessels diffuse, 200–250/mm², mainly in clusters, sometimes in tangential or radial multiples of 2–4, occasionally solitary, often forming, together with vasicentric tracheids, a pattern of tangential bands; angular in cross-section, tangential diameter 20–65 μm, radial diameter up to 100 μm, walls 1–2 μm thick. Vessel member length 580(300–830) μm. Perforations simple in oblique end walls. Inter-vessel pits alternate to diffuse, rounded, 7–10 μm in diameter, with short slit-like, sometimes elliptic apertures. Vessel-parenchyma and vessel-ray pits large, round to elongate, reticulate to scalariform, sometimes unilaterally compound, simple. Vessel walls occasionally with spiral thickenings. Few vessels with tyloses.
Fibres 760(400–1,050) μm long, medium thick-walled, with small simple pits in radial and tangential walls; partly septate.
Parenchyma scanty paratracheal; in up to 9-celled strands.
Rays 6–8/mm, 1–7(14)-seriate, up to ca. 1.5 (rarely up to 6) mm high; nearly

homocellular to heterocellular, composed of procumbent central cells and slightly procumbent, square, sometimes upright marginal cells.
Crystals not observed.

ASCLEPIADACEAE

The two genera from Israel, *Calotropis* and *Periploca,* have many wood anatomical characters in common with the other woody representatives of this family: vessels with vestured pits, fibres with minutely bordered pits, scanty paratracheal parenchyma and heterocellular rays (cf. Metcalfe & Chalk, 1950). The genera differ in degree of vessel grouping, inter-vessel pit size, and presence or absence of laticifers in the rays (see the descriptions).

Calotropis procera (Ait.) Ait. fil.
Plate 13A and B

Shrub; evergreen; desert plains; S. Negev, Lower Jordan Valley, Dead Sea area, Arava Valley, E. Ammon, Moav, Edom; Sudanian element extending into the E. Saharo-Arabian region.

Growth rings absent or very faint. Vessels diffuse, 14–35/mm², mainly in radial multiples of 2–7 or clusters, rarely solitary (ca. 6%); angular to rounded in cross-section (mostly radially elongated), tangential diameter 30–170 μm, radial diameter up to 250 μm, walls ca. 2 μm thick. Vessel member length 200(100–250) μm. Perforations simple in transverse or slightly oblique end walls. Inter-vessel pits vestured, alternate, rounded, ca. 7 μm in diameter, with slit-like to elliptic apertures. Vessel-parenchyma and vessel-ray pits half-bordered to almost simple, round to slit-like in varying directions.
Fibres 710(450–900) μm long, very thin-walled, with minutely bordered pits more numerous in radial than in tangential walls.
Parenchyma scanty paratracheal and diffuse apotracheal, in 2–4-celled strands.
Rays 7–13/mm, 1–4(6)-seriate, up to 25 cells high; heterocellular, composed of procumbent central cells and square and upright cells in a varying number of marginal rows.
Laticifers present in some of the rays.
Crystals not observed.

Archaeological records: Herodion (Judean Desert); 1st century A.D. (N. Liphschitz & Y. Waisel).

Periploca aphylla Decne.
Plate 13C and D

Shrub; rocky deserts; Judean Desert, Upper and Lower Jordan Valley, Dead Sea area, Gilead, Ammon, Moav; Sudanian element extending into the E. Saharo-Arabian region.

Growth rings faint. Vessels diffuse, 3–40/mm^2, mostly solitary, sometimes in pairs, sometimes in a radial or oblique pattern; rounded to angular in cross-section, diameter 15–80 μm, walls 2–3 μm thick. Vessel member length 310(230–480) μm. Perforations simple in oblique end walls. Inter-vessel and vessel-tracheid pits vestured, alternate, rounded, 3–4 μm in diameter, with slit-like apertures. Vessel-parenchyma and vessel-ray pits similar but half-bordered. Tyloses sometimes present. Vasicentric tracheids present.
Fibres 700(490–830) μm, medium thick-walled; with small bordered pits in radial and tangential walls.
Parenchyma scanty paratracheal and diffuse apotracheal; in 2–5-celled strands.
Rays 13–17/mm, mostly uniseriate, some 2–3(4)-seriate, up to ca. 20 cells high; heterocellular, composed of square, upright and some weakly procumbent cells; many of the procumbent cells of the multiseriate rays are vertically and horizontally chambered crystalliferous. Perforated ray cells with simple perforations occasionally present.
Crystals solitary or in aggregates, prismatic, in chambered ray cells and axial parenchyma.

AVICENNIACEAE

Contrary to the treatment in *Flora Palaestina,* we do not include *Avicennia* in the Verbenaceae, but follow several authors who have recognized the monogeneric family Avicenniaceae (e.g., Airy Shaw, 1966). In its wood anatomy *Avicennia* differs from the Verbenaceae in the presence of included phloem and sclereid bands and a number of other qualitative as well as quantitative wood anatomical features (cf. Metcalfe & Chalk, 1950). Airy Shaw's comment (1966) that Avicenniaceae are related to Salvadoraceae is of interest and finds support in the shared presence of included phloem in the wood of those two families. It should be stressed, however, that included phloem by itself is not a strong enough indication of natural affinity, because it occurs in many obviously unrelated families, apparently as a result of parallel evolution.

Avicennia marina (Forssk.) Vierh.
Plate 14A–C

Shrub or tree; evergreen; muddy tidal seashore, in sea water, forming mangrove vegetation; Gulf of Elat (Aqaba), Sinai; tropical seashores and banks of tidal streams and rivers.

Wood with included phloem of the concentric type, and strongly interlocked grain. Vessels mainly in radial multiples of 2–9, occasionally solitary or in clusters; 30–70/mm^2 in the xylem layer; rounded, tangential diameter 25–100 μm, radial diameter up to 100 μm, walls ca. 8 μm thick. Vessel member length 210 (140–300) μm. Perforations simple in mostly horizontal end walls. Inter-vessel pits alternate, rounded, ca. 2–3 μm in diameter, with slit-like, sometimes coalescent apertures. Vessel-parenchyma and vessel-ray pits similar but half-bordered.

Included phloem groups solitary, round to oval in cross-section, embedded in broad concentric bands of lignified conjunctive parenchyma on the inside of sclereid bands. Sclereids rectangular, in 1–2(3)-seriate, occasionally coalescing bands within conjunctive parenchyma bands.
Fibres 550(360–820) μm long, thick- to very thick-walled; with minutely bordered pits mainly in radial walls.
Parenchyma scanty paratracheal and conjunctive in continuous wide bands, in 2–4-celled strands. The sclereid bands divide the conjunctive parenchyma into an inner region that includes the phloem groups and an outer region.
Rays 6–9/mm, 1–6 cells wide, up to 30 cells high; heterocellular, composed of square, upright and infrequent procumbent cells mixed together, many cells crystalliferous.
Crystals small, rectangular to almost spindle-shaped, in varying numbers in most ray cells, and infrequently as druses in ray cells.

BORAGINACEAE

Cordia sinensis Lam. [*Sebestena sanguinea* (Forssk.) Friesen]
Plate 14D and E

Tree or shrub; evergreen; oases in deserts; Dead Sea area; Sudanian element.

Growth rings distinct. Vessels diffuse, 20–25/mm², in multiples of 2–3, and small clusters including some narrow vessels, occasionally solitary; rounded in cross-section, tangential diameter 20–200 μm, radial diameter up to 240 μm, walls 4–6 μm thick. Vessel member length 230(100–280) μm. Perforations simple in horizontal to slightly oblique end walls. Inter-vessel pits alternate, round, 5–7 μm in diameter, with short slit-like sometimes coalescent apertures. Vessel-parenchyma pits and vessel-ray pits similar, sometimes elongate, half-bordered. Many vessels with tyloses.
Fibres 880(610–1,180) μm long, thick- to very thick-walled, with minutely bordered pits mainly in radial walls.
Parenchyma in bands 2–5 cells wide, or only locally confluent, alternating with fibre bands; fusiform and in 2–4-celled strands; storied; rarely crystalliferous.
Rays 2–4/mm, 3–6-seriate, up to ca. 50 cells high; heterocellular, composed of square, procumbent and upright cells, with prominent sheath cells; many cells crystalliferous.
Crystals prismatic, one or more in many of the ray cells, rarely in axial parenchyma; crystal sand infrequent in ray and axial parenchyma cells.

CAPPARIDACEAE

Of the three genera of the Capparidaceae occurring in Israel, *Capparis* and *Maerua* are wood anatomically very similar and apparently closely related. *Cleome*

stands out on account of its included phloem and deviating rays. Solereder (1908) and Metcalfe & Chalk (1950) referred to a number of other genera with included phloem. Airy Shaw (1966) treated *Cleome* with three other genera in the family Cleomaceae. His delimitation of Cleomaceae and Capparidaceae is not in line with the presence or absence of included phloem reported in the literature. However, a reappraisal of family or subfamily boundaries, taking into account the presence or absence of included phloem, seems in order.
Solereder (1908), citing work by Radlkofer & DeBary, also recorded included phloem for *Maerua* (*M. oblongifolia* and *M. uniflora,* a synonym of *M. crassifolia*). Most probably these early observations (also incorporated in Metcalfe & Chalk's surveys of 1950 and 1983) are based on wrongly identified material, because included phloem is definitely absent from authenticated specimens of *M. crassipes* and from another *Maerua* species studied for comparison (*M. kaessneri* Gilg et Bened. from Kenya).

Capparis cartilaginea Decne.
Plate 15A

Shrub; ravines and rocks of hot deserts; Central Negev, Lower Jordan Valley, Arava area; Sudanian element.

Growth rings indistinct. Vessels diffuse; rounded to weakly angular in cross-section; tending to be of two size classes: the wider ones ca. 40/mm², solitary (10% or less), in radial multiples of up to 8(12), occasionally in clusters, tangential diameter 40–135 μm, radial diameter up to 130 μm, walls 2–4 μm thick; the narrow vessels (diameter 25–40 μm) in clusters or multiples with the wider vessels (sometimes intergrading with vascular tracheids) or solitary or in short multiples. Vessel member length 170(130–250) μm. Perforations simple in oblique to transverse end walls. Inter-vessel pits very weakly vestured, alternate, rounded to polygonal, 4–6 μm in diameter, with slit-like, rarely coalescent apertures. Vessel-parenchyma and vessel-ray pits similar but half-bordered.
Fibres 390(280–500) μm long, thin- to medium thick-walled, with numerous simple pits in radial and tangential walls.
Parenchyma scanty paratracheal to vasicentric; fusiform or in 2–3-celled strands; weakly storied.
Rays ca. 5/mm, (1)2–4-seriate, up to 29(38) cells high; usually heterocellular with square to upright, sometimes weakly procumbent, marginal, and procumbent central cells; sheath cells occasionally present.
Crystals not observed.

Capparis decidua (Forssk.) Edgew.
Plate 15B–D

Shrub; wadis and oases in hot deserts among savanna-like vegetation; Upper Jordan area; Sudanian element extending into other Palaeotropical regions.

Wood Anatomy

Growth rings absent or very faint. Vessels diffuse, rounded in cross-section, tending to be of two size classes: the wider ones ca. 20/mm^2, solitary (ca. 30%), in radial multiples or in clusters of 2–5, tangential diameter 40–110 μm, radial diameter up to 140 μm, walls 3–6 μm thick; the narrow vessels (diameter 20–40 μm) in clusters or multiples with the wider vessels (often intergrading with vascular tracheids), or solitary or in short multiples. Vessel member length 170(120–210) μm. Perforations simple in transverse to oblique end walls. Intervessel pits vestured, alternate, rounded to polygonal, 4–6 μm in diameter, with round to slit-like apertures. Vessel-parenchyma and vessel-ray pits similar but half-bordered. Gummy contents in many vessels. Sclerotic tyloses (probably of traumatic origin) in some of the vessels.
Fibres 490(410–710) μm long, medium thick-walled, with simple pits more numerous in radial than in tangential walls.
Parenchyma vasicentric, fusiform and in 2–5-celled strands; storied.
Rays 7/mm, (1)2–3(4)-seriate, ca. 10–24(50) cells high; homocellular to weakly heterocellular, with upright to weakly procumbent marginal cells and square to procumbent central cells.
Crystals not observed, but "ghosts" or integuments of solitary crystals are present in rays, probably belonging to the water-soluble crystals dissolved during microtechnical processing as reported by Miller (1978).

Capparis ovata Desf.
Plate 16A and B

Hemicryptophytic shrub; stony places and abandoned fields in non-typical Mediterranean sites or steppes and deserts; widely distributed; arid East Mediterranean, Irano-Turanian and Saharo-Arabian element.

Growth rings faint to fairly distinct. Vessels diffuse, rounded to weakly angular in cross-section, tending to be of two size classes: the wider ones ca. 40/mm^2, solitary (25%), in short radial multiples of 2–6, and in clusters (forming a denser row at the beginning of growth ring), tangential diameter up to 115 μm, radial diameter up to 150 μm, walls 2–4 μm thick; the narrow vessels (diameter 25–40 μm) in clusters or multiples with the wider vessels (sometimes intergrading with vascular tracheids) or solitary or in short multiples. Vessel member length 180(130–250) μm. Perforations simple in oblique to transverse end walls. Intervessel pits very faintly vestured, alternate, rounded to polygonal, 4–6 μm in diameter, with slit-like, rarely coalescent apertures. Vessel-parenchyma and vessel-ray pits similar but half-bordered. Some vessels with gummy contents.
Fibres 380(280–550) μm long, medium thick-walled, with numerous simple pits in radial and tangential walls.
Parenchyma scanty paratracheal to vasicentric, fusiform or in 2–5-celled strands.
Rays ca. 5/mm, 4–6-seriate, up to 48 cells (1.5 mm) high, weakly heterocellular, composed of procumbent, sometimes square, rarely upright cells; sheath cells rare.
Crystals not observed.

Capparis spinosa L.
Plate 16C and D

Var. *aegyptiaca*: low shrub; on walls, fences and in grassland, often also in the vicinity of human dwellings; widely distributed in lowland and montane areas; Sharon Plain, Philistean Plain, Upper and Lower Galilee, Mt. Carmel, Esdraelon Plain, Mt. Gilboa, Samaria, Judean Mts., W. and C. Negev, Gilead, Ammon; Mediterranean element.
Var. *aravensis*: shrub from wadis and cliffs in hot deserts; Judean Desert, E. Negev, Dead Sea area; East Sudanian element, probably endemic.
The two varieties are very similar in their wood anatomy; the quantitative data in Table 4.4 are based on var. *aegyptiaca.*

Growth rings faint to fairly distinct. Vessels diffuse, rounded to weakly angular in cross-section, tending to be of two size classes: the wider ones ca. 40/mm^2, solitary (ca. 10%) and in radial multiples of 2–6(8) or more rarely in clusters, tangential diameter 40–110 μm, radial diameter up to 145 μm, walls 2–4 μm thick; the narrow vessels (diameter 25–40 μm) in clusters or multiples with the wider vessels (sometimes intergrading with vascular tracheids) or solitary or in short multiples. Vessel member length 170(130–250) μm. Perforations simple in oblique to transverse end walls. Inter-vessel pits very faintly vestured, alternate, rounded to polygonal, 4–6 μm in diameter, with slit-like, rarely coalescent apertures. Vessel-parenchyma and vessel-ray pits similar but half-bordered.
Fibres 340(270–460) μm long, thin- to medium thick-walled, with numerous simple pits in radial and tangential walls.
Parenchyma scanty paratracheal to vasicentric, fusiform or in 2–5-celled strands; weakly storied.
Rays ca. 5/mm, 1–4-seriate, (1)8–18(35) cells high, usually heterocellular, with square to upright (sometimes weakly procumbent) marginal and procumbent central cells; sheath cells present.
Crystals not observed but "ghosts" or integuments of solitary crystals are present in rays, probably belonging to the water-soluble crystals dissolved during microtechnical processing as reported by Miller (1978).

Characteristics of the Genus Capparis

Vessels diffuse, tending to be of two size classes: the wider ones ca. 20–40/mm^2, solitary (not more than 30%), and in radial multiples or clusters; the narrow vessels (intergrading with vascular tracheids) in clusters or multiples with the wider vessels or solitary or in short multiples. Perforations simple in transverse to oblique end walls. Inter-vessel pits alternate, (faintly) vestured, rounded to polygonal, 4–6 μm in diameter, with slit-like, rarely coalescent apertures. Vessel-parenchyma and vessel-ray pits similar but half-bordered. Fibres mostly medium thick-walled, with numerous simple pits. Parenchyma scanty paratracheal or vasicentric.

Wood Anatomy

Differences between the Capparis Species (Table 4.4)

C. cartilaginea stands out on account of its striking radial pattern of vessels, caused by the long multiples and clusters.
C. decidua has the shortest vessel multiples and highest percentage of solitary vessels. It also stands out on account of its relatively abundant vasicentric parenchyma and paucity of tangential wall pits in the fibres.
C. ovata and *C. spinosa* are more or less intermediate between *C. decidua* and *C. cartilaginea* in vessel distribution pattern. Of the limited material studied, *C. ovata* had the widest and tallest rays, but it must be doubted whether this can be used as a reliable character to separate it from the other species in the region.

Cleome droserifolia (Forssk.) Del.
Plate 17A

Small shrub; hot deserts, rocks and wadis; Judean Desert, Lower Jordan Valley, Dead Sea area, Arava Valley; East Sudanian element.

Wood with included phloem of the concentric type. Growth rings very faint. Vessels 50–70/mm^2, solitary (ca. 15%) and in mostly radial multiples of 2–7, rarely in small clusters, together forming a radial pattern; rounded to angular in cross-section, 30–80 μm in tangential diameter, walls 2–3 μm thick. Vessel member length 130(100–180) μm. Perforations simple in horizontal to slightly oblique end walls. Inter-vessel pits weakly vestured, alternate, round to polygonal, 3–5 μm in diameter, with slit-like, sometimes coalescent apertures. Vessel-parenchyma and vessel-ray pits similar but half-bordered. Some vessels with gummy contents.
Included phloem in groups, 60–250 μm in tangential diameter, together with conjunctive parenchyma arranged in concentric bands.
Fibres 290(190–390) μm long, thin- to medium thick-walled, with almost simple pits more numerous in radial than in tangential walls.
Parenchyma paratracheal, ranging from scanty to confluent, fusiform and intergrading with fibres or in 2-celled strands; conjunctive parenchyma in concentric bands outside the phloem groups — the central and sometimes also peripheral region of each band consisting of sclerified cells, the remainder being unlignified.
Rays poorly delimited, 3–5/mm, 1–5 cells wide, often in loose aggregates together with cells intermediate between fusiform parenchyma and wide-lumened fibres; heterocellular, composed of mainly square to upright cells and some procumbent ones.
Crystals not observed.

Maerua crassifolia Forssk.
Plate 17 B–D; Table 4.4

Tree or shrub; wadis and oases; Dead Sea area; Sudanian element.

Growth rings faint. Vessels diffuse, rounded in cross-section, tending to be of

Table 4.4

Survey of some quantitative wood anatomical characters in *Capparis* and *Maerua* (not to be used as diagnostic features).

	Number of Vessels per multiple	% of solitary vessels	Vessel member length (average) (μm)	Fibre length (average) (μm)	Ray width (no. of cells)	Maximum ray height (no. of cells)
C. cartilaginea	2–8(12)	10 or less	160	390	1–4	30
C. decidua	2–3(5)	35	180	500	1–3(4)	40
C. ovata	2–5(6)	25	200	380	(1)3–6	50
C. spinosa	2–6(8)	20	170	340	1–4(5)	30
Maerua crassifolia	2–8	40	140	380	1–3	45

two size classes; the wider ones ca. 40/mm^2, solitary (30%), in radial multiples or clusters of 2–8, tangential diameter 40–80 μm, radial diameter up to 90 μm, walls 4–7 μm thick; the narrow vessels (diameter 10–25 μm) often clustered or in multiples with the wider vessels (intergrading with vascular tracheids) or solitary or in short multiples. Vessel member length 140(50–360) μm. Perforations simple in oblique to transverse end walls. Inter-vessel pits vestured, alternate, 4–5 μm in diameter, with slit-like, sometimes coalescent apertures. Vessel-parenchyma and vessel-ray pits similar but half-bordered. Some vessels with gummy contents.
Fibres 380(250–550) μm long, thin- to medium thick-walled, with simple pits more numerous in radial than in tangential walls.
Parenchyma scanty paratracheal, in 2–3-celled strands.
Rays ca. 4/mm, 1–3-seriate, (1)10–30(44) cells high; weakly heterocellular, composed of weakly procumbent, square and weakly upright cells.
Crystals numerous, prismatic, solitary in ray cells; water-soluble, only observed as "ghosts" or integuments (cf. Miller, 1978).

CAPRIFOLIACEAE

The two genera of the Caprifoliaceae in Israel, *Lonicera* and *Viburnum*, have many wood anatomical features in common with each other and with other genera of the family (cf. Metcalfe & Chalk, 1950). They differ mainly in the vessel perforations, which are exclusively scalariform in *Viburnum*, and mostly simple in *Lonicera*.

Lonicera etrusca Santi

Plate 18

Climbing shrub; rocky places in maquis and garigue; Sharon Plain, Upper and Lower Galilee, Mt. Carmel, Mt. Gilboa, Samaria, Judean Mts., Golan, Gilead, Edom; Mediterranean element.

Growth rings distinct. Vessels diffuse, ca. 250–300/mm^2, mostly solitary (ca. 65–80%) or in radial to oblique multiples of 2 or 3, angular in cross-section, tangential diameter 20–50 μm, radial diameter up to 65 μm, walls 2 μm thick. Vessel member length 500(150–950) μm. Perforations mostly simple in very oblique end walls, rarely scalariform with up to 15 bars. Inter-vessel pits (obscurely vestured?) alternate to opposite, rounded, polygonal or oblong, 6–10 μm in diameter or 12×6 μm, with slit-like, often crossed apertures. Vessel-parenchyma and vessel-ray pits similar but half-bordered. Vessel walls with spiral to annular thickenings.
Fibres 780(590–930) μm long, thin- to medium thick-walled, with numerous distinctly bordered pits (diameter ca. 6 μm) in radial and tangential walls, mostly with fine spiral to annular thickenings.
Parenchyma very scanty diffuse or paratracheal, in 2–4-celled strands.

Rays ca. 15/mm, 1–3(4)-seriate, of various lengths up to more than 1 mm high, heterocellular (Kribs' heterogeneous type I).
Crystals not observed.

Viburnum tinus L.
Plate 19

Shrub; evergreen; shady maquis; Upper Galilee, Mt. Tabor, Mt. Carmel; Mediterranean element.

Growth rings distinct. Vessels diffuse, 120–170/mm², mostly solitary, rarely in radial to tangential multiples of 2–3, angular in cross-section, tangential diameter 25–70 μm, radial diameter up to 80 μm, walls ca. 2 μm thick. Vessel member length 990(550–1,440) μm. Perforations exclusively scalariform, with 10–40 bars. Inter-vessel pits (obscurely vestured?) largely restricted to some overlapping vessel tails, opposite to scalariform, rounded to oblong, 6 μm to 15 × 5 μm in diameter, with slit-like, mostly crossed, apertures. Vessel-parenchyma and vessel-ray pits similar but half-bordered, sometimes unilaterally compound. Part of the vessel tails with spiral or annular thickenings.
Fibres 1,360(720–1,760) μm long, medium thick- to thick-walled, with numerous distinctly bordered pits (diameter 6–8 μm) in radial and tangential walls, often with fine annular or spiral thickenings.
Parenchyma sparse, scanty paratracheal and diffuse apotracheal in 2–8-celled strands.
Rays 8–12/mm, 1–3(4)-seriate, up to 20(50) cells high, heterocellular (Kribs' heterogeneous types I–II), with occasional sheath cells.
Crystals not observed.

Archaeological record: Ein Gedi (inlay on a coffin); 1st century A. D. (A. Fahn).

CARYOPHYLLACEAE

Gymnocarpos decandrum Forssk. [*G. fruticosum* (Vahl) Pers.]
Plate 20A and B

Half-shrub; stony slopes and hammadas; Judean Desert, N., C. and S. Negev, Lower Jordan Valley, Dead Sea area, Arava Valley, Moav, Edom; Saharo-Arabian element.

Growth rings faint to distinct. Vessels diffuse, ca. 200/mm², mostly solitary, also in short tangential to radial multiples or clusters of 2 or 3, rounded in cross-section, tangential diameter 14–45 μm, radial diameter up to 60 μm, walls 4–6 μm thick. Vessel member length 210(130–310) μm. Perforations simple in horizontal to oblique end walls. Inter-vessel pits alternate to opposite, sometimes rather widely spaced, rounded, 5–7 μm in diameter, with slit-like, sometimes

coalescent, apertures. Vessel-fibre pits similar but with crossed apertures. Vessel-parenchyma and vessel-ray pits rounded, crowded, half-bordered.
Fibres 420(270–570) μm long, thick-walled, with very numerous, distinctly bordered pits (diameter 5–7 μm) in radial and tangential walls.
Parenchyma scanty paratracheal, infrequently diffuse apotracheal, and in uniseriate marginal bands; in 2–3-celled strands.
Rays up to 25/mm, 1(2)-seriate, 1–2(3) cells high; cells fusiform with overlapping tips as seen in tangential section, entirely consisting of tall upright cells.
Crystals not observed.

CHENOPODIACEAE

This mainly herbaceous family has numerous woody representatives in the arid flora of the Middle East. All genera have included phloem, and most have storied structure and fusiform parenchyma. The precise arrangement of the phloem groups (concentric or tending to foraminate), the presence, delimitation, or absence of rays, and the crystal complement offer possibilities to identify a number of species or species groups (see the key).

Archaeological records: Timna (S. Negev), around 2000 B. C. Judean Desert; Iron Age (A. Fahn; E. Werker).

Aellenia lancifolia (Boiss.) Ulbrich
Plate 20C and D

Perennial; stony and gypsiferous ground in steppes and deserts; Judean Desert, N. and C. Negev, Lower Jordan Valley, Dead Sea area, Ammon, Edom; Irano-Turanian element.

Wood with included phloem of the concentric type. Growth rings faint to distinct. Vessels solitary, in radial multiples of up to 6 and in clusters, forming a radial pattern inside the phloem groups; rounded in cross-section, ca. 90–130/mm², of two size classes: the wider vessels (20–45 μm in tangential diameter, walls 2–4 μm thick) mixed with narrow ones intergrading with vascular tracheids (diameter 15–20 μm). Vessel member length 100(60–150) μm. Perforations simple in oblique to horizontal end walls. Inter-vessel pits alternate to opposite, rounded to oblong, 4–5 μm in diameter, often with coalescent apertures. Vessel-parenchyma and vessel-ray pits similar but half-bordered. Many vessels with gummy contents.
Included phloem in narrow, often coalescent groups, rounded or tangentially elongated (tangential diameter 65–170 μm), forming long concentric bands together with conjunctive parenchyma.
Fibres 280(210–340) μm long, medium thick- to very thick-walled with simple pits in radial walls; in tangential bands 30–45 cells wide.
Parenchyma scanty paratracheal and conjunctive in tangential bands ca. 8 cells

wide, linked by rays and radial parenchyma strips, fusiform or in strands of various cell numbers; storied; many cells crystalliferous.
Rays distinct, 6–9/mm, 1–16-seriate, of various heights, heterocellular, of square to upright cells; many cells crystalliferous.
Crystals prismatic, in aggregates (sometimes giving the impression of crystal sand), of various sizes and shapes, numerous in ray cells and conjunctive parenchyma, especially at the inner margins of the fibre bands.

Anabasis articulata (Forssk.) Moq.
Plate 21A

Dwarf shrub (chamaephyte); stony, gravelly and sandy deserts; Judean Desert, Negev, Lower Jordan Valley, Dead Sea area, Arava Valley, Sinai, deserts of Ammon, Moav and Edom; mainly a Saharo-Arabian element.

Wood with included phloem of the concentric type. Growth rings not observed.
Vessels in clusters on the inside of the included phloem groups; rounded to angular in cross-section, of two size classes: the wider vessels (ca. 60/mm^2, tangential diameter 15–45 μm, walls ca. 4 μm thick) mixed with narrow ones intergrading with vascular tracheids (diameter ca. 10 μm). Vessel member length 120(80–150) μm. Perforations simple in horizontal to oblique end walls. Inter-vessel pits alternate to opposite, rounded, diameter 3–4 μm, with slit-like, sometimes coalescent apertures. Vessel-parenchyma and vessel-ray pits similar but half-bordered. Vessel elements storied together with the parenchyma and included phloem elements. Some vessels with gummy contents.
Included phloem groups forming together with conjunctive parenchyma long concentric arcs or bands.
Fibres 450(290–540) μm long, very thick-walled with simple pits in radial walls; in tangential bands 30–40 cells wide; some crystalliferous.
Parenchyma scanty paratracheal, and conjunctive in long bands or arcs, ca. 4 cells wide, mostly fusiform; many cells large, chambered crystalliferous mainly where bordering on the inner margins of fibre bands; storied together with vessel members and included phloem elements.
Rays distinct, 4–8/mm, uniseriate, 1(2–5) cells high; cells fusiform in tangential section; 1-cell high rays with square or slightly procumbent cells; taller rays heterocellular with upright cells as well; some crystalliferous.
Crystals solitary, prismatic, very large, sometimes elongate, in chambered parenchyma and ray cells and in fibres.

Anabasis setifera Moq.
Plate 21B

Perennial; hot desert, moist salines and wadi beds; often on gypsiferous ground; Judean Desert, Negev, Lower Jordan Valley, Dead Sea area, Arava Valley, Sinai, Moav, Edom; mainly Saharo-Arabian element.

Wood with included phloem of the concentric type. Vessels in radial multiples and clusters on the inside of the included phloem groups, rarely diffuse in fibrous ground tissue, rounded in cross-section, of two size classes: the wider ones (ca. 80/mm^2, tangential diameter 20–50 μm, radial diameter up to 70 μm, walls 4–8 μm thick), mixed with narrow vessels intergrading with vascular tracheids (diameter 10–15 μm). Vessel member length 70(40–100) μm. Perforations simple in transverse to oblique end walls. Inter-vessel pits alternate to opposite or fairly widely spaced and diffuse, rounded, diameter ca. 5 μm, with slit-like, occasionally coalescent apertures. Vessel-parenchyma and vessel-ray pits similar or oblong, with reduced borders to almost simple. Vessel members weakly storied together with the parenchyma and included phloem elements.
Included phloem in frequently coalescent groups, forming, together with unlignified conjunctive parenchyma, long discontinuous tangential bands.
Fibres 340(260–420) μm long, thick- to very thick-walled with numerous simple pits in radial and tangential walls.
Parenchyma scanty paratracheal and conjunctive around the phloem groups; fusiform and in 2–3-celled strands; weakly storied.
Rays irregular but distinct, 9–15/mm, 1–3-seriate, 1–5(8) cells high, of upright to square (rarely also weakly procumbent) cells.
Crystals not observed.

Anabasis syriaca Iljin (*A. haussknechtii* Bge.)
Plate 21C

Perennial (hemicryptophyte). Var. *syriaca*: steppes; heavy alluvial loess or loess-like soil, often on cultivated land or otherwise disturbed ground; N. and C. Negev. Var. *zoharyi*: as above; Negev, Moav, Edom; Irano-Turanian element.

Wood with included phloem of the concentric type. Growth rings faint. Vessels in clusters, sometimes in 2–6-celled multiples, occasionally solitary, mostly inside the included phloem groups; rounded to angular in cross-section, of two size classes: the wider ones (30–85/mm^2, tangential diameter 15–80 μm, walls 2–5 μm thick) mixed with narrow vessels intergrading with vascular tracheids (tangential diameter 7–15 μm). Vessel member length 90(50–120) μm. Perforations simple in horizontal to oblique end walls. Inter-vessel pits alternate, rounded, 3–5 μm in diameter, with elliptic to slit-like apertures. Vessel-parenchyma pits round, half-bordered, occasionally gash-like. Vessel elements storied together with the parenchyma and included phloem elements.
Included phloem groups often coalescent, mostly forming, together with unlignified conjunctive parenchyma, concentric bands of various lengths.
Fibres 290(220–350) μm long, medium thick- to thick-walled, with simple pits mainly in radial walls; in bands, 20–40 cells wide.
Parenchyma scanty paratracheal and conjunctive unlignified in bands 4–10 cells wide and in radial strips; fusiform; many cells crystalliferous.
Rays hardly distinguishable from radial strips of parenchyma, 0–3/mm, 1–7 cells

wide, of various heights (from a few cells to more than 1 mm); part of the cells crystalliferous.
Crystals prismatic, of various sizes, and irregularly shaped, in aggregates (sometimes giving the impression of crystal sand), in parenchyma and ray cells.

Arthrocnemum macrostachyum (Moric.) Moris et Delponte [*A. glaucum* (Del.) Ung.]
Plate 22A

Shrub; inundated salines, banks of saline water bodies; Acco Plain, Sharon Plain, N. Negev, Lower Jordan Valley, Dead Sea area, Edom; mainly Mediterranean and Saharo-Arabian element.

Wood with included phloem of the concentric type. Growth rings faint to distinct. Vessels ca. 130–180/mm², solitary, in radial or tangential multiples of 2–3, or in clusters on the inside of the phloem groups; rounded to angular in cross-section; tangential diameter 15–45 μm, walls 3–5 μm thick, the narrow ones intergrading with vascular tracheids. Vessel member length 100(70–150) μm. Perforations simple in horizontal to somewhat oblique end walls. Inter-vessel pits alternate, 3–4 μm in diameter, with slit-like, often coalescent apertures. Vessel-parenchyma pits similar but half-bordered. Vessel elements storied together with conjunctive parenchyma and included phloem elements.
Included phloem in distinct, very small groups, rounded (tangential diameter 10–50 μm), forming, together with lignified conjunctive parenchyma, more or less concentric bands.
Fibres 200(140–250) μm long, medium thick- to very thick-walled, with numerous simple pits in radial and tangential walls; in bands.
Parenchyma scanty paratracheal, fusiform or in 2-celled strands, and conjunctive in 4–6-seriate bands, storied together with vessel members and included phloem elements; radial parenchyma strips absent.
Rays absent.
Crystals rarely observed, in irregular aggregates in axial parenchyma.

Arthrocnemum perenne (Mill.) Moss

Shrub; coastal marshes; Acco Plain, Sharon Plain; Mediterranean element.

Wood with included phloem of the foraminate to somewhat concentric type. Growth rings faint. Vessels ca. 90/mm², solitary (ca. 25%), in variously directed multiples of 2–3 and in small clusters, on the inside of the phloem groups; angular to rounded in cross-section, tangential diameter 15–45 μm, walls ca. 4 μm thick, the narrow vessels intergrading with vascular tracheids. Vessel member length 190(120–270) μm. Perforations simple in horizontal end walls. Inter-vessel pits alternate, rounded, 4–5 μm in diameter, with slit-like apertures. Vessel-parenchyma pits similar but half-bordered. Vessel elements storied together with parenchyma and included phloem elements. Some vessels with gummy contents.

Included phloem in distinct small groups (tangential diameter 35–80 μm).
Fibres 260(180–340) μm long, medium thick- to thick-walled, with numerous simple pits in radial and tangential walls.
Parenchyma scanty paratracheal, and conjunctive around phloem strands, sometimes forming short tangential bands 3–5 cells wide; fusiform, storied together with vessel and included phloem elements. Radial parenchyma strips absent.
Rays absent.
Crystals not observed.

Atriplex halimus L.
Plate 22B

Shrub. Var. *halimus*: salines, wadi beds and sandy soils; Sharon Plain, Philistean Plain, Judean Mts., Judean Desert, Negev, Upper and Lower Jordan Valley. Var. *schweinfurthii*: wadi beds and sandy soils; Sharon Plain, Philistean Plain, Judean Desert, S. Negev, Upper and Lower Jordan Valley, Dead Sea area, Arava Valley, Sinai. Mediterranean and Saharo-Arabian element.

Wood with included phloem of the foraminate to concentric type. Growth rings faint. Vessels in radial multiples of up to 8 and clusters on the inside of the included phloem groups, rarely diffuse in fibrous ground tissue; rounded to angular in cross-section, of two size classes: the wider ones (ca. 50/mm^2, 25–100 μm in tangential diameter, up to 160 μm in radial diameter, walls 3–6 μm thick) mixed with narrow vessels intergrading with vascular tracheids (15–20 μm in diameter). Vessel member length 160(90–210) μm. Perforations simple in transverse to oblique end walls. Inter-vessel pits alternate, sometimes rather widely spaced, round, 6–9 μm in diameter, with slit-like, sometimes coalescent apertures. Vessel-parenchyma pits similar but half-bordered. Vessel elements storied together with parenchyma and included phloem elements.
Included phloem in distinct, frequently coalescent groups (tangential diameter ca. 50–200 μm), occasionally forming short, more or less tangential bands together with unlignified conjunctive parenchyma.
Fibres 360(210–470) μm long, medium thick-walled, with numerous almost simple pits in tangential and radial walls.
Parenchyma mostly conjunctive, forming poorly defined arcs together with the phloem groups outside the vessel multiples, variously directed extensions of parenchyma arcs often in contact with other arcs, also scanty paratracheal, mostly fusiform, partly in 2–5-celled strands, and then often crystalliferous, usually chambered, sometimes resembling upright ray cells, mostly storied together with vessel members and included phloem elements.
Rays typically absent, but some narrow to broad (up to 10 cells wide) radial parts of conjunctive parenchyma tending to have ray-like structure, and consisting of upright to weakly procumbent cells.
Crystals prismatic, of various sizes and shapes, solitary or in aggregates in parenchyma cells.

Trabeculae in a file found in vessel members and fibres.

Archaeological records: Timna mines (S. Negev); late Bronze, early Iron Age (E. Werker).

Chenolea arabica Boiss.
Plate 22C

Dwarf-shrub; steppes and deserts, especially on gypsiferous ground; Judean Desert, Negev, Lower Jordan Valley, Arava Valley, Ammon, Moav, Edom; East Saharo-Arabian element.

Wood with included phloem. Vessels solitary, in radial multiples of 2–10 or in clusters, forming, together with vascular tracheids, a radial pattern within radially elongated groups of fibres which are embedded in unlignified parenchymatous ground tissue; rounded, tangential diameter 10–40 μm, the narrow ones intergrading with vascular tracheids, walls 2–4 μm thick. Vessel member length 120(90–140) μm. Perforations simple in horizontal end walls. Inter-vessel pits alternate, round to oblong, diameter 4–5 μm, with slit-like, sometimes coalescent apertures. Vessel-parenchyma pits half-bordered, round to oval. Some vessels with gummy substances. Vessel elements weakly storied.
Included phloem in distinct arc-like groups (tangential diameter 180–450 μm), each forming, together with xylem, a radial band embedded in parenchymatous tissue.
Fibres 330(220–410) μm long, thick- to very thick-walled, with simple pits in radial and tangential walls; in long radial strips.
Parenchyma scanty paratracheal, and in tangential (marginal?) bands 1 cell wide, fusiform or in 2-celled strands, and abundant as ground tissue widely separating the radial strips of lignified vascular tissue.
Rays absent but ground parenchyma tissue between lignified vascular strips giving impression of wide rays; composed of square, slightly upright or slightly procumbent cells; including crystalliferous ones.
Crystals mostly prismatic, variously shaped and sized, in groups (some aggregates of minute crystals giving the impression of crystal sand) in the parenchyma cells.

Halogeton alopecuroides (Del.) Moq.
Plate 22D

Dwarf-shrub. Var. *alopecuroides*: steppes and saline soil; Judean Desert, C. and S. Negev, Lower Jordan Valley, Arava Valley, Sinai, deserts of Ammon, Moav and Edom. Var. *papilosus*: steppes; Central Negev. Saharo-Arabian element.

Wood with included phloem of the foraminate to concentric type. Growth rings faint. Vessels in clusters inside the included phloem groups, rounded to angular in cross-section, of two size classes: the wider vessels ca. 50/mm² (25–60 μm in tangential diameter, walls 2–4 μm thick) mixed with narrow ones intergrading

with vascular tracheids (diameter ca. 10 μm). Vessel member length 120(50–160) μm. Perforations simple in horizontal to oblique end walls. Inter-vessel pits alternate to opposite, round to oblong, with slit-like, often coalescent apertures, ca. 6 μm in diameter. Vessel-parenchyma pits round, half-bordered. Many vessels with gummy contents.
Included phloem in distinct groups (tangential diameter 100–300 μm), often coalescent, partly surrounded by unlignified conjunctive parenchyma, sometimes merging with lignified conjunctive parenchyma, together forming bands of various lengths.
Fibres 300(240–390) μm long, thick-walled, with simple pits in radial and tangential walls.
Parenchyma scanty paratracheal, and conjunctive unlignified around phloem groups and lignified, in more or less tangential bands, 2–3 cells wide; fusiform or in 2-celled strands; many-chambered crystalliferous.
Rays absent.
Crystals prismatic, of various sizes, in aggregates or solitary, in chambered parenchyma cells.

Haloxylon persicum Bge.
Plate 23A

Shrub or tree; sandy wadis and wadis crossing a saline depression; Arava Valley, Dead Sea area, Edom, Sinai; Irano-Turanian element.

Wood with included phloem of the concentric type. Growth rings not observed. Vessels in variously sized clusters inside the included phloem groups; rounded to angular in cross-section, of two size classes: the wider ones (30–40/mm^2, 50–95 μm in tangential diameter, walls 5–8 μm thick) mixed with narrow ones intergrading with vascular tracheids (diameter 10–20 μm). Vessel member length 110(80–160) μm. Perforations simple in horizontal to oblique end walls. Inter-vessel pits alternate to opposite, round to oblong, ca. 6 μm in diameter, with slit-like, often coalescent apertures. Vessel-parenchyma and vessel-ray pits round, half-bordered.
Included phloem in distinct groups (tangential diameter 150–220 μm) sometimes coalescent, usually arranged in tangential bands interrupted by rays.
Fibres 380(160–600) μm long, thick-walled, with sparse simple pits in radial walls; some crystalliferous.
Parenchyma scanty paratracheal and conjunctive in a narrow layer around the phloem groups; some cells chambered crystalliferous.
Rays distinct, 3–5/mm, 1–18-seriate, up to more than 70 cells high; heterocellular, with square, procumbent and upright cells mixed together; many crystalliferous, some horizontally or longitudinally chambered.
Crystals of various sizes, prismatic, variously shaped, in aggregates or solitary; in ray cells and in some parenchyma cells and fibres.

Archaeological records: Arava Valley, S. Negev and N. E. Sinai; ranging from 2000–1000 B. C. to the 6–8th centuries A. D. (E. Werker; N. Liphschitz & Y. Waisel).

Hammada negevensis Iljin et Zoh.
Plate 23B

Dwarf-shrub; stony deserts, often on somewhat gypsiferous soils; Central Negev, Arava Valley, Sinai; East Saharo-Arabian element; endemic.

Wood with included phloem of the foraminate type. Growth rings faint. Vessels in radial multiples of up to 6 and in clusters, occasionally solitary, inside included phloem groups; rounded to angular in cross-section; of two size classes: the wider ones (80–100/mm², diameter 20–55 μm, walls 3–5 μm thick) mixed with narrow ones intergrading with vascular tracheids (diameter ca. 10 μm). Vessel member length 140(100–200) μm. Perforations simple in oblique end walls. Inter-vessel pits alternate, rounded, 3–4 μm in diameter, with short slit-like apertures. Vessel-parenchyma pits similar but half-bordered. Vessel elements storied together with the parenchyma and included phloem elements.
Included phloem groups round to elliptic (tangential diameter 65–130 μm), occasionally coalescent.
Fibres 270(190–380) μm long, thick-walled, with simple pits more numerous in radial than in tangential walls; some crystalliferous.
Parenchyma sparse, very scanty paratracheal, and unlignified conjunctive (extensions of lignified conjunctive, paenchyma absent or very inconspicuous); fusiform; some chambered crystalliferous; storied together with vessel members and included phloem elements.
Rays absent.
Crystals prismatic, of various shapes and sizes, mostly in aggregates, in some chambered parenchyma cells and in a few fibres.

Archaeological record of the genus: Timna (S. Negev); 2000 B. C. (E. Werker).

Hammada salicornica (Moq.) Iljin

Half-shrub; sandy ground, debris of sandstone and granite in hotter deserts; C. and S. Negev, Dead Sea area, Arava Valley, Edom; East Sudanian element.

Wood with included phloem of the foraminate type. Growth rings faint. Vessels mainly in clusters inside the phloem groups; rounded in cross-section; of two size classes: the wider vessels (ca. 50/mm², diameter 25–60 μm, walls ca. 4 μm thick) mixed with narrow ones intergrading with vascular tracheids (diameter ca. 10 μm). Vessel member length 100(60–140) μm. Perforations simple in horizontal to oblique end walls. Inter-vessel pits alternate, rounded, ca. 3 μm in diameter, with slit-like apertures. Vessel-parenchyma pits similar but half-bordered. Vessel elements storied together with the parenchyma and included phloem elements.
Included phloem groups elliptic to crescent-shaped (tangential diameter 100–230 μm).

Fibres 290(160–370) μm long, very thick-walled, with simple pits more numerous in radial than in tangential walls; some crystalliferous.
Parenchyma sparse, scanty paratracheal, unlignified conjunctive associated with phloem groups, occasionally with short lignified tangential extensions; some cells chambered crystalliferous (scattered between fibres, forming more or less tangential rows); storied together with vessel members and included phloem elements.
Rays absent.
Crystals prismatic, of various shapes and sizes, mostly in aggregates, in chambered parenchyma cells and fibres.

Hammada scoparia (Pomel) Iljin
Plate 23C and D

Shrub; steppes and deserts, mainly on alluvial loess soil, sometimes in cultivated depressions; W., N. and C. Negev, Lower Jordan Valley, Dead Sea area, desert parts of Gilead, Ammon, Moav and Edom; Saharo-Arabian and Irano-Turanian element.

Wood with included phloem of the foraminate type. Growth rings faint. Vessels in clusters inside included phloem groups; rounded to angular in cross-section; of two size classes: the wider vessels (ca. 100/mm^2, 20–42 μm in tangential diameter, walls ca. 4 μm thick) mixed with narrow ones intergrading with vascular tracheids (diameter ca. 10 μm). Vessel member length 70(45–90) μm. Perforations simple in horizontal to slightly oblique end walls. Inter-vessel pits alternate, rounded, ca. 5 μm in diameter, with slit-like, often coalescent, apertures. Vessel-parenchyma pits similar but half-bordered. Vessel elements storied together with the parenchyma and included phloem. Many vessels with gummy contents.
Included phloem groups round to crescent-shaped (tangential diameter 80–160 μm), occasionally 2–5 groups linked by lignified conjunctive parenchyma; fusiform and in 2-celled strands.
Fibres 210(120–270) μm long, thick- to very thick-walled, with simple pits more numerous in radial than in tangential walls.
Parenchyma sparse, scanty paratracheal, and unlignified conjunctive occasionally with lignified extensions linking 2–5 phloem groups; fusiform; storied together with vessel members and included phloem; solitary large cells with sphaerocrystals rarely present.
Rays absent.
Crystals infrequent, prismatic, of various shapes and sizes and as sphaerocrystals, in parenchyma cells.

Noaea mucronata (Forssk.) Aschers. et Schweinf.
Plate 24A

Low shrub; steppes, mainly on grey, calcareous soil; Judean Mts., Judean Desert, N. C. and S. Negev, Lower Jordan Valley, Sinai, deserts of Gilead, Ammon, Moav and Edom; Irano-Turanian element.

Wood with included phloem of the foraminate to concentric type. Growth rings very faint. Vessels, together with vascular tracheids and parenchyma cells, in clusters inside included phloem groups; rounded to angular in cross-section, of two size classes: the wider vessels (ca. 150–210/mm², tangential diameter 15–35 μm, radial diameter up to 45 μm, walls ca. 3 μm thick) mixed with narrow vessels intergrading with vascular tracheids (diameter ca. 10 μm). Vessel member length 130(100–170) μm. Perforations simple in horizontal to oblique end walls. Inter-vessel pits alternate, rounded, ca. 5 μm in diameter, with slit-like apertures. Vessel-parenchyma pits similar but half-bordered. Some of the narrow vessels with faint spiral thickenings. Vessel members storied together with parenchyma and included phloem elements.
Included phloem in distinct, sometimes coalescent groups, usually crescent-shaped (tangential diameter 50–180 μm), sometimes forming tangential, oblique or irregular bands of varying lengths together with unlignified conjunctive parenchyma.
Fibres 310(220–580) μm long, medium thick- to thick-walled, with simple pits more numerous in radial than in tangential walls.
Parenchyma scanty paratracheal and conjunctive; fusiform, sometimes in 2-celled strands; storied together with vessel members and included phloem elements.
Rays absent.
Crystals not observed.

Salsola baryosma (Roem. et Schult.) Dandy
Plate 24B

Shrub. Var. *baryosma*: saline, ruderal and segetal sites; Central Negev, Lower Jordan Valley, Dead Sea area, Arava Valley. Var. *viridis*: as above; Arava Valley. Sudanian element.

Wood with included phloem of the concentric type. Growth rings distinct. Vessels in small clusters or in multiples of 2–6, together with vascular tracheids inside included phloem groups; rounded to angular in cross-section; of two size classes: the wider vessels (ca. 80–95/mm², tangential diameter 20–45 μm, radial diameter up to 50 μm, walls 2–3 μm thick) mixed with narrow ones intergrading with vascular tracheids (diameter 10–20 μm). Vessel member length 140(60-190) μm. Perforations simple in oblique to horizontal end walls. Inter-vessel pits alternate, rounded, ca. 6 μm in diameter, with elliptic apertures. Vessel-parenchyma and vessel-ray pits round, half-bordered. Vessel elements storied together with the parenchyma and included phloem elements.
Included phloem in distinct, sometimes coalescent groups, round to elliptic (tangential diameter 20–110 μm); connected by lignified conjunctive parenchyma to form concentric bands.
Fibres 310(230–460) μm long, thick-walled, with simple pits mainly in radial walls; some chambered crystalliferous.

Parenchyma scanty paratracheal and conjunctive lignified, in tangential bands 2–5 cells wide; mostly fusiform, sometimes in 2-celled strands, some chambered crystalliferous; storied together with vessel members and included phloem elements.
Rays absent or indistinguishable from radial strips of conjunctive parenchyma, 2–6-seriate, 2–30 cells (more than 1 mm) high, composed of upright cells, some crystalliferous.
Crystals prismatic, of various sizes and shapes, solitary or in aggregates, in scattered chambered fibres and in conjunctive chambered parenchyma cells, especially where bordering on the fibres.

Salsola tetrandra Forssk.
Plate 24C

Dwarf-shrub; deserts; dry saline soils; Judean Desert, W. and C. Negev, Lower Jordan Valley, Dead Sea area, Arava Valley, Sinai, deserts of Ammon, Moav and Edom; Saharo-Arabian element.

Wood with included phloem of the foraminate to concentric type. Growth rings present. Vessels mostly in clusters together with vascular tracheids, parenchyma cells and fibres, inside included phloem groups, occasionally solitary vessels are embedded in fibrous tissue not in connection with a phloem group; rounded to angular in cross-section, of two size classes: the wider vessels (ca. 70–80/mm^2, tangential diameter 20–45 μm, radial diameter up to 70 μm, walls 4–7 μm thick) mixed with narrow vessels intergrading with vascular tracheids (diameter 15–20 μm). Vessel member length 120(80–160) μm. Perforations simple in horizontal to oblique end walls. Inter-vessel pits alternate, rounded to oblong, 4–7 μm in diameter, with slit-like, often coalescent apertures. Vessel-parenchyma and vessel-ray pits similar but half-bordered. Some narrow vessels with fine spiral thickenings. Vessel elements storied together with the parenchyma and included phloem elements.
Included phloem in distinct, sometimes coalescent groups, rounded to crescent-shaped (tangential diameter 100–300 μm), occasionally linked by lignified, crystalliferous, partly thick-walled conjunctive parenchyma to form tangential bands of varying lengths.
Fibres 350(200–400) μm long, medium thick- to very thick-walled, with numerous simple pits in radial and tangential walls; crystalliferous fibres and irregular islands of fibres with thinner walls scattered in the bulk of thick-walled fibres.
Parenchyma sparse, scanty paratracheal, and conjunctive lignified around each phloem group and sometimes in bands 2–5 cells wide; fusiform or in 2-celled strands, many crystalliferous; storied together with vessel members and included phloem elements.
Rays uncommon, 1–3-seriate, 1–5(8) cells high, heterocellular, with upright and square cells, many crystalliferous.

Crystals abundant, small, prismatic, in aggregates, in parenchyma cells bordering on the fibres, in ray cells and in chambered fibres.

Archaeological records: Timna (S. Negev); late Bronze, early Iron Age (E. Werker).

Salsola vermiculata L.

Plate 24D

Shrubby perennial; calcareous stony steppes and also somewhat saline soils; Judean Desert, N. Negev, Upper and Lower Galilee, Dead Sea area, Arava Valley, deserts of Ammon, Moav and Edom; Saharo-Arabian and Irano-Turanian element.

Wood with included phloem of the concentric type. Growth rings distinct. Vessels in clusters, inside included phloem groups; rounded in cross-section, of two size classes: the wider vessels (ca. 70/mm^2, tangential diameter 20–35 μm, walls 3–4 μm thick) mixed with narrow ones intergrading with vascular tracheids (diameter ca. 10 μm). Vessel member length 100(80–120) μm. Perforations simple in horizontal to oblique end walls. Inter-vessel pits alternate, rounded, with slit-like apertures, 4–7 μm in diameter. Vessel-parenchyma pits round to oblong, half-bordered. Vessel elements storied together with the parenchyma and included phloem elements.

Included phloem in distinct, sometimes coalescent groups, rounded (tangential diameter 60–130 μm); forming concentric bands together with mostly lignified conjunctive parenchyma.

Fibres 250(190–290) μm long, medium thick- to thick-walled, with slightly bordered pits more numerous in radial than in tangential walls; some chambered crystalliferous.

Parenchyma scanty paratracheal and conjunctive in broad concentric bands, 4–7 cells wide, linked by radial, short parenchyma strips (2–4 cells wide); mostly fusiform, sometimes in 2-celled strands, partly chambered crystalliferous; storied together with vessel members and included phloem elements.

Rays absent but radial strips of conjunctive parenchyma narrow and ray-like, 2–6-seriate, up to 10 cells high, composed of upright, square and irregularly shaped cells.

Crystals prismatic, of various sizes, solitary or in small aggregates, in scattered chambered fibres, and in chambered conjunctive parenchyma especially where bordering on the fibres.

Seidlitzia rosmarinus Bge.

Low shrub; saline soils in hot deserts; Negev, Lower Jordan Valley, Dead Sea area, Arava Valley, deserts of Edom; East Saharo-Arabian element.

Wood with included phloem of the foraminate to concentric type. Growth rings distinct. Vessels in small clusters or radial multiples of up to 4 inside included phloem groups; rounded to angular in cross-section; of two size classes: the wider vessels (ca. 80–100/mm^2, tangential diameter 22–45 μm, walls 4 μm thick) mixed with narrow ones intergrading with vascular tracheids (diameter ca. 10 μm). Vessel member length 90(60–110) μm. Perforations simple in horizontal to oblique end walls. Inter-vessel pits alternate to opposite, rounded, 4–6 μm in diameter, with slit-like apertures. Vessel-parenchyma and vessel-ray pits similar but half-bordered. Many vessels with gummy contents. Vessel members storied together with parenchyma and phloem elements.

Included phloem in round, elliptic to crescent-shaped groups (tangential diameter 90–325 μm), mostly single, occasionally coalescent, often forming tangential bands or arcs of various lengths together with lignified or unlignified conjunctive parenchyma.

Fibres 250(140–330) μm long, thick- to very thick-walled, with simple pits mainly in radial walls; few-chambered crystalliferous.

Parenchyma scanty paratracheal, in marginal bands 1 cell wide and conjunctive lignified or unlignified, in bands or arcs 2–8 cells wide, linked by irregular radial parenchyma strips; fusiform or in 2-celled strands. Many cells chambered crystalliferous; storied together with vessel members and included phloem elements.

Rays vaguely delimited from radial strips of conjunctive parenchyma, often short; 3–6/mm, 1–10-seriate, 1–30 cells (more than 1 mm) high; cells upright, some longitudinally chambered crystalliferous.

Crystals prismatic, of various sizes and shapes, mostly in aggregates, some solitary, in parenchyma and ray cells, occasionally in fibres.

Suaeda asphaltica (Boiss.) Boiss.
Plate 25A

Dwarf-shrub; deserts, frequently on gypsiferous soils and steep slopes; Judean Desert, Central Negev, Lower Jordan Valley, Dead Sea area, deserts of Moav and Edom; Saharo-Arabian element.

Wood with included phloem of the concentric type. Growth rings present. Vessels solitary (ca. 20%), in radial or sometimes tangential multiples of 2–3(4), or in clusters, inside included phloem groups; rounded to angular in cross-section, of two size classes: the wider vessels (ca. 60/mm^2, tangential diameter 25–50 μm, radial diameter up to 65 μm, walls 4–7 μm thick) mixed with narrow ones intergrading with vascular tracheids (diameter ca. 13 μm). Vessel member length 80(50–110) μm. Perforations simple in horizontal to oblique end walls. Inter-vessel pits alternate, rounded, ca. 5 μm in diameter, with slit-like, sometimes coalescent apertures. Vessel-parenchyma pits half-bordered, round to oblong. Vessel elements storied together with the parenchyma and included phloem elements.

Included phloem in distinct groups, round to elliptic [tangential diameter 20–100(160) μm], forming irregular concentric bands together with lignified conjunctive parenchyma.

Fibres 240(160–350) μm long, medium thick- to thick-walled, with numerous simple pits in radial and tangential walls.
Parenchyma scanty paratracheal and in conjunctive lignified bands 2–4 cells wide, sometimes linked by radial parenchyma strips; mostly fusiform, sometimes in 2-celled strands, partly chambered crystalliferous; storied together with vessel members and included phloem elements.
Rays absent but radial strips of conjunctive parenchyma occasionally present, composed of upright, square and weakly procumbent cells.
Crystals prismatic, large, sometimes solitary, mostly in aggregates with small ones, mostly in chambered conjunctive parenchyma, especially where bordering on the fibre bands.

Suaeda fruticosa Forssk.
Plate 25B

Shrub; saline marshes in hot deserts; Judean Desert; Lower Jordan Valley, Dead Sea area, Arava Valley, deserts of Moav; mainly Sudanian element.

Wood with included phloem of the concentric type. Vessels mostly in clusters, sometimes in radial multiples of 2–5 inside the included phloem groups; rounded or angular in cross-section; of two size classes: the wider vessels (ca. 50/mm^2, tangential diameter 30–90 μm, radial diameter up to 110 μm, walls 4–6 μm thick) mixed with narrow vessels intergrading with vascular tracheids (diameter 15–25 μm). Vessel member length 130(90–160) μm. Perforations simple in horizontal to oblique end walls. Inter-vessel pits alternate, rounded, 4–5 μm in diameter, with slit-like, sometimes coalescent apertures. Vessel-parenchyma and vessel-ray pits similar but half-bordered. Some vessels with gummy contents. Vessel elements storied together with parenchyma and included phloem elements.
Included phloem in distinct groups, round to elliptic (tangential diameter 40–150 μm), sometimes coalescent; forming concentric bands together with lignified conjunctive parenchyma.
Fibres 310(230–380) μm long, medium thick-walled, with numerous simple pits more numerous in radial than in tangential walls.
Parenchyma scanty paratracheal and conjunctive lignified in broad concentric bands 4–5 cells wide, linked by numerous ray-like or irregular radial parenchyma strips; mostly fusiform, sometimes in 2–3-celled strands, partly chambered crystalliferous; storied together with the vessel members and included phloem elements.
Rays indistinguishable from long radial strips of conjunctive parenchyma, ca. 1–2/mm, (1)6–10(17)-seriate and up to or over 3 mm high, composed of upright to square cells.
Crystals prismatic, mostly large and solitary or in aggregates together with small crystals, in chambered conjunctive parenchyma, especially where bordering on the fibres, occasionally in non-chambered parenchyma and ray cells.

Suaeda monoica Forssk.
Plate 25C

Tall shrub; wet salines; Lower Jordan Valley, Dead Sea area, Arava Valley, Sinai; Sudanian element.

Wood with included phloem of the concentric type. Vessels rarely solitary, mostly in radial multiples of 2–4 or in clusters inside the included phloem groups, rounded or angular in cross-section, of two size classes: the wider vessels [ca. 50/mm^2, tangential diameter 20–60 μm, walls 4–6(10) μm thick] mixed with narrow ones intergrading with vascular tracheids (diameter ca. 15 μm). Vessel member length 120(70–160) μm. Perforations simple in horizontal to oblique end walls. Inter-vessel pits alternate to opposite, rounded, 4–5 μm in diameter, with slit-like, sometimes coalescent apertures. Vessel-parenchyma pits similar but half-bordered. Some vessels with gummy contents. Vessel elements storied together with the parenchyma and included phloem elements.
Included phloem in distinct groups, usually round (tangential diameter 40–130 μm), sometimes coalescent; usually forming tangential bands of various lengths together with lignified conjunctive parenchyma.
Fibres 290(210–390) μm long, medium thick-walled, with minutely bordered pits in radial and tangential walls; rarely crystalliferous.
Parenchyma scanty paratracheal and conjunctive lignified in tangential bands 2–5 cells wide and of various lengths, linked by irregular radial parenchyma strips, mostly fusiform, sometimes in 2-celled strands, partly chambered crystalliferous; storied together with vessel members and included phloem elements.
Rays absent but very rare strips of radial conjunctive parenchyma present.
Crystals prismatic, large, mostly solitary, in chambered parenchyma cells, rarely in fibres.

Suaeda palaestina Eig et Zoh.
Plate 25D

Dwarf-shrub; hot desert salines; Lower Jordan Valley, Dead Sea area; Saharo-Arabian and East Sudanian element.

Wood with included phloem of the concentric type. Growth rings distinct. Vessels solitary (ca. 17%), in clusters or in radial multiples of 2–3(4), inside the included phloem groups; rounded or angular in cross-section, of two size classes: the wider vessels (ca. 120/mm^2, tangential diameter 20–35 μm, walls ca. 5 μm thick) mixed with narrow ones intergrading with vascular tracheids (diameter ca. 15 μm). Vessel member length 90(60–130) μm. Perforations simple in horizontal to oblique end walls. Inter-vessel pits alternate, rounded, 3–4 μm in diameter, with slit-like, sometimes coalescent apertures. Vessel-parenchyma pits similar but half-bordered. Most vessels with gummy contents. Vessel elements storied together with the parenchyma and included phloem elements.
Included phloem in very narrow groups, round to elliptic in cross-section (tan-

gential diameter 20–65 μm); forming more or less tangential bands of various lengths together with lignified conjunctive parenchyma.
Fibres 210(110–340) μm long, thick- to very thick-walled, with minutely bordered pits mainly in radial walls.
Parenchyma scanty paratracheal and lignified conjunctive in more or less tangential bands 2–5 cells wide (individually fluctuating in width) and of various lengths, occasionally linked by irregular radial parenchyma strips; fusiform or in 2-celled strands; partly chambered crystalliferous; storied together with the vessel members and included phloem elements.
Rays absent.
Crystals prismatic, large and solitary or often in aggregates together with small ones, in chambered conjunctive parenchyma mainly where bordering on the fibres.

Archaeological records: Timna (S. Negev); 1300 B. C. (E. Werker).

CISTACEAE

Cistus creticus L.
Plate 26A

Dwarf-shrub; batha, garigue and devastated maquis; on sandy clay, calcareous sandstone, terra rossa and rendzina, frequently on stony ground; Coastal Galilee, Acco Plain, Sharon Plain, Philistean Plain, Upper and Lower Galilee, Mt. Carmel, Samaria, Shefela, Judean Mts., Judean Desert, Dan Valley, Golan, Gilead, Ammon, Moav; Mediterranean element.

Growth rings faint to distinct. Vessels diffuse or wood with a slight tendency towards semi-ring-porosity. Vessels ca. 170/mm^2, almost exclusively solitary, rounded in cross-section, 20–60 μm in tangential diameter, radial diameter up to 70 μm, walls ca. 3 μm thick. Vessel member length 290(150–390) μm. Perforations simple in mainly oblique end walls. Inter-vessel pits not observed. Vessel-fibre pits diffuse, fully bordered, weakly vestured, rounded, diameter 5–6 μm, with slit-like, partly crossed, sometimes coalescent apertures. Vessel-parenchyma and vessel-ray pits similar but half-bordered. Irregular, coarse—yet faint—spiral thickenings present in some of the vessels. Gummy contents often present.
Fibres 410(270–550) μm long, thick- to very thick-walled, with numerous distinctly bordered, obscurely vestured pits (diameter 4–5 μm) in radial and tangential walls; sometimes with faint annular to spiral thickenings.
Parenchyma scanty paratracheal and diffuse apotracheal and as infrequent marginal cells; fusiform or in 2–4-celled strands.
Rays ca. 10/mm, 1–4-seriate, up to 45 cells high; heterocellular (Kribs' heterogeneous type II), but central cells of multiseriates only weakly procumbent to square.
Crystals not observed.

Cistus salvifolius L.
Plate 26B–D

Dwarf-shrub; batha, garigue and devastated maquis; on sandy clay, calcareous sandstone, terra rossa and rendzina, frequently on stony ground; Coastal Galilee, Acco Plain, Sharon Plain, Philistean Plain, Upper and Lower Galilee, Mt. Carmel, Mt. Gilboa, Samaria, Shefela, Judean Mts., Upper Jordan Valley, Golan, Gilead; Mediterranean element.

Growth rings faint to distinct. Vessels diffuse or wood with a slight tendency towards semi-ring-porosity. Vessels ca. 250–300/mm^2, almost exclusively solitary, rounded in cross-section, tangential diameter 15–50 μm, radial diameter up to 60 μm, walls ca. 3 μm thick. Vessel member length 300(210–550) μm. Perforations simple in horizontal to oblique end walls. Inter-vessel pits not observed. Vessel-fibre pits diffuse, fully bordered, obscurely vestured, rounded, 5 μm in diameter, with slit-like, mostly crossed, sometimes coalescent apertures. Vessel-parenchyma and vessel-ray pits similar but half-bordered. Irregular, coarse—yet faint—spiral thickenings present in some of the vessels. Gummy contents often present.
Fibres 460(250–620) μm long, thick- to very thick-walled, with numerous distinctly bordered, sometimes weakly vestured pits (diameter ca. 4 μm) in radial and tangential walls, sometimes with faint annular to spiral thickenings.
Parenchyma scanty paratracheal and diffuse apotracheal and as infrequent marginal cells; fusiform or in 2-celled strands.
Rays ca. 10/mm, 1–4-seriate, up to 40(60) cells high; heterocellular (Kribs' heterogeneous types I–II), but central cells of multiseriates only square to weakly procumbent, with occasional sheath cells.
Crystals infrequent, solitary, prismatic, in ray cells.

Characteristics of the Genus Cistus

Growth rings faint to distinct. Vessels mainly diffuse (or wood with a slight tendency towards semi-ring-porosity); almost exclusively solitary, small and numerous. Perforations simple. Vessel wall pits obscurely vestured. Faint spiral thickenings occasionally present. Fibres thick- to very thick-walled, with numerous bordered, sometimes obscurely vestured pits in the radial and tangential walls, sometimes with faint annular to spiral thickenings. Parenchyma scanty paratracheal and diffuse apotracheal and as infrequent marginal cells. Rays ca. 10/mm, 1–4-seriate, heterocellular but central cells of multiseriates only square to weakly procumbent.

Separation of Cistus Species

The two species are very similar in their wood anatomy. Further testing is required to establish whether the presence of crystals in *C. salvifolius* and their absence in *C. creticus* are really diagnostic.

COMPOSITAE

The three woody genera of the Compositae in Israel, *Artemisia, Inula* and *Pluchea,* share many wood anatomical characters and are well within the range described for the whole family (cf. Carlquist, 1966 and literature cited there). *Inula* differs from the other two genera by its exceptionally wide rays. The wood anatomical differences between *Artemisia* and *Pluchea* are only slight (see descriptions).

Artemisia arborescens L.

Plate 27A–C

Shrub; steep rocky slopes; Coastal Galilee, Acco Plain, Sharon Plain, Upper and Lower Galilee, Mt. Carmel, Gilead(?), Ammon; Mediterranean element.

Growth rings absent or very faint. Vessels diffuse, ca. 130/mm^2, solitary (ca. 45%) or in radial to tangential multiples of 2–6(8) or in clusters, including some vascular tracheids, rounded to angular in cross-section, tangential diameter 15–60 μm, radial diameter up to 70 μm, walls 2–6 μm thick. Vessel member length 210(130–330) μm. Perforations simple in transverse to oblique end walls. Inter-vessel pits alternate, rounded to polygonal, 4–6 μm in diameter, with slit-like, mostly crossed, sometimes coalescent apertures. Vessel-parenchyma and vessel-ray pits similar but half-bordered; weakly storied together with parenchyma cells and fibres.
Fibres 360(240–480) μm, thin- to medium thick-walled, with simple pits mainly in radial walls; weakly storied.
Parenchyma scanty paratracheal, fusiform and in 2-celled strands; weakly storied together with vessel members.
Rays ca. 7/mm, 1–4(5)-seriate, of various heights up to 1(2) mm, composed of procumbent and square cells in the multiseriate parts of the rays (including pronounced sheath cells) and square to upright marginal cells.
Crystals not observed.

Artemisia monosperma Del.

Plate 27D

Chamaephyte; maritime and semi-stabilized desert sands; Acco Plain, Sharon Plain, Philistean Plain, W. and N. Negev, Arava Valley, N. Sinai, Edom; Saharo-Arabian element extending along the East Mediterranean coast.

Growth rings faint. Vessels diffuse, in a radial pattern, 90–120/mm^2, rarely solitary (ca. 10%), mostly in radial multiples and clusters of up to 15 elements (often including some parenchyma cells, vascular tracheids or fibres); angular to rounded in cross-section, tangential diameter 20–70 μm, radial diameter up to 80 μm, walls 2–3 μm thick. Vessel member length 180(110–270) μm. Perforations simple in oblique end walls. Inter-vessel pits alternate to almost scalariform, round to oblong, diameter 4–6 μm or up to 10 × 4 μm, with slit-like, partly crossed, sometimes coalescent apertures. Vessel-parenchyma and vessel-ray pits similar but half-bordered. Vessel members somewhat storied. Some vessels with faint spiral sculpturings, some with gummy contents.

Fibres 340(220–430) μm long, thin- to medium thick-walled, with numerous simple pits in radial and tangential walls.

Parenchyma scanty paratracheal and marginal apotracheal; in 2-celled strands; the marginal parenchyma developing into interxylary cork.

Rays 5–8/mm, 1–5(8)-seriate, up to 37 cells (more than 1 mm) high, heterocellular, composed of square central, but mostly upright marginal and sheath cells. Suberization of rays present.

Interxylary cork develops from marginal parenchyma of each growth ring, in 1–2-seriate bands.

Crystals not observed.

Characteristics of the Genus Artemisia

Growth rings absent to faint. Vessels diffuse. Perforations simple. Fibres thin- to medium thick-walled with simple pits. Multiseriate rays up to 1 mm or more in height. Otherwise the two species differ significantly in their wood anatomy.

Key to Artemisia Species

Fibres with pits mainly in radial walls; marginal parenchyma absent. **A. arborescens**

Fibres with numerous pits in radial and tangential walls; marginal parenchyma which develops into interxylary cork present. **A. monosperma**

Inula crithmoides L.

Plate 28A

Shrub, woody at the base; saline marshes and rocks exposed to sea spray; Coastal Galilee, Acco Plain, coast of Carmel, Sharon Plain, Esdraelon Plain, Central Negev, Upper Jordan Valley, Beit Shean Valley, Lower Jordan Valley, Dead Sea area; Mediterranean element.

Growth rings faint. Vessels diffuse; angular to rounded in cross-section, of two size classes: the wider vessels ca. 45/mm^2 (tangential diameter 20–50 μm, radial diameter up to 60 μm, walls 2–4 μm thick), seemingly solitary (ca. 40%) or in mostly radial multiples of 2–4 or small clusters, often associated with groups of few to numerous very narrow vessels and vascular tracheids (ca. 15 μm in diameter), which are very difficult to distinguish from fibres in cross-section. Vessel member length 120(80–230) μm. Perforations simple in horizontal to oblique end walls. Inter-vessel pits mostly alternate, rounded, rarely oblong, 4–5(6) μm in diameter, with slit-like apertures. Vessel-parenchyma and vessel-ray pits similar but half-bordered. Faint spiral thickenings present in some of the vessels. Some vessels with gummy contents.

Fibres 300(190–440) μm long, medium thick-walled, with simple pits somewhat more numerous in radial than in tangential walls.
Parenchyma scanty paratracheal to vasicentric, fusiform or in 2–3-celled strands.
Rays ca. 3/mm, mostly very wide and tall (4–20-seriate and up to or over 2 mm high), infrequently 1–3-seriate and low; heterocellular, composed of procumbent, square and upright cells.
Crystals not observed.

Inula viscosa (L.) Ait.
Plate 28B–D; Fig. 8A

Chamaephyte or perennial herb, woody at base; marly soils on hillsides, often newly disturbed, alluvial soils, marshes and banks of water courses; Acco Plain, coast of Carmel, Sharon Plain, Philistean Plain, Samaria, Shefela, Judean Mts., Judean Desert, Dan Valley, Hula Plain, Upper Jordan Valley, Dead Sea area, Golan, Gilead, Ammon, Moav, Edom; Mediterranean element.

Growth rings faint to fairly distinct. Wood semi-ring-porous or vessels diffuse. Vessels angular to rounded in cross-section, of two size classes: the wider vessels ca. 60–70/mm^2 (tangential diameter 25–80 μm, radial diameter up to 100 μm, walls 2–3 μm thick), seemingly solitary (ca. 50%) or in radial multiples of 2–6 or small clusters, often associated with groups of few to numerous narrow vessels and vascular tracheids (15–20 μm in diameter), especially in the latewood. Vessel member length 190(110–250) μm. Perforations simple in mostly oblique end walls. Inter-vessel pits mostly alternate, rounded, rarely oblong, 4–6(9) μm in diameter, with slit-like, sometimes coalescent apertures. Vessel-parenchyma and vessel-ray pits similar but half-bordered. Some vessels with gummy contents.
Fibres 340(210–450) μm long, thin- to medium thick-walled, with numerous simple pits mainly in radial walls or also fairly frequent in tangential walls, rarely with conspicuous spiral thickenings.
Parenchyma scanty paratracheal, fusiform or in 2-celled strands.
Rays 2–3/mm, mostly very wide and high, 4–15(20)-seriate and up to (2)4 mm high, with margins tending to merge with axial parenchyma, infrequently 1–3-seriate and low; heterocellular, mainly composed of upright and square cells with a small proportion of weakly procumbent cells.
Crystals not observed.
Trabeculae of Sanio continuous over several elements (including wide and narrow vessel members, fibres and axial parenchyma cells as well as ray cells) in a radial file, fairly frequent in one specimen.

Characteristics of the Genus Inula

Vessels of two size classes: the wider ones seemingly solitary and in radial multiples often associated with narrow vessels and vascular tracheids. Perforations simple. Fibres with simple pits. Parenchyma scanty paratracheal. Rays mostly very wide and high; heterocellular.

Separation of Inula Species

In *I. crithmoides* the small vessels are less distinct in cross-section, and the fibres have more pits in the tangential walls than in *I. viscosa*. Whether these differences are diagnostic must be tested in more (possibly also more mature) specimens.

Pluchea dioscoridis (L.) DC. [*Conyza dioscoridis* (L.) Desf.]
Plate 29A and B

Shrub; banks of streams; Sharon Plain, Philistean Plain, Lower Galilee, Esdraelon Plain, Judean Desert, Upper Jordan Valley, Beit Shean Valley, Golan; East Sudanian and Saharo-Arabian element.

Growth rings absent or very faint. Vessels diffuse, 20–60/mm^2, solitary (ca. 25%) or in radial multiples of 2–5(7), rarely clustered, round to angular in cross-section, tangential diameter 25–80 μm, radial diameter up to 100 μm, walls 2–5 μm thick. Vessel member length 290(140–400) μm. Perforations simple in oblique end walls. Inter-vessel pits mostly alternate, rounded, 3–6 μm in diameter, with slit-like apertures. Vessel-parenchyma and vessel-ray pits similar but half-bordered.
Fibres 620(460–900) μm long, thin- to medium thick-walled, with simple pits more numerous in radial than in tangential walls.
Parenchyma vasicentric or scanty paratracheal, in 2–4-celled strands.
Rays 4–6/mm, (1)4–7(8)-seriate, mostly up to 1–2(5) mm high; heterocellular, mainly composed of upright cells, but with some square to procumbent cells; some aggregate; some cells occasionally crystalliferous.
Crystals solitary, prismatic, infrequent in ray cells.

CRUCIFERAE

Zilla spinosa (L.) Prantl
Plate 29C and D

Perennial. Var. *spinosa*: mainly sandy deserts; Judean Desert, C. and S. Negev, Lower Jordan Valley, Arava Valley, deserts of Moav and Edom. Var. *microcarpa*: mainly sandy desert; C. and S. Negev, Dead Sea area, Arava Valley, Sinai, desert of Edom. Saharo-Arabian element.

Growth rings faint to distinct. Vessels diffuse or wood with a tendency to semi-ring-porosity. Vessels ca. 160/mm^2 (excluding very narrow elements), mainly in radial multiples of 2–6(15) or clusters, often including very narrow vessels and

vascular tracheids, more rarely solitary (ca. 15%), angular to rounded in cross-section, tangential diameter 15–70 μm, radial diameter up to 100 μm, walls 2–3 μm thick. Vessel member length 140(80–200) μm. Perforations simple in mostly oblique end walls. Inter-vessel pits vestured, mostly alternate, rounded to polygonal or slightly oblong, 5–8 μm in diameter, with slit-like apertures. Vessel-parenchyma and vessel-ray pits similar but half-bordered. Vessel members storied together with adjacent short fibres and parenchyma cells. Some vessels with gummy contents.
Fibres 270(140–380) μm long, medium thick-walled, with numerous simple pits in radial and tangential walls.
Parenchyma scanty paratracheal (often difficult to discern in the vessel multiples); mostly fusiform, occasionally in 2-celled strands.
Rays ca. 4/mm, (1)3–5(6)-seriate, up to 1.5(3) mm high; heterocellular, composed mainly of square to upright cells (including some sheath cells) and some very weakly procumbent cells.
Crystals not observed.

ELAEAGNACEAE

Elaeagnus angustifolia L.
Plate 30A and B

Shrub or tree; deciduous; cultivated and probably escaped from cultivation; especially on sand dunes of coastal Galilee and Acco Plain; Mediterranean element.

Growth rings distinct. Wood ring-porous. Earlywood vessels embedded in ground tissue of tracheids, ca. 50/mm^2, solitary (ca. 35%) or in tangential to radial multiples of 2–4(5) or small clusters, rounded in cross-section, tangential diameter 30–130 μm, radial diameter up to 150 μm, walls 2–4 μm thick. Latewood vessels solitary or in radial to tangential pairs, and embedded in intergrading tracheids near the growth ring boundaries; rounded to angular in cross-section, tangential diameter 20–60 μm, radial diameter up to 80 μm. Vessel member length 180(110–230) μm. Perforations simple in oblique to transverse end walls (vestigial bars once observed in a very narrow vessel). Inter-vessel pits alternate, rounded, ca. 4–7 μm in diameter, with slit-like apertures. Vessel-tracheid pits similar but diffuse, and often with crossed apertures. Vessel-parenchyma and vessel-ray pits more or less simple, round to gash-like, ca. 4 μm in diameter or up to 4 × 11 μm. Vessel walls with faint spiral thickenings in wide vessels and distinct spirals in narrow ones. Thin-walled tyloses present in many vessels.
Vasicentric tracheids (sometimes intergrading with narrow vessels, i.e., vascular tracheids) forming ground tissue in earlywood and latest formed latewood, usually also present around other latewood vessels; with distinct spiral thickenings.

Wood Anatomy

Fibres 580(340–800) μm long, medium thick-walled, with rather infrequent distinctly bordered pits (diameter ca. 5 μm) more numerous in radial than in tangential walls; mostly without, sometimes with faint spiral thickenings.
Parenchyma apotracheal diffuse or diffuse-in-aggregates, rarely touching the vessels, fusiform or in 2-celled strands; with a tendency to be storied.
Rays ca. 5/mm, (1)3–7(9)-seriate, up to 35(58) cells (0.8 mm) high; homocellular, composed of procumbent cells only (with the exception of some square to weakly upright cells in low uniseriates).
Crystals not observed.

ERICACEAE

Arbutus andrachne L.

Plate 30C and D

Shrub or tree; evergreen; maquis and forest, mainly on marly soil; Upper and Lower Galilee, Mt. Tabor, Mt. Carmel, Samaria, Judean Mts., Gilead, Ammon; East Mediterranean element.

Growth rings distinct. Wood diffuse to weakly semi-ring-porous. Vessels 100–130/mm², mostly in radial multiples of 2–6(12), more rarely solitary (10–20%) or in clusters; angular to rounded in cross-section, tangential diameter 25–80 μm, radial diameter up to 120 μm, walls 2–3 μm thick. Narrow vessels sometimes intergrading with vascular tracheids. Vessel member length 420(230–660) μm. Perforations mainly simple, partly scalariform, especially in the small vessel members, with 1–4(6) bars or even reticulate, in oblique end walls. Inter-vessel pits opposite to alternate, rounded, 5(9) μm in diameter, with slit-like apertures. Vessel-parenchyma and vessel-ray pits similar but half-bordered. Vessels with coarse and distinct spiral thickenings.
Fibres 610(350–830) μm long, thin- to medium thick-walled, mostly with simple pits confined to radial walls, partly septate and without spiral thickenings (libriform fibres); partly with distinctly bordered pits in radial and tangential walls and with spiral thickenings (fibre-tracheids) often associated with vessel multiples, especially in the latewood; rarely chambered crystalliferous.
Parenchyma very sparse, scanty paratracheal and diffuse apotracheal, in 2–5-celled strands; sometimes chambered crystalliferous.
Rays ca. 8/mm, 1–4(5)-seriate, 1–25(43) cells high; heterocellular and tending to be of two distinct sizes (Kribs' heterogeneous type II); some of the upright cells chambered crystalliferous.
Crystals absent or prismatic, solitary or in aggregates together with minute ones in chambered axial parenchyma (in 3–6-celled chains) and upright chambered ray cells, rarely in chambered fibres.

Archaeological records: A powder box from the "Cave of Letters", Judean Desert; 1st century A.D. (A. Fahn).

EUPHORBIACEAE

The large, cosmopolitan family of the Euphorbiaceae is represented in Israel by two genera with woody species, viz. *Euphorbia* and *Ricinus*, exhibiting only a small part of the large wood anatomical range of variation in the family (cf. Metcalfe & Chalk, 1950). The differences in wood anatomy between *Ricinus* and *Euphorbia* are possibly partly related to the different habit and ecology of the species; partly they probably reflect the different taxonomic alliances of the genera which belong to different tribes in the subfamily Euphorbioideae (=Crotonoideae) according to Airy Shaw (1966).
Ray size and fibre wall thickness are reliable technical characters for distinction of the genera on the basis of wood anatomy (see the descriptions).

Euphorbia hierosolymitana Boiss.
Plate 31A and B

Dwarf shrub. Var. *hierosolymitana*: batha and garigue; Philistean Plain, Upper Galilee, Mt. Carmel, Esdraelon Plain, Mt. Gilboa, Samaria, Judean Mts., C. Negev, Golan, Gilead, Ammon, Moav, Edom. Var. *ramanensis*: rocky ground; Judean Desert, N. and C. Negev. East Mediterranean element with slight extensions into adjacent Irano-Turanian territories.

Growth rings faint. Vessels diffuse, 125–170/mm^2, mostly in radially arranged clusters, some in radial multiples of up to 8, or solitary; angular to rounded in cross-section; tangential diameter 15–50 μm, radial diameter up to 70 μm, walls 3–4 μm thick. Vessel member length 250(140–370) μm. Perforations simple in oblique end walls. Inter-vessel pits alternate, rounded to oblong, occasionally partly scalariform, 5–10 μm in horizontal diameter, with slit-like apertures. Vessel-parenchyma and vessel-ray pits similar but half-bordered to simple.
Fibres 630(440–830) μm long, medium thick-walled, with simple pits in radial walls.
Parenchyma scanty paratracheal and abundantly diffuse apotracheal or diffuse-in-aggregates; in 2–3-celled strands.
Rays 10–16/mm; of two distinct sizes, 1–6-seriate and 10–30-seriate; of various heights, up to more than 1 mm; occasionally compound; heterocellular (Kribs' heterogeneous types I–III); ray cells increasingly procumbent towards the centre; in multiseriates, in contrast to uniseriates, walls of most cells unlignified; many of the cells crystalliferous.
Laticifers, some very wide, with thick walls present in many of the wide rays.
Crystals prismatic, in aggregates of large and very small ones, in many horizontally and some vertically chambered ray cells.

Ricinus communis L.
Plate 31C and D

Shrub; subspontaneous in the hotter parts of the region, in the Dead Sea area, in wadis, at roadsides and in waste places; Sharon Plain, Philistean Plain, Esdraelon Plain, Samaria, Judean Mts., W. and N. Negev, Hula Plain, Beit Shean Valley, Moav; Sudanian and East African element.

Growth rings fairly distinct. Vessels diffuse; 10–50/mm², frequency higher in early- than in latewood; solitary and in radial multiples of up to 9 (multiples more frequent in early- than latewood), sometimes in radially arranged clusters, some radial multiples appear double as a result of overlapping end walls; rounded to angular in cross-section, tangential diameter 30–190 μm, radial diameter up to 200 μm; walls 3–5 μm thick. Vessel member length 540(320–840) μm. Perforations simple in oblique end walls. Inter-vessel pits alternate, rounded to polygonal, 10–12 μm in diameter, with short, oval to slit-like apertures. Vessel-parenchyma and vessel-ray pits large, simple or with reduced borders, round to oval.
Fibres 960(460–1,370) μm long; thin-walled; with simple pits in radial walls.
Parenchyma scanty paratracheal and diffuse apotracheal; in 2–5-celled strands.
Rays 7–9/mm; 1–5(6) cells wide, up to 40 cells high; heterocellular, composed of square and upright cells, a few also with procumbent central cells; some cells crystalliferous.
Crystals prismatic and in irregular clusters or aggregates in chambered or nonchambered ray cells.

FAGACEAE

Quercus boissieri Reut. (*Q. infectoria* Oliv.)
Plate 32A and B; Fig. 5D

Small tree; deciduous; forest and maquis, usually preferring higher altitudes or sites with colder winters. Var. *boissieri*: Upper Galilee, Mt. Carmel, Samaria, Judean Mts., Gilead. Var. *latifolia*: Upper and Lower Galilee, Mt. Carmel, Samaria, Judean Mts., Gilead. East Mediterranean element.

Growth rings distinct. Wood ring-porous. Vessels almost exclusively solitary, rarely in pairs, forming a flame-like or dendritic pattern together with parenchyma cells and vasicentric tracheids; 5–75/mm², earlywood vessels rounded, latewood vessels angular to slightly rounded in cross-section, tangential diameter 30–200 μm, radial diameter up to 250 μm, walls 2–3 μm thick. Vessel member length 450 (250–590) μm. Perforations simple in horizontal to oblique end walls. Vessel-tracheid pits widely spaced or alternate, round, 7–9 μm in diameter, with slit-like or oval apertures. Vessel-parenchyma and vessel-ray pits large and simple or with reduced borders, round to horizontally or vertically elongate.
Vasicentric tracheids abundant, forming the ground tissue associated with the vessels together with axial parenchyma.
Fibres 1,020(670–1,380) μm long, thick- to very thick-walled, with minutely bordered pits mainly in radial walls; sometimes gelatinous.

Parenchyma scanty paratracheal and apotracheal diffuse, diffuse-in-aggregates and in 1-seriate tangential bands; in many-celled strands; often chambered crystalliferous.
Rays ca. 10/mm, of two distinct sizes, 1(2)-seriate rays, 2–11(18) cells high, and multiseriates 24–40 cells wide and up to 11 mm high; typically homocellular, but some upright and square cells scattered in the multiseriate rays; multiseriates partly aggregate; some cells chambered crystalliferous.
Crystals solitary, prismatic, in chambered axial parenchyma and ray cells; some crystalliferous cells enlarged and/or sclerified.

Archaeological records: Golan Heights; 3rd millennium B.C. (N. Liphschitz & Y. Waisel).

Quercus calliprinos Webb
Plate 32C

Tree or shrub; evergreen; maquis and forest; on terra rossa, rendzina, sandy clay and also in rock fissures, from sea level to 1,200 m; Sharon Plain, Upper and Lower Galilee, Mt. Carmel, Samaria, Shefela, Judean Mts., Dan Valley, Gilead, Ammon, Edom; East Mediterranean element.

Growth rings fairly distinct. Vessels diffuse, almost exclusively solitary, occasionally in pairs, forming, together with vasicentric tracheids and parenchyma cells, an irregular flame-like or dendritic pattern; 5–20/mm^2; rounded in cross-section, tangential diameter 25–130 μm, radial diameter up to 160 μm, walls ca. 5 μm thick. Vessel member length 360(140–600) μm. Perforations simple in mainly oblique end walls. Vessel-tracheid pits round, diffuse, opposite or alternate, ca. 7 μm in diameter, with slit-like apertures. Vessel-parenchyma pits simple or with reduced borders, round to elongate. Vessel-ray pits large and simple, mainly vertically elongate.
Vasicentric tracheids abundant, foring the ground tissue associated with the vessels together with axial parenchyma.
Fibres 810(580–980) μm long, thick- to very thick-walled, sometimes gelatinous; with simple to minutely bordered pits more numerous in radial than in tangential walls.
Parenchyma very scanty paratracheal and apotracheal diffuse, diffuse-in-aggregates and in wavy 1(2)-seriate tangential bands; in 2- to many-celled strands; often chambered crystalliferous.
Rays 10–13/mm, of two distinct sizes, 1(2)-seriates up to 17 cells high and multiseriates 14–25 cells wide and up to more than 8 mm high; typically homocellular but some weakly procumbent, square or upright cells scattered in multiseriates; multiseriates partly aggregate; some cells chambered crystalliferous.
Crystals solitary, prismatic, in chambered axial parenchyma and ray cells; some crystalliferous cells enlarged and/or sclerified.

Quercus ithaburensis Decne.
Plate 32D

Tree; deciduous; forest and forest remnants on various soils (rendzina, terra rossa, basalt, sandy soils), up to 500 m above sea level; Coastal Galilee, Şamaria, Dan Valley, Hula Plain, Upper Jordan Valley, Gilead; East Mediterranean element.

Growth rings fairly distinct. Wood ring- to semi-ring-porous. Vessels almost exclusively solitary, occasionally in pairs, forming a radial, sometimes oblique or dendritic pattern; 3–12/mm^2, rounded in cross-section, tangential diameter 50–200 μm, radial diameter up to 250 μm, walls 3–5 μm thick. Vessel member length 360(260–540) μm. Perforations simple in horizontal to oblique end walls. Vessel-tracheid pits rounded, widely spaced, alternate or opposite, ca. 6 μm in diameter, with slit-like apertures. Vessel-parenchyma pits simple or with reduced borders, horizontally elongate. Vessel-ray pits large and simple, round to elongate in various directions. Tyloses abundant.
Vasicentric tracheids surrounding wide and narrow vessels difficult to distinguish from fibres in cross-section.
Fibres 880(360–1,750) μm long, thick- to very thick-walled, with minutely bordered pits more numerous in radial than in tangential walls; partly gelatinous.
Parenchyma very scanty paratracheal and apotracheal diffuse, diffuse-in-aggregates and in short 1-seriate tangential bands; in many-celled strands; many cells chambered crystalliferous.
Rays ca. 11/mm, of two distinct sizes, 1(2)-seriate rays 2–10(20) cells high and multiseriates up to 30 cells or more in width and up to 11 mm high; typically homocellular; many cells crystalliferous; multiseriates often compound or aggregate.
Crystals solitary, prismatic, in chambered or non-chambered ray and parenchyma cells; many of the crystalliferous cells enlarged and sclerified.

Archaeological records: Six locations; Coastal Plain, Beit Shean, Negev; ranging from the Chalcolithic Period to the Arabic Period (N. Liphschitz & Y. Waisel).

Quercus libani Oliv.
Plate 33A

Tree; deciduous; Mt. Hermon; East Mediterranean and West Irano-Turanian element.

Growth rings fairly distinct. Wood ring-porous. Vessels almost exclusively solitary, occasionally in pairs, earlywood vessels in tangential rows or in groups, and narrow latewood vessels forming a radial, sometimes slightly dendritic pattern together with vasicentric tracheids and parenchyma; 18–35/mm^2, rounded in cross-section, diameter 20–170 μm, walls 3–4 μm thick. Vessel member length 340(190–480) μm. Perforations simple in horizontal to oblique end walls. Vessel-tracheid pits alternate or widely spaced, rounded, 6–7 μm in diameter, with slit-like apertures. Vessel-parenchyma and vessel-ray pits large and simple or with reduced borders, round or elongate in various directions. Tyloses sometimes present.

Wood Anatomical Descriptions

Vasicentric tracheids abundant, forming ground tissue associated with the vessels, but difficult to distinguish from fibres in the latewood in cross-section.
Fibres 940(570–1,340) μm long, thick- to very thick-walled; with minutely bordered pits more numerous in radial than in tangential walls; partly gelatinous.
Parenchyma scanty paratracheal and apotracheal diffuse, diffuse-in-aggregates and in short, more or less tangential 1(2)-seriate bands; in long strands; infrequently chambered crystalliferous.
Rays 7–13/mm, of two distinct sizes, uniseriates up to 15 cells high and multiseriates 14–28 cells wide and up to ca. 2.5 mm high; typically homocellular, but with some square or upright cells scattered in the multiseriates; multiseriates partly aggregate or compound; some cells crystalliferous.
Crystals solitary, prismatic, in ordinary or chambered ray cells and chambered axial parenchyma; some crystalliferous cells enlarged and/or sclerified.

Characteristics of the Genus Quercus

Vessels almost exclusively solitary, in a radial, flame-like or dendritic pattern. Perforations simple. Vasicentric tracheids present. Large and simple vessel-parenchyma and vessel-ray pits present. Fibres thick- to very thick-walled, with minutely bordered pits. Parenchyma scanty paratracheal, and apotracheal diffuse, diffuse-in-aggregates and in short narrow tangential bands. Rays of two distinct sizes, uniseriates and very wide and high multiseriates; the latter partly compound. Crystals solitary, prismatic, in chambered and non-chambered ray and parenchyma cells.

Key to Quercus Species

1. Wood diffuse-porous; vasicentric tracheids often distinct in cross-section. **Q. calliprinos**
– Wood ring-porous, sometimes semi-ring-porous; and if diffuse, vasicentric tracheids difficult to distinguish from fibres in cross-section 2
2. Latewood vessels often numerous and angular to slightly rounded. **Q. boissieri**
– Latewood vessels infrequent, mostly rounded. **Q. libani**
Q. ithaburensis

JUGLANDACEAE

Juglans regia L.
Plate 33B–D

Cultivated fruit tree; deciduous; native in southeast Europe, Himalayas and China.

Wood Anatomy

Growth rings distinct. Wood diffuse- to semi-ring-porous. Vessels 2–20/mm², mostly solitary (70%), partly in radial multiples of 2–4(5); rounded to angular in cross-section; tangential diameter 40–150(280) μm, radial diameter up to 175(360) μm, walls 3–6 μm thick. Vessel member length 470(400–590) μm. Perforations simple in oblique end walls. Inter-vessel pits alternate, polygonal to rounded, 7–12 μm in diameter, with slit-like apertures. Vessel-parenchyma and vessel-ray pits opposite to alternate, half-bordered, with wide, gash-like to slit-like, partly extended apertures. Thin-walled tyloses frequent.
Fibres 900(550–1,180) μm long, mostly medium thick-walled, and thick-walled and flattened in narrow latewood zones; with minutely bordered pits in radial walls; sometimes gelatinous.
Parenchyma abundant, vasicentric and apotracheal diffuse and diffuse-in-aggregates to 1-seriate tangential bands, and in 1-seriate, discontinuous marginal bands; in 4–8-celled strands.
Rays (4)6–8/mm, 1–3(5)-seriate, up to ca. 35 cells high; homocellular, composed of procumbent cells only, or heterocellular with procumbent central cells and one row of square marginal cells; infrequently crystalliferous.
Crystals absent or infrequent, solitary, prismatic, in ray cells.

LABIATAE

The genera of the Labiatae with woody species in Israel, *Coridothymus, Prasium, Rosmarinus, Salvia* and *Teucrium*, have many wood anatomical characters in common: numerous narrow vessels, spiral vessel wall thickenings, libriform fibres, scanty paratracheal parenchyma and heterocellular rays with a high proportion of upright cells. These features have also been reported for other shrubby Labiatae by Metcalfe & Chalk (1950). In the taxa from Israel, *Coridothymus* stands out somewhat on account of its narrow rays.

Coridothymus capitatus (L.) Reichenb. fil.
Plate 34A

Chamaephyte; batha on rocky ground and compact soil; Coastal Galilee, Acco Plain, coast of Carmel, Sharon Plain, Philistean Plain, Upper and Lower Galilee, Mt. Carmel, Samaria, Shefela, Judean Mts., Judean Desert, Golan, Gilead, Ammon, Moav; Mediterranean element.

Growth rings distinct, boundaries wavy (lobed stem). Wood diffuse- to semi-ring-porous. Vessels ca. 250–300/mm², solitary, in usually radial multiples of 2–4(7) and in small clusters, associated with numerous vascular tracheids (difficult to distinguish from fibres or narrow vessels in cross-section); rounded to angular in cross-section, diameter 10–65 μm, walls 1.5–2 μm thick. Vessel member length 160(100–200) μm. Perforations simple in horizontal to very oblique end walls. Inter-vessel pits alternate, rounded to polygonal, ca. 4 μm in diameter, with slit-like, sometimes coalescent apertures. Vessel-parenchyma and vessel-ray

pits similar but half-bordered or simple. Narrow vessels and vascular tracheids with faint spiral thickenings.
Fibres 340(250–570) μm long, medium thick- to thick-walled, with simple pits more numerous in radial than in tangential walls.
Parenchyma sparse, scanty paratracheal, in 2–3-celled strands.
Rays inconspicuous in tangential section, clear in cross-section but short; 1(3)-seriate, 1 to several tall upright cells high.
Crystals not observed.

Prasium majus L.
Plate 34B

Climbing shrub; maquis, garigue and shady hedges; Coastal Galilee, Sharon Plain, Philistean Plain, Upper and Lower Galilee, Mt. Carmel, Esdraelon Plain, Mt. Gilboa, Samaria, Judean Mts., N. Negev, Upper and Lower Jordan Valley, Golan, Gilead, Ammon, Moav; Mediterranean element.

Growth rings distinct. Wood semi-ring- to ring-porous. Vessels 160–220/mm^2, solitary (ca. 15%), in tangential (predominantly at the beginning of growth ring) to radial multiples of 2–5(7) or small clusters; forming ground tissue in latest formed latewood together with vascular tracheids (only visible in radial section); rounded to angular in cross-section, tangential diameter 20–70 μm, radial diameter up to 90 μm, walls ca. 3–5 μm thick. Vessel member length 220(150–270) μm. Perforations simple in mostly oblique end walls. Inter-vessel pits alternate to opposite, round to polygonal, ca. 5 μm in diameter, with slit-like, sometimes crossed, often coalescent apertures. Vessel-parenchyma and vessel-ray pits similar but half-bordered or simple. Spiral sculpturing (thickenings and/or grooves?) mainly restricted to some of the narrow vessels and vascular tracheids.
Fibres 360(240–490) μm long, medium thick-walled, with simple pits, more numerous in radial than in tangential walls.
Parenchyma sparse, scanty paratracheal and marginal; fusiform and in 2-celled strands.
Rays 6–8/mm, 1–4(5)-seriate, 1–45 cells (more than 1 mm) high, mainly composed of upright cells, with some square or very weakly procumbent cells in the central part of the multiseriates.
Masses of fine crystalline material in most of the ray cells and some in the axial parenchyma cells.

Rosmarinus officinalis L.
Plate 34C

Cultivated shrub; Mediterranean element.

Growth rings distinct. Wood ring- to semi-ring-porous. Vessels 170–180/mm^2, mainly in loose clusters and tangential multiples of 2–3(4), associated with numerous vascular tracheids (difficult to distinguish from fibres or very narrow

vessels in cross-section), partly forming a dendritic pattern; rounded to angular in cross-section, tangential diameter 13–50 μm, radial diameter up to 80 μm, walls 2–4 μm thick. Vessel member length 220(150–270) μm. Perforations simple in oblique end walls. Inter-vessel pits alternate, rounded to polygonal, with slit-like, sometimes coalescent apertures, 4–5 μm in diameter. Vessel-parenchyma and vessel-ray pits similar but half-bordered. Most vessels and vascular tracheids with distinct spiral thickenings; some vessels with gummy contents. Tyloses in some vessels.
Fibres 470(270–610) μm long; thick- to very thick-walled; with minutely bordered pits mainly in radial walls.
Parenchyma very sparse, mainly restricted to some cells associated with the vessels and vascular tracheids; mainly fusiform.
Rays 3–5/mm, 1–4-seriate, up to ca. 27 cells high; heterocellular, composed of square, upright and weakly procumbent cells.
Crystals not observed.

Salvia fruticosa Mill. (*S. triloba* L. fil.)
Plate 34D

Shrub; garigue and maquis; Coastal Galilee, coast of Carmel, Mt. Gilboa, Samaria, Judean Mts.; mainly East Mediterranean element.

Growth rings distinct. Wood semi-ring-porous. Vessels ca. 150/mm², solitary (ca. 50%) or in tangential to radial multiples of 2(7) predominantly wide vessels, or in small clusters; narrow vessels forming ground tissue in latest formed latewood together with vascular tracheids (only visible in radial section); rounded to slightly angular in cross-section, tangential diameter 15–80 μm, radial diameter up to 100 μm, walls 2–5 μm thick. Vessel member length 200(130–270) μm. Perforations simple in mainly oblique end walls. Inter-vessel pits mostly alternate, rounded, ca. 5 μm in diameter, with slit-like, sometimes crossed apertures. Vessel-parenchyma and vessel-ray pits similar but half-bordered or simple, occasionally elongate. Narrow vessels and vascular tracheids with very densely spaced spiral thickenings. Wider vessels sometimes with faint grooves. Some vessels with thin-walled tyloses.
Fibres 400(250–590) μm long, medium thick-walled, with simple pits mainly in radial walls.
Parenchyma sparse, scanty paratracheal and marginal; mostly in 2-celled strands, sometimes fusiform.
Rays ca. 10/mm, 1–3(5)-seriate, 1–20(24) cells high; heterocellular, mainly composed of upright and square cells, with (mostly weakly) procumbent cells in central part of the multiseriates.
Crystals not observed.
Trabeculae of Sanio continuous over several elements (including vessel members and fibres) in a radial file, observed in one specimen.

Teucrium creticum L.
Plate 35A

Chamaephyte; garigue and batha; Upper and Lower Galilee, Mt. Carmel, Samaria, Judean Mts.; East Mediterranean element.

Growth rings distinct. Vessels diffuse but often a narrow region with distinctly smaller ones present at the end of the growth rings; ca. 160/mm^2; in clusters and radial and tangential multiples of 2–5, associated with vascular tracheids, some solitary, occasionally tending to form tangential bands; angular to rounded in cross-section, tangential diameter 20–60 μm, radial diameter up to 100 μm, walls 1.5–3 μm thick. Vessel member length 250(160–350) μm. Perforations simple in oblique end walls. Inter-vessel pits alternate, rounded to polygonal, ca. 5 μm in diameter, with slit-like apertures. Vessel-parenchyma and vessel-ray pits similar but half-bordered or simple. Some vessel elements and vascular tracheids with spiral thickenings.
Fibres 470(310–630) μm long, medium thick- to thick-walled, with simple or minutely bordered pits more numerous in radial than in tangential walls.
Parenchyma very scanty paratracheal; in 2-celled strands.
Rays 8–12/mm; 1–3(4)-seriate, 1–24 cells high; weakly heterocellular, mainly composed of upright cells with some square and occasionally weakly procumbent cells.
Crystals not observed.

LAURACEAE

Laurus nobilis L.
Plate 35C and D; Fig. 8F

Tree; evergreen; maquis; Upper and Lower Galilee, Mt. Carmel, Samaria, Judean Mts., Dan Valley; Mediterranean element.

Growth rings distinct. Vessels diffuse, 15–50/mm^2, solitary (14–35%), in radial multiples of 2–6(11), occasionally in small clusters; rounded to somewhat angular in cross-section, tangential diameter 40–90 μm, radial diameter up to 100 μm, walls 3.5 μm thick. Vessel member length 360(140–520) μm. Perforations simple in oblique end walls. Inter-vessel pits alternate, round to elliptic or polygonal, 5–7 μm in diameter, with round to slit-like, often coalescent apertures. Vessel-parenchyma and vessel-ray pits round to elongate, half-bordered or with strongly reduced borders to simple.
Fibres 640(340–840) μm long, thin- to medium thick-walled, with simple to minutely bordered pits, more numerous in radial than in tangential walls; partly septate.
Parenchyma scanty paratracheal, in 2–5-celled strands.
Rays 10–12/mm, (1)2–3-seriates up to 6 cells high, multiseriates up to 30 or more

cells high; mostly heterocellular, with 1 row of square to upright marginal cells (Kribs' heterogeneous type II); many cells crystalliferous; oil cells infrequently present.
Crystals abundant as styloids, one to several per ray cell.
Oil cells occasionally present in rays.

LEGUMINOSAE

The division into three subfamilies (or even into separate families, as adopted in *Flora Palaestina*) is clearly reflected in the wood anatomy of the trees and shrubs of Israel, while all genera of the Leguminosae share vestured inter-vessel pits, libriform fibres and paratracheal parenchyma.
The two genera of the Caesalpinioideae, *Ceratonia* and *Cercis,* share a basically diffuse vessel distribution and vasicentric plus marginal parenchyma, but can be differentiated according to vessel frequency and a number of other quantitative and qualitative features (see the descriptions).
Acacia and *Prosopis,* the two genera of the Mimosoideae, stand out on account of their wide, confluent parenchyma bands. Within *Acacia*, *A. albida* differs markedly from the other species on account of its narrow rays. This character could be used to advocate the reinstatement of the genus *Faidherbia* for this species (cf. Robbertse et al., 1980).
The eight genera of the Papilionoideae (=Faboideae) in Israel, with the exception of *Ononis,* all share an oblique to dendritic pattern of normal and very narrow vessels together with paratracheal parenchyma, spiral thickenings in at least the narrow vessels and vascular tracheids, and a storied structure. Although *Ononis* lacks the typical vessel distribution pattern and spiral thickenings, it shares the two vessel size classes and the storied structure with the other Papilionoideae.
It should be noted that the coincidence of wood anatomical grouping with subfamily delimitation in the genera of Israel and adjacent regions, does not hold true if Leguminosae from other parts of the world are considered (cf. Metcalfe & Chalk, 1950). In particular, the tropical Papilionoideae differ widely in their wood anatomy from the Mediterranean genera.

CAESALPINIOIDEAE

Ceratonia siliqua L.
Plate 36

Tree; evergreen; maquis, especially in lower altitudes; Coastal Galilee, Acco Plain, Sharon Plain, Philistean Plain, Upper and Lower Galilee, Mt. Carmel, Mt. Gilboa, Samaria, Shefela, Judean Mts., Gilead, C. Negev, Edom; Mediterranean element.

Growth rings faint to fairly distinct. Vessels diffuse, ca. 14/mm², solitary (ca. 25%) and in radial multiples of 2–4(8), rounded in cross-section, tangential diameter 25–155 μm, radial diameter up to 180 μm, walls ca. 4 μm thick. Vessel member length 260(150–440) μm. Perforations simple in horizontal to slightly oblique end walls. Inter-vessel pits vestured, opposite or alternate, rounded or polygonal, ca. 6–8 μm in diameter, with slit-like, occasionally coalescent apertures. Vessel-parenchyma and vessel-ray pits similar but half-bordered, sometimes unilaterally compound. Some vessels with gummy contents.
Fibres 830(440–1,240) μm long, medium thick- to thick-walled, with simple pits mainly in radial walls; septate, rarely gelatinous.
Parenchyma mainly paratracheal, vasicentric to aliform, rarely scanty, also very scanty diffuse apotracheal and marginal; in (2)4(8)-celled strands; part of the cells chambered crystalliferous (some non-crystalliferous cells bordering on the vessels with radial partition walls).
Rays ca. 9/mm, 1–4-seriate, up to 25 cells high; heterocellular (Kribs' heterogeneous types II and III).
Crystals usually solitary in short strands of chambered parenchyma cells, rarely also in marginal ray cells.

Archaeological records: Apollonia (coastal plain); early Arabic to Crusader Period (N. Liphschitz & Y. Waisel).

Cercis siliquastrum L.
Plate 37A and B

Tree or shrub; deciduous; maquis; Upper and Lower Galilee, Mt. Carmel, Samaria, Judean Mts.; East and North Mediterranean element.

Growth rings distinct. Wood diffuse- to semi-ring-porous. Vessels ca. 75/mm² in earlywood, ca. 110/mm² in latewood, solitary (20% in earlywood, 10% in latewood) and in radial multiples of 2–7 or clusters (especially in the latewood); rounded to faintly angular in cross-section, tangential diameter 50–110 μm in earlywood, 30–70 μm in latewood, radial diameter up to 130 μm, walls 2–4 μm thick. Vessel member length 240(210–280) μm. Perforations simple in transverse to slightly oblique end walls. Inter-vessel pits vestured, alternate (to opposite), rounded or polygonal, 6–7 μm in diameter, with slit-like apertures. Vessel-parenchyma and vessel-ray pits similar but half-bordered. Spiral thickenings present. Sometimes weakly storied together with the parenchyma. Many vessels with gummy contents.
Fibres 960(410–1,340) μm long, medium thick- to thick-walled, with simple pits mainly in radial walls; partly gelatinous.
Parenchyma scanty paratracheal to vasicentric and scanty diffuse apotracheal and in narrow marginal (initial) bands; in (2)4-celled strands; weakly storied, partly chambered crystalliferous.
Rays ca. 10/mm, 1–4-seriate, mostly ca. 8–16 cells high, but sometimes up to

about twice as high (up to 32 cells); homocellular and entirely composed of procumbent cells or weakly heterocellular and with one marginal row of square, occasionally weakly upright cells; lower rays with a tendency for storying.
Crystals prismatic, solitary or in pairs or groups of three, in chambered parenchyma cells.

MIMOSOIDEAE

Acacia albida Del.
Plate 37C and D

Tree; deciduous; alluvial and sandy soils, banks of wadis and basalt ground; Sharon Plain, Philistean Plain, Lower Galilee, Shefela, Upper Jordan Valley, Beit Shean Valley, Lower Jordan Valley, Dead Sea area; Sudanian element.

Growth rings absent. Vessels diffuse, ca. 4/mm^2, solitary (ca. 60%) or in radial multiples of 2–5 (or small clusters), rounded in cross-section, tangential diameter 50–200 μm, radial diameter up to 260 μm, walls 4–8 μm thick. Vessel member length 230(170–280) μm. Perforations simple in transverse to oblique end walls. Inter-vessel pits vestured, mostly alternate, rounded or polygonal, ca. 7–10 μm in diameter, with slit-like, fairly often coalescent apertures. Vessel-parenchyma and vessel-ray pits similar but half-bordered. Some vessels with gummy contents.
Fibres in more or less tangential bands alternating with the parenchyma, 1,060 (760–1,400) μm long, medium thick- to very thick-walled, with simple pits mainly in radial walls; partly gelatinous.
Parenchyma in wide, confluent, paratracheal bands varying in thickness (up to 25 cells wide); storied; occasionally fusiform, mostly in 2-celled strands, strands bordering on the vessels sometimes up to 4-celled; cells bordering on the fibres sometimes chambered crystalliferous.
Rays ca. 16/mm, mainly uniseriate, sometimes with small biseriate portions, up to 12(30) cells high; homocellular, entirely composed of procumbent cells; weakly storied.
Crystals solitary in chambered parenchyma cells, in short chains bordering on the fibres, and rarely in chambered ray cells.

Acacia gerrardii Benth. subspecies **negevensis** Zoh.
Plate 38

Shrub or tree; evergreen; wadis, sandy and pebbly ground; N., C. and S. Negev, Arava Valley, Edom; less heat-demanding than *A. raddiana* and *A. tortilis*; ascends in the S. Negev up to 800 m above sea level; Sudanian and East African element.

Growth rings faint to fairly distinct. Vessels diffuse, ca. 5/mm^2, solitary (ca. 40%) and in radial multiples of 2–4 or small clusters, rounded in cross-section,

tangential diameter 40–190 μm, radial diameter up to 220 μm, walls ca. 8 μm thick. Vessel member length 160(100–230) μm. Perforations simple in transverse to oblique end walls. Inter-vessel pits vestured, alternate, rounded or polygonal, 6–8 μm in diameter, with slit-like, often coalescent apertures. Vessel-parenchyma and vessel-ray pits similar but half-bordered. Many vessels with gummy contents. Spiral thickenings occasionally present.
Fibres in more or less tangential bands alternating with the parenchyma, 940 (400–1,350) μm long, medium thick- to thick-walled, with simple pits mainly in radial walls; occasionally gelatinous, rarely septate.
Parenchyma abundant and predominantly in paratracheal, confluent bands of varying thickness (up to 35 cells wide), also scanty diffuse apotracheal and marginal in 1–2-seriate bands; rarely fusiform, mostly in 2-celled strands; occasionally chambered crystalliferous (especially the diffuse and marginal strands, and the confluent strands where bordering on the fibres).
Rays ca. 5–6/mm, 1–6-seriate, generally up to 40(100) cells high; homocellular, composed of strongly procumbent cells.
Crystals usually solitary, prismatic, in chambered parenchyma cells.

Acacia raddiana Savi

Plate 39A and B; Fig. 4D

Small tree; evergreen; hot desert wadis, depressions and oases; W., N. and C. Negev, Lower Jordan Valley, Dead Sea area, Arava Valley, deserts of Edom, Sinai; Sudanian element.

Growth rings absent to fairly distinct. Vessels diffuse, ca. 8/mm², solitary (ca. 40%) and in radial multiples of 2–4 or in clusters including some narrow vessels; rounded in cross-section, tangential diameter 50–190 μm, radial diameter up to 190 μm, walls ca. 6–8 μm thick. Vessel member length 250(110–450) μm. Perforations simple in transverse to oblique end walls. Inter-vessel pits vestured, mostly alternate, round to oval or polygonal, ca. 8 μm in diameter, with slit-like, fairly often coalescent apertures. Vessel-parenchyma and vessel-ray pits similar but half-bordered. Spiral thickenings occasionally present in narrow vessels.
Fibres in more or less tangential bands alternating with the parenchyma, 1,320 (670–1,790) μm long, thick- to very thick-walled, with simple pits mainly in radial walls; partly gelatinous, very rarely septate.
Parenchyma abundant paratracheal, mainly in long confluent bands (up to 25 cells wide), also aliform and scanty diffuse apotracheal; fusiform and in 2-celled strands; partly chambered crystalliferous (especially the diffuse parenchyma).
Rays ca. 5/mm, 1–6(7)-seriate, tending to be of two distinct sizes: low uniseriates generally 2–10 cells high, and (2)3–6(7)-seriates up to 60(100) cells high; all rays homocellular, composed of strongly procumbent cells.
Crystals solitary, prismatic, in short chambered parenchyma cells, chambers often enlarged and sclerified, within and on the boundary of fibre bands.

Wood Anatomy

Acacia tortilis (Forssk.) Hayne

Small tree; evergreen; hot desert wadis; Dead Sea area, Arava Valley and adjoining wadis, Sinai; Sudanian element.

Growth rings absent to fairly distinct. Vessels diffuse, ca. 5/mm^2, solitary (75%), or in radial multiples of 2–4 or in clusters including some narrow vessels; rounded in cross-section, tangential diameter 40–180(250) μm, radial diameter up to 200 μm, walls 4–8 μm thick. Vessel member length 210(100–400) μm. Perforations simple in horizontal to slightly oblique end walls. Inter-vessel pits vestured, mostly alternate, sometimes opposite, rounded or polygonal, ca. 8 μm in diameter, often with coalescent apertures. Vessel-parenchyma and vessel-ray pits similar but half-bordered.

Fibres more or less in tangential bands alternating with the parenchyma, 840 (350–1,400) μm long, thick- to very thick-walled, with simple pits mainly in radial walls; often gelatinous.

Parenchyma abundant paratracheal, sometimes aliform, mostly confluent in fairly continuous tangential bands of up to about 30 cells wide; also scanty diffuse apotracheal, and marginal in discontinuous 1(2)-seriate bands; fusiform or in 2-celled strands; partly chambered crystalliferous (especially the diffuse parenchyma).

Rays ca. 5/mm, 1–6-seriate, tending to be of two distinct sizes: low, 1–2-seriate and 2–12 cells high, and 3–6-seriate and up to 35(100) cells high; homocellular, composed of strongly procumbent cells.

Crystals solitary, prismatic, in short chambered parenchyma cells, chambers partly enlarged and sclerified.

Characteristics of the Genus Acacia

Vessels diffuse, infrequent, solitary and in radial multiples of 2–4 or in clusters. Perforations simple. Inter-vessel pits vestured, mainly alternate. Fibres in more or less tangential bands alternating with parenchyma bands; with simple pits mainly in the radial walls; partly gelatinous. Rays homocellular, composed of procumbent cells. Crystals solitary, prismatic, in chambered parenchyma cells.

Key to Acacia Species

1. Rays mainly uniseriate; parenchyma cells storied. **A. albida**
- Rays 1–6-seriate; parenchyma non-storied 2
2. Chambered crystalliferous parenchyma cells partly enlarged and sclerified; fibres thick- to very thick-walled. **A. raddiana**
   A. tortilis
- Enlarged sclerified crystalliferous parenchyma cells absent; fibres medium thick- to thick-walled. **A. gerrardii**

Archaeological records: *Acacia albida* identified in Beit Shean Valley; Chalcolithic

Period (N. Liphschitz & Y. Waisel). Wood belonging to the group *A. raddiana, A. tortilis* and *A. gerrardii* identified in 16 locations in the Negev, Dead Sea area and Sinai; periods ranging from 2000 B. C. to 7th century A. D. (E. Werker; A. Fahn; A. Fahn & E. Zamski; N. Liphschitz & Y. Waisel).

Prosopis farcta (Banks et Sol.) Macbride

Plate 39C and D

Shrub or dwarf shrub; alluvial soils, among crops; saline ground and river banks; Coastal Galilee, Acco Plain, Sharon Plain, Philistean Plain, Upper and Lower Galilee, Esdraelon Plain, Samaria, Shefela, coastal and N. Negev, Dan Valley, Hula Plain, Upper Jordan Valley, Beit Shean Valley, Lower Jordan Valley, Dead Sea area, Moav; W. Irano-Turanian element.

Growth rings distinct. Vessels diffuse, ca. $32/mm^2$, solitary (ca. 25%) and in radial to oblique multiples or clusters of 2–7(11), often including and sometimes entirely composed of very narrow vessel elements; rounded in cross-section, tangential diameter 20–140 μm, radial diameter up to 180 μm, walls 5–7 μm thick. Vessel member length 190(100–280) μm. Perforations simple in horizontal to slightly oblique end walls. Inter-vessel pits vestured, mostly alternate, rounded to polygonal, 5–8 μm in diameter, with slit-like, sometimes coalescent apertures. Vessel-parenchyma and vessel-ray pits similar but half-bordered. Some vessels with gummy contents.

Fibres 620(210–870) μm long, mostly medium thick-walled, with simple pits mainly in radial walls; partly gelatinous, occasionally septate.

Parenchyma paratracheal, abundant, mostly in short to long confluent bands (up to 13 cells wide), also scanty diffuse apotracheal and in initial marginal bands (but then mostly continuous with the paratracheal parenchyma); fusiform or in 2-celled strands, strands bordering on the vessels sometimes up to 5-celled; partly chambered crystalliferous, especially on the growth ring boundary.

Rays ca. 8/mm, 1–4(5)-seriate, (1)6–40(75) cells high; mostly homocellular and entirely composed of procumbent cells, rarely with some square marginal cells.

Crystals solitary, prismatic, in narrow, long chambered parenchyma chains (unidentified sphaerocrystalline masses present in addition in untreated hand sections).

PAPILIONOIDEAE

Anagyris foetida L.

Plate 40C

Shrub; primarily in open semi-steppe vegetation, also in sites of devastated Mediterranean vegetation, never in maquis or garigue; Acco Plain, Sharon Plain, Philistean Plain, Upper and Lower Galilee, Esdraelon Plain, Mt. Gilboa, Judean Mts., N. Negev, Ammon; Mediterranean and West Irano-Turanian element.

Growth rings distinct. Wood diffuse- to weakly semi-ring-porous. Vessels in a dendritic or oblique pattern together with the paratracheal parenchyma, ca. 75/mm², mostly in clusters (rarely solitary); rounded to angular in cross-section, of different size classes: wider vessels (tangential diameter 20–90 μm, radial diameter up to 100 μm, walls ca. 3 μm thick), and very narrow vessels intergrading with vascular tracheids (diameter ca. 16 μm) mostly surrounding the wider ones. Vessel member length 140(110–190) μm. Perforations simple in transverse to oblique end walls. Inter-vessel pits vestured, mostly alternate, rounded to polygonal, 6–10 μm in diameter, with slit-like, often coalescent apertures. Vessel-parenchyma and vessel-ray pits similar but half-bordered. Some vessels with gummy contents. Smaller vessels with fine, sometimes inconspicuous, spiral thickenings. Small vessels storied together with the parenchyma.
Fibres 690(570–850) μm long, thick- to very thick-walled, with simple pits mainly in radial walls; partly gelatinous.
Parenchyma fairly abundant paratracheal, forming tangential or oblique and dendritic pattern with the vessels, also as narrow marginal, mainly initial, bands; mostly fusiform, also in 2-celled strands; storied together with the vessel members.
Rays ca. 6/mm, 1–4-seriate, 10–20(40) cells high; homocellular to heterocellular, composed of weakly procumbent, square and sometimes weakly upright central cells and square to procumbent marginal cells.
Crystals not observed.

Calycotome villosa (Poir.) Link.
Plate 40D

Shrub; maquis, batha and garigue; Coastal Galilee, Acco Plain, Sharon Plain, Philistean Plain, Upper and Lower Galilee, Mt. Carmel, Mt. Gilboa, Samaria, Shefela, Judean Mts., W. Judean Desert, Upper Jordan Valley, Gilead, Ammon; Mediterranean element.

Growth rings distinct. Wood semi-ring-porous. Vessels in an oblique to dendritic pattern together with the paratracheal parenchyma, ca. 140/mm², mostly in clusters including some very narrow vessels (intergrading with vascular tracheids) and radial to oblique multiples of 2–4, rarely (ca. 10%) solitary, rounded (or slightly angular) in cross-section, tangential diameter of earlywood vessels 25–70 μm, of latewood vessels 20–40 μm, radial diameter up to 80 μm, walls 3–4 μm thick. Vessel member length 140(120–180) μm. Perforations simple in oblique to transverse end walls. Inter-vessel pits vestured, alternate, opposite or diffuse (pattern often obscured by coarse spiral thickenings), rounded, ca. 6 μm in diameter, with slit-like apertures. Vessel-parenchyma and vessel-ray pits similar but half-bordered. All vessels with mostly coarse spiral thickenings. Gummy contents often present. Vessel members storied together with the paratracheal parenchyma.
Fibres 600(100–790) μm long, thick- to very thick-walled, with simple pits mainly in radial walls; partly gelatinous.

Parenchyma mainly paratracheal, forming an oblique to dendritic pattern together with the vessels, also scanty diffuse apotracheal and in narrow marginal bands; fusiform or in 2-celled strands; storied together with the vessel members.
Rays ca. 9/mm, 1–3(4)-seriate, (1)6–14(23) cells high; heterocellular, composed of (mostly weakly) procumbent central cells and square to upright marginal cells (sometimes marginal cells also weakly procumbent; uniseriate rays sometimes composed of square to upright cells only).
Crystals not observed.

Colutea cilicica Boiss. et Bal.
Plate 41A and B

Shrub; Mt. Carmel(?), at the foot of Mt. Hermon; East Mediterranean element.

Growth rings distinct. Wood ring- to semi-ring-porous. Vessels sometimes forming a very faint oblique to dendritic pattern together with the paratracheal parenchyma; rounded to weakly angular in cross-section; of two size classes: wider vessels ca. 38/mm^2 [solitary or in radial to tangential multiples of 2–3(4), tangential diameter 40–100 μm in the earlywood, 20–60 μm in the latewood, radial diameter up to 120 μm, walls 2–4 μm thick], and very narrow vessels intergrading with vascular tracheids (diameter ca. 15 μm) often clustered with the wider vessels. Perforations simple in transverse to slightly oblique end walls. Vessel member length 120(100–160) μm. Inter-vessel pits vestured, mostly alternate, polygonal to rounded, 6–8 μm in diameter, with slit-like apertures. Vessel-parenchyma and vessel-ray pits similar but half-bordered. Most vessels (some wide vessels excepted) with fine spiral thickenings. Vessel members storied together with the paratracheal parenchyma.
Fibres 640(380–850) μm long, thick- to very thick-walled, with simple pits mainly in radial walls; frequently gelatinous.
Parenchyma mainly paratracheal, vasicentric to confluent, in latewood tending to form a weak oblique to dendritic pattern together with the vessels; also scanty diffuse apotracheal and marginal; fusiform or in 2-celled strands; storied together with the vessel members.
Rays ca. 6–8/mm, (1)3–4-seriate, often very tall (over 70 cells or 1 mm high); heterocellular, composed of procumbent to square to weakly upright cells, the latter including both marginal and sheath cells.
Crystals not observed.

Colutea istria Mill.

Shrub; on rocks or walls and in desert wadis; Lower Galilee, Samaria, Judean Mts., Judean Desert, Central Negev, Edom; West Irano-Turanian element.

Growth rings fairly distinct. Wood semi-ring-porous. Vessels forming a faint oblique to dendritic pattern together with the paratracheal parenchyma; rounded to weakly angular in cross-section; of two size classes: wider vessels ca. 30/mm^2

(solitary or in radial to tangential multiples of 2–3, tangential diameter 30–90 μm in the earlywood, 20–50 μm in the latewood, radial diameter up to 120 μm, walls 2–4 μm thick), and very narrow vessels intergrading with vascular tracheids (diameter ca. 15 μm) and often clustered with the wider vessels. Perforations simple in transverse to slightly oblique end walls. Vessel member length 120 (80–160) μm. Inter-vessel pits vestured, alternate, opposite or sometimes scalariform, polygonal to rounded, sometimes oblong, 6–9 μm in diameter, with slit-like, sometimes coalescent apertures. Vessel-parenchyma and vessel-ray pits similar but half-bordered. Narrow vessels with fine spiral thickenings. Vessel members storied together with the paratracheal parenchyma.
Fibres 610(440–770) μm long, thick- to very thick-walled, with simple pits mainly in radial walls; frequently gelatinous.
Parenchyma mainly paratracheal, vasicentric to confluent, in latewood tending to form an oblique or dendritic pattern together with the vessels; also scanty diffuse apotracheal and marginal; fusiform; storied together with the vessel members.
Rays ca. 6/mm, (1)3–6-seriate, (1)12–35(55) cells high (often over 1 mm high); heterocellular, composed of procumbent, square and upright cells, the latter including sheath and/or marginal cells.
Crystals not observed.

Separation of Colutea Species

The two species are very similar in wood anatomy; the minor differences in the material studied by us (see descriptions) are probably not diagnostic.

Genista fasselata Decne.
Plate 41C and D

Shrub; mainly in maquis and pine forests; Mt. Carmel and adjacent coastal hills; East Mediterranean element.

Growth rings faint to distinct. Wood diffuse- to semi-ring-porous. Vessels rounded in cross-section, of different size classes: larger vessels (ca. 70/mm^2, tangential diameter 50–90 μm in earlywood, radial diameter up to 120 μm, walls 3–4 μm, diameter in latewood 25–50 μm) mostly surrounded by numerous very narrow vessels intergrading with vascular tracheids (diameter ca. 20 μm), together forming an oblique to dendritic pattern with the paratracheal parenchyma, sometimes solitary. Vessel member length 150(90–310) μm. Perforations simple in transverse to oblique end walls. Inter-vessel pits faintly vestured, mostly alternate, rounded, 4–7 μm in diameter, with slit-like apertures. Vessel-parenchyma and vessel-ray pits similar but half-bordered. Vessels with prominent spiral thickenings. Gummy contents present in part of the vessels. Vessel members storied together with the parenchyma.
Fibres 660(440–880) μm long, thick- to very thick-walled, with simple pits mainly in radial walls; partly gelatinous.

Parenchyma predominantly paratracheal forming an oblique to dendritic pattern together with the vessels, also scanty diffuse apotracheal and in narrow marginal bands; usually fusiform, sometimes in 2-celled strands; storied together with the vessel members.
Rays ca. 10/mm, 1–3(4)-seriate, (1)8–16(46) cells high; heterocellular to homocellular, with procumbent central cells and weakly procumbent, square to upright marginal cells.
Crystals not observed.

Gonocytisus pterocladus (Boiss.) Sp.
Plate 42A and B

Shrub; maquis and garigue; Upper Galilee; East Mediterranean element.

Growth rings distinct. Wood semi-ring-porous. Vessels rounded to angular in cross-section, of two different size classes: wider vessels (10–20/mm^2, mainly in clusters or in short radial to tangential multiples, rarely solitary, tangential diameter 25–100 μm, radial diameter up to 120 μm, walls ca. 2–3 μm thick) usually surrounded by very narrow vessels intergrading with vascular tracheids (diameter 10–20 μm), together with paratracheal parenchyma forming tangential bands on the growth ring boundaries and in faint oblique to dendritic pattern in the latewood. Vessel member length 130(100–160) μm. Perforations simple in oblique to transverse end walls. Inter-vessel pits vestured, mostly alternate, rounded to polygonal, ca. 7 μm in diameter, with slit-like, sometimes coalescent apertures. Vessel-parenchyma and vessel-ray pits similar but half-bordered. All vessels with spiral thickenings. Some vessels with gummy contents. Vessel members storied together with the paratracheal parenchyma.
Fibres 750(660–970) μm long, thick- to very thick-walled, with simple pits mainly in radial walls; partly gelatinous, rarely septate.
Parenchyma predominantly paratracheal, contributing to the tangential, oblique or dendritic pattern in cross-section, and marginal (as part of the paratracheal parenchyma) and scanty diffuse apotracheal, diffuse-in-aggregates and in narrow, short tangential bands; mostly fusiform, rarely in 2-celled strands; storied together with the vessel members.
Rays ca. 6/mm, (1)3–5(6)-seriate, (1)12–45 cells high; heterocellular, composed of procumbent central cells and weakly procumbent to square and upright marginal cells; sheath cells sometimes present.
Crystals not observed.

Ononis natrix L.
Plate 42C and D

Half-shrub; 3 subspecies; batha; Upper and Lower Galilee, Mt. Carmel, Esdraelon Plain, Mt. Gilboa, Samaria, Sharon Plain, Philistean Plain, Judean Mts., Judean Desert, Dan Valley, Hula Plain, Lower Jordan Valley, Dead Sea area, Arava Valley, Moav, Edom; Mediterranean element with extensions into the Saharo-Arabian territories.

Growth rings faint to distinct. Vessels diffuse, rounded to angular in cross-section, of two size classes: wider vessels ca. 30–40/mm², solitary and in radial to tangential multiples of 1–4 and small clusters (tangential diameter 20–60 μm, radial diameter up to 80 μm, walls 2–3 μm thick), and very narrow vessels intergrading with vascular tracheids (diameter 10–20 μm), often associated with the wider vessels. Vessel member length 90(70–130) μm. Perforations simple in oblique to transverse end walls. Inter-vessel pits vestured, mostly alternate, round to polygonal or oblong, 6–12 μm in diameter, with slit-like to rather wide apertures. Vessel-parenchyma and vessel-ray pits similar but half-bordered and often with slightly to strongly reduced borders. Vessel members weakly storied together with the parenchyma.

Fibres 550(370–680) μm long, thick- to very thick-walled, with simple pits mainly in radial walls; partly gelatinous.

Parenchyma mainly paratracheal, scanty or vasicentric to confluent, also scanty diffuse apotracheal and marginal; mostly fusiform, but also in 2–5-celled strands, weakly storied together with the vessel members.

Rays ca. 5–6/mm, (1)3–10(16)-seriate, uni- and biseriates very low, multiseriates usually high [often over 1 mm or 40(75) cells]; heterocellular, composed mainly of square to weakly procumbent central cells and square to upright marginal and sheath cells.

Crystals solitary, prismatic, in ordinary or in chambered ray cells.

Retama raetam (Forssk.) Webb

Plate 43A and B

Shrub. Var. *raetam*: mainly in deserts, sandy, rocky and gravelly ground; Judean Desert, E., C. and S. Negev, Upper and Lower Jordan Valley, Arava Valley, Sinai, deserts east of Ammon, Moav, Edom. Var. *sarcocarpa*: sands and sandy soils; Acco Plain, Sharon Plain, Philistean Plain, W. Negev. Saharo-Arabian element.

Growth rings indistinct to fairly distinct. Wood diffuse-porous. Vessels in an oblique to dendritic pattern together with the paratracheal parenchyma; rounded in cross-section, of two different size classes: the wider vessels [often clustered, sometimes solitary, 50–75(160)/mm², tangential diameter 30–100 μm, radial diameter up to 110 μm, walls ca. 3–5 μm thick] mostly embedded in few to numerous very narrow vessels intergrading with vascular tracheids (diameter 15–20 μm). Vessel member length 130(100–160) μm. Perforations simple in oblique to transverse end walls. Inter-vessel pits vestured, mostly alternate, rounded to polygonal, 5–8 μm in diameter, with slit-like, sometimes coalescent apertures. Vessel-parenchyma and vessel-ray pits similar but half-bordered. Spiral thickenings present, especially coarse in narrow vessels. Some vessels with gummy contents. Vessel members storied together with axial parenchyma.

Fibres 540(270–1,050) μm long, thick- to very thick-walled, with simple pits mainly in radial walls; partly gelatinous.

Parenchyma paratracheal, forming an oblique to dendritic pattern with the ves-

sels, also apotracheal, in irregular or tangential 1–2-seriate bands, diffuse and diffuse-in-aggregates, and marginal, the latter often in touch with the paratracheal parenchyma; fusiform, rarely in 2-celled strands; storied together with the vessel members.
Rays ca. 10–13/mm, 1–4-seriate, (1)4–20(37) cells high; heterocellular, composed of procumbent and square central cells and square to upright marginal cells.
Crystals not observed.

Archaeological records: 11 locations; coastal plain, Negev, Dead Sea area, Sinai; late Bronze Age to Arabic Period (N. Liphschitz & Y. Waisel; A. Fahn & E. Werker; E. Werker).

Spartium junceum L.
Plate 43C and D

Shrub or small tree; maquis; Upper and Lower Galilee, Mt. Carmel, Samaria, Judean Mts., Dan Valley; Mediterranean element.

Growth rings distinct. Wood ring- to semi-ring-porous. Vessels in an oblique to dendritic pattern together with the paratracheal parenchyma, rounded to angular in cross-section, of two different size classes: wider vessels ca. 35/mm² (tangential diameter 40–100 μm in earlywood, radial diameter up to 120 μm, walls 2–3 μm thick; in latewood tangential diameter ca. 30–50 μm), and numerous very narrow vessels intergrading with vascular tracheids (diameter 15–20 μm), mostly surrounding the wider ones, especially in the latewood; wide vessels rarely solitary in earlywood. Vessel member length 150(120–170) μm. Perforations simple in transverse to oblique end walls. Inter-vessel pits vestured, mostly alternate, rounded to somewhat polygonal, sometimes oblong, ca. 5–7(10) μm in diameter, with slit-like apertures. Vessel-parenchyma and vessel-ray pits similar but half-bordered. All vessels with prominent spiral thickenings. Some vessels with gummy contents. Vessel members storied together with the paratracheal parenchyma.
Fibres 1,040(700–1,350) μm long, medium thick- to thick-walled, with infrequent simple pits; partly gelatinous.
Parenchyma predominantly paratracheal, vasicentric in the earlywood, forming an oblique to dendritic pattern with the vessels in the latewood, also scanty diffuse apotracheal and in narrow marginal bands; fusiform (very rarely in 2–3-celled strands); storied with the vessel members.
Rays ca. 6/mm, (1)2–5-seriate, (1)8–16(35) cells high; heterocellular to homocellular, composed of procumbent central and weakly procumbent or square to upright marginal cells.
Crystals not observed.

LORANTHACEAE

The hemi-parasitic species of *Loranthus* and *Viscum* in Israel differ widely in their wood anatomy. *Viscum* has a very unusual wood anatomy with inconspic-

uous vessels and a very high ratio of fibre to vessel member length (7.0). Moreover, the two genera differ in their type of fibre tissue (libriform in *Viscum*, fibre-tracheids in *Loranthus*), and *Loranthus* is outstanding because of the stone cells in its multiseriate rays.
It would be interesting to establish whether the large differences between *Viscum* and *Loranthus* coincide with the delimitation of Loranthaceae *sensu stricto* and Viscaceae of some authors (e.g. Airy Shaw, 1966; Dahlgren, 1980; Cronquist, 1981). The wood anatomical data summarized by Metcalfe & Chalk (1950) are not indicative of a perfect coincidence, but critical updating of our anatomical knowledge of the family(ies), using authenticated specimens, is required.

Loranthus acaciae Zucc.
Plate 44A and B

Perennial hemi-parasite; Judean Desert, N. Negev, Lower Jordan Valley, Dead Sea area, Arava Valley, Edom; parasitic mostly on *Acacia* and *Ziziphus*, also on *Punica, Tamarix, Atriplex* etc.; East Sudanian element.

Growth rings very faint. Vessels diffuse, 50–100/mm^2, in variously sized, poorly defined clusters and variously directed multiples of up to 6 and infrequently solitary; rounded to angular in cross-section, tangential diameter 20–65 μm, radial diameter up to 100 μm, walls 3–5 μm thick. Vessel member length 150 (100–210) μm. Perforations simple in mainly horizontal end walls. Inter-vessel pits alternate, round to gash-like and scalariform, 5–6 μm in diameter, with slit-like, sometimes coalescent apertures. Vessel-parenchyma and vessel-ray pits of two types: mostly similar to inter-vessel pits but half-bordered, occurring together with large and simple ones (10 μm or more). Vessel members storied together with parenchyma cells.
Fibres 290(150–420) μm long, thick-walled, with distinctly bordered pits in radial and tangential walls.
Parenchyma abundant, scanty paratracheal to vasicentric and diffuse apotracheal, mostly fusiform; storied together with vessel members.
Rays 2–3/mm, 5–20 cells wide, of various heights (up to 3 mm or more); heterocellular, composed of square, weakly procumbent and some upright cells; some cells crystalliferous; large groups of crystalliferous stone cells present in the centre of the rays.
Crystals solitary, prismatic, in all stone cells and in some unsclerified ray cells.

Viscum cruciatum Sieb.
Plate 44C and D

Perennial hemi-parasite; Philistean Plain, Upper and Lower Galilee, Samaria, Judean Mts., Gilead, Ammon, Moav, Edom; parasitic mainly on *Olea*, also on *Rhamnus, Crataegus, Rubus, Amygdalus, Robinia, Citrus*, etc.; Mediterranean element.

Growth rings very faint. Vessels diffuse; hardly distinguishable in cross-section from parenchyma cells, indistinctly arranged in radial multiples of up to 14 and in clusters; rounded to angular in cross-section, diameter ca. 30 μm, walls 4–6 μm thick. Vessel member length 80(60–130) μm. Perforations simple in oblique to horizontal end walls. Inter-vessel pits alternate to opposite, round to polygonal, 3–4 μm in diameter, with slit-like, very often coalescent apertures, creating a scalariform appearance. Vessel-parenchyma and vessel-ray pits half-bordered to simple, round, gash-like to scalariform. Vessel members storied together with parenchyma cells.
Fibres 560(350–680) μm long, thick- to very thick-walled, mostly gelatinous, with very inconspicuous and infrequent simple pits.
Parenchyma scanty paratracheal and diffuse apotracheal, mostly fusiform; storied together with vessel members.
Rays 3–5/mm, very poorly delimited, up to ca. 5 cells wide, of various heights (up to ca. 0.5 mm); heterocellular, composed of procumbent, square and upright cells. Some cells crystalliferous.
Crystals solitary, prismatic, in some of the ray cells.

MALVACEAE

Abutilon and *Hibiscus* have many wood anatomical characters in common with each other as well as with other woody Malvaceae: vessels partly solitary, partly in rather long radial multiples, libriform fibres, paratracheal and marginal parenchyma and heterocellular rays (cf. Metcalfe & Chalk, 1950). The differences between the two genera, recorded in the descriptions (mainly in ray structure), are only slight.

Abutilon fruticosum Guill. et Perr.

Shrub; wadis and oases in hot deserts; E. and S. Negev, Dead Sea area, Arava Valley; Sudanian element.

Growth rings distinct. Vessels diffuse, 70–120/mm^2, solitary (20–37%), in radial multiples of 2–3(5), sometimes in small clusters, rounded in cross-section, tangential diameter 13–70 μm, radial diameter up to 80 μm, walls 2–4 μm thick. Vessel member length 140(90–200) μm. Perforations simple in horizontal to oblique end walls. Inter-vessel pits alternate to diffuse, rounded, 5–6 μm in diameter, with slit-like, occasionally coalescent apertures. Vessel-parenchyma and vessel-ray pits similar but half-bordered. Some vessels with gummy contents.
Fibres 580(380–870) μm long, medium thick- to very thick-walled, with simple, sometimes minutely bordered pits mainly in radial walls.
Parenchyma scanty paratracheal, sometimes aliform or confluent, in short irregular tangential bands 2–4 cells wide and in narrow, discontinuous marginal bands; fusiform and in up to 5-celled strands; some cells crystalliferous.

Rays 5–8/mm, 1–7 cells wide, uniseriates up to 6 cells high, multiseriates up to 50 cells (ca. 1 mm) high; heterocellular, composed of square, upright and weakly procumbent central cells and square to upright marginal cells; many cells crystalliferous.
Crystals prismatic, mainly solitary, sometimes several per cell; also in irregular aggregates or as clusters and druses, in ray cells.
Trabeculae found in ray cells and vessels close to a disturbance in the wood structure.

Abutilon pannosum (Forst. fil.) Schlecht.
Plate 45A and B

Shrub; oases in hot deserts; on cultivated land and near water; Lower Jordan Valley, Dead Sea area; Tropical element.

Growth rings distinct. Vessels diffuse, 30–55/mm², solitary (ca. 25%) and in radial multiples of 2–5(9), occasionally in clusters; rounded in cross-section, tangential diameter 30–95 μm, radial diameter up to 120 μm, walls 4–6 μm thick. Vessel member length 190(130–250) μm. Perforations simple in horizontal to oblique end walls. Inter-vessel pits alternate, rounded, ca. 6 μm in diameter, with short slit-like, occasionally coalescent apertures. Vessel-parenchyma and vessel-ray pits similar but half-bordered.
Fibres 860(570–1,260) μm long, medium thick- to thick-walled; with simple pits more numerous in radial than in tangential walls.
Parenchyma paratracheal, mostly confluent forming bands alternating with fibre bands, sometimes scanty to vasicentric or weakly aliform, and in marginal bands (the latter sometimes indistinct from confluent paratracheal parenchyma); fusiform and in 2–4-celled strands.
Rays 4–8/mm, 1–4(7) cells wide, uniseriates up to 5 cells high, multiseriates up to ca. 60 cells (1.5 mm) high;. heterocellular, composed of weakly procumbent, square and upright cells.
Crystals solitary, prismatic, infrequent in ray cells.

Characteristics of the Genus Abutilon

Growth rings distinct. Vessels diffuse; solitary and in radial multiples, sometimes in small clusters. Perforations simple. Inter-vessel pits mostly alternate. Vessel-parenchyma and vessel-ray pits half-bordered. Fibres with simple or minutely bordered pits. Parenchyma paratracheal, sometimes aliform or confluent and marginal. Rays uni- to multiseriate; heterocellular.

Separation of Abutilon Species

In the material studied *A. fruticosum* shows a variable parenchyma distribution within a single stem; this makes it doubtful whether the two species can be separated on the basis of parenchyma distribution. *A. pannosum* with its distinct tangential parenchyma bands and *A. fruticosum* might appear to overlap if more specimens were studied.

Hibiscus micranthus L. fil.
Plate 45C and D

Shrub; hot deserts, in wadis or rock fissures; S. Negev, Lower Jordan Valley, Dead Sea area, Arava Valley; element from tropical regions of Asia and Africa.

Growth rings distinct. Vessels diffuse, with a weak tendency to form a radial pattern; ca. 100/mm²; mostly solitary (ca. 70%), in radial multiples of 2–5 and in small clusters; rounded in cross-section, tangential diameter 20–70 μm, radial diameter up to 70 μm, walls 3–5 μm thick. Vessel member length 140(80–330) μm. Perforations simple in horizontal to oblique end walls. Inter-vessel pits alternate to diffuse, rounded, 5–6 μm in diameter, with slit-like apertures. Vessel-parenchyma and vessel-ray pits similar but half-bordered. Many vessels with gummy contents.
Fibres 400(210–640) μm long, medium thick- to thick-walled; with simple to minutely bordered pits in radial walls.
Parenchyma scanty paratracheal to vasicentric, sometimes aliform, and in 1–3-seriate marginal (initial) bands; thick-walled; fusiform or in 2(3)-celled strands.
Rays 5–8/mm, 1–8 cells wide, very variable in shape, height and type of cells; some aggregate, some compound; not always well delimited; heterocellular, composed of variously arranged procumbent, square and upright cells; some cells occasionally crystalliferous.
Crystals occasionally in aggregates in ray cells.
Trabeculae found in a double file in vessels and fibres of a much twisted branch.

MELIACEAE

Melia azedarach L.
Plate 46A–C

Tree; deciduous; cultivated or escaped from cultivation; native in southwest Asia.

Growth rings distinct. Wood ring- to semi-ring-porous. Vessels rounded to angular in cross-section, of two different size classes: wider vessels 9–13/mm², solitary or in multiples of 2(3) (tangential diameter 100–330 μm, radial diameter up to 370 μm, walls 4–8 μm thick), and small ones (diameter 20–80 μm) in clusters or multiples together with 1–2 large vessels in the earlywood and intermediate wood, or in clusters of numerous exclusively narrow vessels in the latest formed latewood, the latter sometimes in an oblique to tangential or occasionally dendritic pattern. Vessel member length 340(170–430) μm. Perforations simple in horizontal to oblique end walls. Inter-vessel pits alternate, rounded to polygonal, 5–7 μm in diameter, with slit-like apertures. Vessel-parenchyma and vessel-ray

pits similar but half-bordered. Most vessels, except for the largest ones, with fine, very dense, spiral thickenings. Some vessels with gummy contents.
Fibres 820(510–1,040) μm long, thin- to thick-walled, with simple pits more numerous in radial than in tangential walls.
Parenchyma abundant, vasicentric or forming ground tissue in the earlywood together with wide and narrow vessels, and marginal; in 2–4-celled strands; tending to be storied; sometimes chambered crystalliferous.
Rays 3–4/mm, (1)4–8 cells wide, up to 50 cells high; homocellular to heterocellular, composed mostly of procumbent cells, sometimes also of square or upright marginal cells.
Crystals solitary, prismatic, in chambered parenchyma.

MENISPERMACEAE

Cocculus pendulus (J. R. et G. Forst.) Diels
Plate 46D and E

Woody climber on rocks, cliffs and trees; E. and S. Negev, Lower Jordan Valley, Arava Valley, Edom; Sudanian element, also native in other Palaeotropical regions.

Wood with included phloem of the concentric type. Vessels diffuse, 16–36/mm², solitary, rarely in pairs; of various sizes, rounded in cross-section, tangential diameter 25–200 μm, radial diameter up to 230 μm, walls up to 6 μm thick. Vessel member length 270(90–390) μm. Perforations simple, in horizontal to oblique end walls. Inter-vessel pits infrequent, alternate, rounded, 6–7 μm in diameter, with slit-like apertures. Vessel-fibre pits similar but more widely spaced. Vessel-ray pits half-bordered and similar to inter-vessel pits or simple and scalariform.
Included phloem in distinct, semicircular or crescent-shaped, occasionally coalescent groups (tangential diameter 250–450 μm) arranged in concentric rings.
Fibres 710(380–980) μm, medium thick-walled, with numerous distinctly bordered pits in radial and tangential walls.
Sclereids isodiametric or radially elongated, in concentric bands between the non-sclerified conjunctive parenchyma bands and the outer border of the phloem groups, intruding centripetally between the latter and slightly into rays.
Parenchyma scanty, diffuse apotracheal, in 2-celled strands and conjunctive in tangential bands inside the xylem and outside the sclereid bands, up to ca. 5 cells wide, with 1–2 outer rows of lignified and inner non-lignified cells.
Rays 2–4/mm, 6–20 cells wide, very infrequently 1–2-seriate, up to 4 mm or more in height; heterocellular, composed of procumbent and square cells, sclerified where intersecting sclereid bands and non-lignified where intersecting non-sclerified conjunctive parenchyma; usually discontinuous from one vascular band to another; some cells crystalliferous.
Crystals minute, variously shaped, prismatic and elongate, abundant in ray cells.

MORACEAE

Ficus and *Morus* differ markedly in their wood anatomy: *Ficus* is diffuse-porous and has wide tangential parenchyma bands, while *Morus* has ring-porous wood with latewood vessels in a tangential or oblique pattern. The differences are partly related to the different taxonomic affinities of the genera (they belong to different tribes of the Moraceae), partly to their phytogeographical background: *Ficus* is largely a tropical genus, while *Morus alba* is a species of warm temperate to subtropical regions.

Ficus carica L.
Plate 47A and B

Cultivated fruit tree, deciduous, one uncultivated variety native in Upper Galilee, Mt. Gilboa, Samaria; Mediterranean and Irano-Turanian element.

Growth rings distinct. Vessels diffuse, 8–20/mm^2, solitary (5–20%) and in radial multiples of 2–4(7), occasionally in clusters; rounded to angular in cross-section, tangential diameter 40–180 μm, radial diameter up to 220 μm, walls 4–10 μm thick. Vessel member length 280(160–360) μm. Perforations simple in oblique to transverse end walls. Inter-vessel pits alternate to opposite, round to polygonal, 4–6 μm in diameter, with slit-like, very often coalescent apertures. Vessel-parenchyma pits similar or much longer, elliptic. Vessel-ray pits similar but simple to half-bordered. Tyloses present.
Fibres 870(620–1,160) μm long, very thick-walled, gelatinous; in wide bands alternating with parenchyma bands; with simple to minutely bordered pits mainly in radial walls.
Parenchyma in 2–6-seriate tangential bands alternating with fibre bands, vasicentric and marginal; fusiform and in 2–8-celled strands; many cells chambered crystalliferous.
Rays 6–8/mm, 1–4 cells wide, up to ca. 45 cells (0.8 mm) high; heterocellular, composed of procumbent central cells and square and upright marginal cells (Kribs' heterogeneous types II–III); some cells crystalliferous.
Crystals prismatic, mainly solitary, large ones in ordinary ray cells, smaller ones in chambered parenchyma cells.

Ficus pseudo-sycomorus Decne.
Plate 47C and D; Fig. 9B

Shrub; deciduous; among rocks in dry, hot creeks; N. and C. Negev, Arava Valley, Edom: East Sudanian element.

Growth rings faint, marked by progressive change in width of parenchyma and fibre bands and poorly delimited marginal parenchyma. Vessels diffuse, ca. 25/mm^2, in radial multiples of 2–6(8), often including both wide and narrow vessels, rarely solitary (4–12%) or in clusters: rounded to angular in cross-section, tangential diameter 35–200 μm, radial diameter up to 180 μm, walls 3–9

μm thick. Vessel member length 270(200–360) μm. Perforations simple in oblique end walls. Inter-vessel pits alternate (to opposite), rounded, 4–7 μm in diameter, with elliptic, very often coalescent apertures. Vessel-parenchyma and vessel-ray pits simple to half-bordered, rounded or angular to scalariform. Tyloses present.
Fibres to 710(340–1,010) μm long, medium thick- to thick-walled, many gelatinous, with simple to minutely bordered pits mainly in radial walls.
Parenchyma in 2–5-seriate tangential bands alternating with fibre bands, vasicentric and marginal; in 2–4-celled strands; many cells chambered, some non-chambered crystalliferous.
Rays 5–8/mm, 1–5 cells wide, up to ca. 80 cells (1.2 mm) high; heterocellular (Kribs' heterogeneous type II); some cells crystalliferous.
Crystals prismatic, mainly solitary, in chambered and non-chambered parenchyma cells and in ray cells.
Laticifers occasionally observed in rays.

Archaeological records: N. E. Sinai, 8–9th century B. C. (N. Liphschitz & Y. Waisel).

Ficus sycomorus L.
Plate 48

Tree, cultivated; evergreen, deciduous in cold winters; Acco Plain, Sharon Plain, Philistean Plain, W. Negev, Lower Jordan Valley; native in Ethiopia and elsewhere in tropical Africa.

Growth rings absent (or very faint), marked only by progressive change in width of parenchyma and fibre bands. Vessels diffuse, 3–7/mm^2, in radial multiples of 2–3(6), sometimes in clusters, rarely solitary (ca. 10%); rounded in cross-section, tangential diameter 90–240 μm, radial diameter up to 300 μm, walls ca. 6 μm thick. Vessel member length 490(320–680) μm. Perforations simple in mostly horizontal end walls. Inter-vessel pits alternate (to opposite), round to polygonal, 7–9 μm in diameter, with slit-like to elliptic apertures. Vessel-parenchyma and vessel-ray pits simple or half-bordered, round, angular to elongate in various directions. Many vessels with tyloses.
Fibres 1,450(910–1,920) μm long, medium thick- to thick-walled, with simple to minutely bordered pits mainly in radial walls.
Parenchyma in wide (6–20-seriate) tangential bands alternating with fibre bands; fusiform and in 2–8-celled strands; rarely crystalliferous.
Rays 3–5/mm, of two sizes: 1–4-seriates up to 14 cells high, and 5–14-seriates up to 1.4 mm high; heterocellular, composed of square, slightly procumbent or upright marginal cells and strongly procumbent central cells (Kribs' heterogeneous types II–III); some cells crystalliferous.
Crystals solitary, prismatic, in ordinary ray and parenchyma cells.
Laticifers occasionally observed in rays.

Archaeological records: N. E. Sinai, 8–9th century B. C. (N. Liphschitz & Y. Waisel); Jerusalem, 6th century B. C. and 1st century A. D. (A. Fahn).

Wood Anatomical Descriptions

Characteristics of the Genus Ficus

Vessels diffuse; frequency not more than 25/mm²; mostly in radial multiples. Perforations simple. Fibres with simple to minutely bordered pits mainly in the radial walls. Parenchyma in wide tangential bands. Rays heterocellular.

*Key to Ficus Species**

Multiseriate rays over 10 cells wide present.	**F. sycomorus**
Multiseriate rays less than 8 cells wide.	**F. carica**
	F. pseudo-sycomorus

Morus alba L.

Plate 49A and B

Cultivated tree; deciduous; native in China.

Growth rings distinct. Wood ring-porous. Vessels 10–125/mm², in clusters and radial multiples of 2–3(4), some large vessels solitary, the groups of small vessels, especially in latewood, often forming a pattern of tangential or discontinuous oblique bands; rounded in cross-section, tangential diameter 20–260 μm, radial diameter up to 350 μm, walls 2–3 μm thick. Vessel member length 190(130–230) μm. Perforations simple in horizontal to oblique end walls. Inter-vessel pits alternate, polygonal to rounded, 8–10 μm in diameter, with short slit-like apertures. Vessel-parenchyma and vessel-ray pits rounded to elliptic, simple to half-bordered. Vessels, except for the widest, with fine, distinct spiral thickenings.
Fibres 840(500–1,020) μm long, medium thick- to very thick-walled; with simple pits mainly in radial walls; many gelatinous.
Parenchyma vasicentric, partly confluent; in 2–5-celled strands.
Rays 3–5/mm, 1–2 but mostly 3–10 cells wide, up to 38 cells high; heterocellular, composed of procumbent central cells and square, irregular and upright marginal cells.
Crystals absent to frequent, prismatic in ray cells.

MORINGACEAE

Moringa peregrina (Forssk.) Fiori (*M. aptera* Gaertn.)

Plate 49C and D

Tree; deciduous; tropical oases; Judean Desert (hot ravines), S. Negev, Lower Jordan Valley, Dead Sea area, Arava Valley, Edom; East Sudanian element.

* *F. carica* and *F. pseudo-sycomorus* are very similar in their wood anatomy. Perhaps *F. carica* can be distinguished from *F. pseudo-sycomorus* (see descriptions) on the basis of the combination of growth ring distinctness, fairly low rays and two distinct size classes of prismatic crystals, but this requires further testing.

Growth rings faint. Vessels diffuse, 3–10/mm², solitary, in mostly radial, sometimes tangential multiples of 2–3(5) and in clusters; rounded in cross-section, tangential diameter 70–200 μm, radial diameter up to 260 μm, walls 4–6 μm thick. Vessel member length 340(190–510) μm. Perforations simple in mainly horizontal end walls. Inter-vessel pits alternate, polygonal, 10–18 μm in diameter, with large, elliptic to slit-like apertures. Vessel-parenchyma and vessel-ray pits rounded to elliptic, half-bordered to simple. Many vessels with tyloses.
Fibres 760(350–1,070) μm long, very thin-walled with wide lumina; with simple pits mainly in radial walls; storied.
Parenchyma paratracheal, vasicentric to aliform; in 2–4-celled strands or fusiform.
Rays 9–11/mm, 1–2(3) cells wide, up to ca. 7 cells high; heterocellular, composed of procumbent central cells and weakly procumbent, square or upright marginal cells; storied; some cells crystalliferous.
Crystals solitary, prismatic or irregularly shaped, in ray cells.

Archaeological records: Arava Valley; 1000 B. C. and 6–8th centuries A. D. (N. Liphschitz & Y. Waisel).

MYRTACEAE

Myrtus communis L.
Plate 50A and B

Shrub; evergreen; maquis and riverine thickets; Upper Galilee, Mt. Carmel, Dan Valley, Upper Jordan Valley, Golan; Mediterranean element.

Growth rings distinct. Vessels diffuse, 100–125/mm², predominantly solitary, occasionally in pairs; angular to rounded in cross-section, tangential diameter 15–40 μm, radial diameter up to 45 μm, walls ca. 1.5 μm thick. Vessel member length 450(270–660) μm. Perforations simple in oblique end walls. Vessel-parenchyma and vessel-ray pits vestured, rounded, 3–4 μm in diameter, simple to half-bordered. Vessels with spiral thickenings.
Fibres 700(450–950) μm long, medium thick- to thick-walled, with distinctly bordered pits in radial and tangential walls.
Parenchyma scanty paratracheal and diffuse apotracheal, in 4–10-celled strands.
Rays ca. 10–12/mm, 1–2(3) cells wide, up to 20 cells high; heterocellular (Kribs' heterogeneous type I).
Crystals not observed.

NYCTAGINACEAE

Commicarpus africanus (Lour.) Dandy (*Boerhavia africana* Lour.)
Plate 50C; Fig. 8E

Perennial; shady places. Var. *africanus*: Philistean Plain, Upper and Lower Galilee, Esdraelon Plain, Judean Desert, N. and C. Negev, Upper Jordan Valley, Beit

Shean Valley, Lower Jordan Valley, Dead Sea area, Arava Valley, Moav, Edom. Var. *viscosus*: S. Negev, Arava Valley. Probably a Sudanian element.

Wood with included phloem of the concentric type. Growth rings absent. Vessels of two size classes: wider vessels 15–35/mm², seemingly solitary or in variously directed multiples of 2–3, rounded in cross-section, tangential diameter 50–160 μm, radial diameter up to 170 μm, walls 5–6 μm thick; narrow vessels clustered with the wider ones, intergrading with vascular tracheids, 15–30 μm in diameter. Vessel member length 100(60–140) μm. Perforations simple in horizontal to oblique end walls. Inter-vessel pits alternate, rounded, 5–6 μm in diameter, with slit-like, sometimes coalescent apertures. Vessel-parenchyma pits similar but half-bordered. Vessel elements storied together with phloem and parenchyma elements.
Included phloem groups distinct, sometimes coalescent, crescent-shaped (tangential diameter 60–600 μm), mostly forming concentric bands of various lengths together with the thin-walled conjunctive parenchyma.
Fibres 330(170–470) μm long, thin- to medium thick-walled, with simple pits mainly in radial walls.
Parenchyma scanty paratracheal, lignified and unlignified conjunctive, often in concentric bands of various lengths, sometimes extending inwards between the fibres fusiform and in 2-celled strands; storied together with vessel and phloem elements. Some conjunctive parenchyma cells crystalliferous.
Rays not seen in cross-section, small ray-like cell configurations observed in tangential view.
Crystals present as raphides in conjunctive parenchyma.

OLEACEAE

The Oleaceae of Israel can easily be differentiated on the basis of wood anatomical features. *Jasminum* is the only genus with distinctly bordered pits in the tangential and radial fibre walls (fibre-tracheids); *Fraxinus* stands out on account of its distinct ring-porosity and homocellular rays; *Olea* is diffuse-porous and has many vessels in radial multiples; *Phillyrea* can be recognized by its dendritic pattern of vessels, vascular tracheids and paratracheal parenchyma. The species from Israel fit neatly into the different wood anatomical groups recognized for the family as a whole by Esser & Van der Westen (1983).

Fraxinus syriaca Boiss.
Plate 51A and B

Tree; deciduous; along rivers and streams; Sharon Plain, Upper Galilee, Mt. Carmel, Hula Plain, Beit Shean Valley, Golan, Gilead; East Mediterranean and West Irano-Turanian element.

Growth rings distinct. Wood ring-porous. Vessels 4–23/mm², solitary (20–60%),

in radial multiples of 2–3(4) or rarely in small clusters; rounded in cross-section, tangential diameter in the earlywood 110–250 μm, radial diameter up to 300 μm, diameter in the latewood ca. 40 μm, walls 5–8 μm thick. Vessel member length 260(210–330) μm. Perforations simple in horizontal to oblique end walls. Intervessel pits alternate, rounded to polygonal, 3–4 μm in diameter, with slit-like, sometimes coalescent apertures. Vessel-parenchyma and vessel-ray pits similar but half-bordered. Many vesssels with tyloses.
Fibres 880(490–1,070) μm long, medium thick- to very thick-walled, with minutely bordered to simple pits mainly in radial walls.
Parenchyma vasicentric in the earlywood, aliform and confluent towards the end of growth ring, and marginal; in up to ca. 3–8-celled strands.
Rays 7–10/mm, 1–4(5) cells wide, up to 20 cells high; homocellular, composed of procumbent cells.
Crystals not observed.

Jasminum fruticans L.
Plate 51C and D

Shrub; Quercetum calliprine maquis, about 400–1,000 m above sea level; Upper Galilee, Mt. Gilboa, Samaria, Judean Mts., Golan, Gilead; Mediterranean and West Irano-Turanian element.

Growth rings distinct. Wood ring- to semi-ring-porous. Vessels ca. 200 –300/mm², mostly solitary, sometimes in tangential multiples of 2–4 mainly at the beginning of growth ring; angular to rounded in cross-section, tangential diameter 15–50 μm, radial diameter up to 60 μm, walls 2–3 μm thick. Vessel member length 220(150–420) μm. Perforations simple in oblique end walls. Intervessel pits alternate, rounded, ca. 5 μm in diameter, with slit-like apertures. Vessel-parenchyma and vessel-ray pits similar but half-bordered. Many vessel members with fine spiral thickenings.
Fibres 390(220–600) μm long, thick- to very thick-walled, with distinctly bordered pits in radial and tangential walls; many with spiral thickenings.
Parenchyma scanty paratracheal, and diffuse apotracheal and scanty marginal (initial); in up to 2–5-celled strands.
Rays 7–9/mm, 1(3) cells wide, up to 20 cells high; heterocellular, with upright and square marginal cells and weakly procumbent central cells.
Crystals not observed.

Olea europaea L.
Plate 52A–C

Tree or shrub; evergreen. Var. *sylvestris*: maquis; Upper Galilee, Golan, Gilead, Ammon; East Mediterranean element.

Growth rings distinct to faint. Vessels diffuse, ca. 80/mm², solitary (ca. 20%), in radial multiples of 2–4(6) or occasionally in clusters; angular to rounded in cross-section, tangential diameter 30–60 μm, radial diameter up to 70 μm, walls 3–6

μm thick. Vessel member length 310(190–380) μm. Perforations simple in horizontal to oblique end walls. Inter-vessel its alternate, rounded, 3–4 μm in diameter. Vessel-parenchyma and vessel-ray pits similar but half-bordered. Some vessels with gummy substances.
Fibres 790(520–1,030) μm long, medium thick- to very thick-walled, with minutely bordered pits in radial walls.
Parenchyma mostly vasicentric, sometimes aliform or confluent, and sparse apotracheal diffuse; fusiform and in 2–4-celled strands.
Rays 8–13/mm, 1–2(3)-seriate, up to 12(20) cells high; heterocellular (Kribs' heterogeneous types I and II); often crystalliferous.
Crystals acicular, minute, numerous in ray cells.

Archaeological records: 17 locations throughout Israel; since very early periods (N. Liphschitz & Y. Waisel; E. Werker & A. Fahn; E. Werker; A. Fahn & E. Zamski).

Phillyrea latifolia L. (*P. media* L.)
Plate 52D

Shrub or small tree; evergreen; maquis and forest; Coastal Galilee, Acco Plain, Sharon Plain, Upper and Lower Galilee, Mt. Carmel, Mt. Gilboa, Samaria, Judean Mts.; Mediterranean element.

Growth rings distinct. Vessels diffuse, 120–200/mm^2, solitary and in short multiples and small clusters, forming a dendritic pattern together with paratracheal parenchyma and vascular tracheids; angular in cross-section, tangential diameter 15–45 μm, radial diameter up to 50 μm, walls 2–3 μm thick. Vessel member length 220(110–350) μm. Perforations simple in oblique end walls. Inter-vessel pits alternate to diffuse, rounded, ca. 4 μm in diameter, with slit-like apertures. Vessel-parenchyma and vessel-ray pits similar but half-bordered. Vascular tracheids and some vessels with fine spiral thickenings. Some vessels with gummy contents.
Fibres 520(280–870) μm long, thick-walled, with simple pits mainly in radial walls.
Parenchyma scanty paratracheal and marginal (initial) apotracheal in 1–4-seriate rows; in 2–5-celled strands; some cells crystalliferous.
Rays 9–14/mm, 1–3-seriate, up to 15(23) cells high; heterocellular (Kribs' heterogeneous types II–III); some cells crystalliferous.
Crystals minute, prismatic, solitary or several per cell, in ray and parenchyma cells.

PLATANACEAE

Platanus orientalis L.
Plate 53A and B

Tree; deciduous; river banks; Upper Galilee, Judean Mts., Dan Valley, Hula Plain, Upper Jordan Valley, Gilead, Ammon; East Mediterranean element.

Growth rings distinct. Vessels diffuse, ca. 75/mm², solitary (ca. 50%) or in tangential to radial multiples or clusters of 2–4(6); angular in cross-section, tangential diameter 40–100 μm, radial diameter up to 130 μm, walls ca. 2 μm thick. Vessel member length 570(390–780) μm. Perforations simple and scalariform in oblique end walls, the latter with up to ca. 20 bars. Inter-vessel pits opposite to diffuse, round to elongate, 6–10 μm in diameter, with slit-like apertures. Vessel-parenchyma and vessel-ray pits similar but half-bordered.
Fibres 1,450(840–1,800) μm long, medium thick-walled, with distinctly bordered pits in radial and tangential walls.
Parenchyma scanty paratracheal, and apotracheal diffuse-in-aggregates and in short uniseriate tangential bands; in 2–8-celled strands.
Rays 2–6/mm, rarely 1(2)-seriate, mostly multiseriate up to 14 cells wide, distending tangentially at the ring boundaries, up to ca. 3 mm high; homocellular, composed almost exclusively of procumbent cells, sometimes with one marginal row of square cells; many chambered crystalliferous.
Crystals solitary, prismatic, abundant in chambered ray cells.

Archaeological records: Jerusalem, 1st century A. D.; scabbard, Crusader Period (A. Fahn; E. Werker & A. Fahn).

POLYGONACEAE

Calligonum comosum L'Hér.
Plate 53C and D

Shrub; sands and sand dunes; W., C. and S. Negev, Dead Sea area, Arava Valley, deserts of Edom; Saharo-Arabian and West Irano-Turanian element.

Growth rings distinct. Wood ring- to semi-ring-porous. Vessels ca. 20–50/mm², in radial, tangential or oblique multiples of 2–6(13), or in clusters, rarely solitary (4%); rounded to angular in cross-section, of two size classes: wide vessels (tangential diameter 40–200 μm, radial diameter up to ca. 300 μm, walls ca. 7 μm thick), and narrow vessels intergrading with vascular tracheids (diameter 20–35 μm) associated with wide ones or clustered together, especially in the latest formed latewood. Vessel member length 150(90–190) μm. Perforations simple in horizontal to oblique end walls. Inter-vessel pits alternate, rounded or polygonal, 4–5 μm in diameter, with slit-like, sometimes coalescent apertures. Vessel-parenchyma and vessel-ray pits similar but half-bordered. Vessel members storied together with parenchyma cells.
Fibres 630(420–810) μm long, medium thick-walled, with simple to minutely bordered pits mainly in radial walls; with starch grains in sap wood.
Parenchyma scanty paratracheal to vasicentric, sometimes aliform to confluent, and marginal (initial); fusiform or in 2-celled strands; storied together with the vessel elements.
Rays 7–9/mm, 1–3(4)-seriate, up to ca. 30 cells high; homocellular, composed of strongly procumbent cells.

Large cells or sacs(?) with unidentified contents present throughout the wood. Crystals not observed.

PUNICACEAE

Punica granatum L.
Plate 54A–C

Cultivated and escaped from cultivation; deciduous; tall shrub or small tree; probably native in Asia Minor.

Growth rings faint to distinct. Vessels diffuse, 45–85/mm², solitary (20–80%) and in radial multiples of 2–4(9); rounded in cross-section, tangential diameter 25–85 μm, radial diameter up to ca. 110 μm, walls 3-6 μm thick. Vessel member length 300(130–530) μm. Perforations simple in oblique end walls. Inter-vessel pits vestured, alternate, round to polygonal, ca. 3–5 μm in diameter, with short oval apertures. Vessel-parenchyma and vessel-ray pits similar but half-bordered. Many vessels with gummy contents.
Fibres 610(300–830) μm long, thin- to medium thick-walled, with simple pits mainly in radial walls; septate, many crystalliferous.
Parenchyma very scanty paratracheal to almost absent, in 2–4-celled strands.
Rays 13–21/mm, 1(2)-seriate, up to 20 cells high; heterocellular, composed of upright and square cells, sometimes of procumbent cells in the biseriate portions.
Crystals prismatic, in chambered fibres, often one per chamber.

RANUNCULACEAE

Clematis cirrhosa L.
Plate 54D and E

Climber; maquis and maquis remnants; Sharon Plain, Philistean Plain, Upper and Lower Galilee, Mt. Carmel, Samaria, Judean Mts.; Mediterranean element.

Growth rings faint to distinct. Wood semi-ring- to ring-porous. Vessels 85–125/mm², of two size classes: wider vessels solitary (ca. 25%), in tangential multiples of 2–4, or in small clusters; rounded to angular in cross-section (diameter 30–160 μm, walls ca. 4 μm thick), mixed together with narrow vessels intergrading with vascular tracheids (diameter ca. 20 μm). Vessel member length 270 (180–340) μm. Perforations simple in horizontal to oblique end walls. Inter-vessel pits alternate to diffuse, elliptic, ca. 7 μm in diameter, with long slit-like, sometimes coalescent apertures. Vessel-parenchyma and vessel-ray pits similar to elongate, half-bordered. Vessels, except for the widest, with very fine spiral thickenings. Vessel members more or less storied together with parenchyma cells and fibres.

Wood Anatomy

Fibres 320(230–420) μm long, medium thick-walled, with numerous simple pits in radial and tangential walls; storied.
Parenchyma scanty paratracheal, fusiform and in 2-celled strands; storied.
Rays 1–4/mm, 2–25 cells wide, poorly delimited from surrounding tissue, the widest rays over 1.5 cm high; heterocellular, composed of upright and square cells and some slightly procumbent cells in the centre.
Crystals not observed.

RESEDACEAE

Ochradenus baccatus Del.
Plate 55

Shrub; hot desert wadis and depressions; Judean Desert, W., N. and C. Negev, Lower Jordan Valley, Dead Sea area, Arava Valley, deserts of Moav and Edom; Sudanian element.

Growth rings distinct. Wood diffuse- to weakly semi-ring-porous. Vessels 42–64/mm^2, mostly in clusters or variously directed multiples of 2–3, rarely solitary (ca. 10%); rounded in cross-section, tangential diameter 20–120 μm (tending to be of two size classes), radial diameter up to 170 μm, walls 3–8 μm thick. Vessel member length 140(100–180) μm. Perforations simple in horizontal to oblique end walls. Inter-vessel pits alternate, round to oval, 4–5 μm in diameter, with slit-like apertures. Vessel-parenchyma and vessel-ray pits similar but half-bordered. Vessel members storied together with parenchyma cells. Some vessels with gummy contents.
Fibres 660(460–920) μm long, thick-walled, with simple to minutely bordered pits mainly in radial walls; weakly storied.
Parenchyma scanty paratracheal or vasicentric, sometimes unilaterally aliform or confluent, and apotracheal in 1–3-seriate marginal bands; fusiform or in 2-celled strands; storied together with the vessel elements.
Rays 3–6/mm, (1)2–4-seriate, up to 20(32) cells high; heterocellular, composed of slightly procumbent, square and upright cells.
Crystals not observed.

RHAMNACEAE

The three genera of the Rhamnaceae in Israel, *Paliurus, Rhamnus* and *Ziziphus,* share such general features as simply perforated vessels with alternate pits, libriform fibres, and weakly heterocellular rays. *Rhamnus* stands out on account of its dendritic vessel pattern and the occurrence of druses in addition to prismatic crystals (N.B. a group of *Rhamnus* species belonging to subgenus *Frangula,* not represented in the flora of Israel, lacks the dendritic vessel pattern, cf. Metcalfe

& Chalk, 1950); *Paliurus* and *Ziziphus* are quite similar in their wood anatomy and can only be differentiated on the basis of a combination of quantitative characters, each of which taken by itself is of doubtful diagnostic value (see the descriptions).

Paliurus spina-christi Mill.
Plate 56A and B

Tree or shrub; deciduous; among shrubs mostly on alluvial soils; Acco Plain, Sharon Plain, Philistean Plain, Upper and Lower Galilee, Samaria, Judean Mts. (cultivated?), Dan Valley, Sinai (mostly cultivated); North and East Mediterranean and West Irano-Turanian element with extensions into South Euro-Siberian territories.

Growth rings distinct. Vessels diffuse, 35–50/mm^2, solitary (ca. 20–50%), in mostly radial multiples of 2–4, rarely in small clusters; rounded in cross-section, tangential diameter 30–100 μm, radial diameter up to 135 μm, walls 4–7 μm thick. Vessel member length 300(100–440) μm. Perforations simple in oblique end walls. Inter-vessel pits alternate, round to polygonal, ca. 5–7 μm in diameter, with short, slit-like, sometimes coalescent apertures. Vessel-parenchyma and vessel-ray pits similar but half-bordered.
Fibres 520(380–690) μm long, thick-walled, very narrow, with minute simple pits in radial walls.
Parenchyma vasicentric, occasionally confluent, and apotracheal in 1-seriate marginal bands; in 4–8-celled strands.
Rays 22–29/mm, mostly 1(2)-seriate, up to ca. 40(70) cells high; weakly heterocellular, composed of weakly procumbent, square and weakly upright cells mixed together; some cells crystalliferous.
Crystals solitary, prismatic, in some of the ray cells.

Rhamnus alaternus L.
Plate 56C and D

Shrub or tree; evergreen; semi-humid maquis; Coastal Galilee, Sharon Plain, Upper Galilee, Mt. Carmel, Samaria, Judean Mts., Dan Valley, Golan; Mediterranean element.

Growth rings distinct. Wood diffuse- to semi-ring-porous. Vessels ca. 150/mm^2, solitary and in short multiples and clusters forming a dendritic pattern together with paratracheal parenchyma and vascular tracheids; rounded to angular in cross-section, tangential diameter 12–40 μm, radial diameter up to 60 μm, walls ca. 2 μm thick. Vessel member length 290(130–380) μm. Perforations simple in oblique end walls. Inter-vessel pits opposite to alternate, rounded, 4–6 μm in diameter, with slit-like, often crossed apertures. Vessel-parenchyma and vessel-ray pits similar but half-bordered. Vessels and vascular tracheids with spiral thickenings.

Fibres 620(360–810) μm long, thick-walled, with numerous simple pits mainly in radial walls.
Parenchyma very scanty paratracheal, in 2–4-celled strands.
Rays 10–15/mm, 1–4-seriate, up to ca. 25 cells high; heterocellular, composed of square to upright or weakly procumbent marginal cells and procumbent or weakly procumbent central cells; many cells crystalliferous.
Crystals prismatic, one to many per cell, sometimes as druses, in ray cells.
Trabeculae file found in one ray.

Rhamnus dispermus Ehrenb.
Plate 57A

Shrub; deciduous; mountains in deserts; stony ground or among rocks; Judean Desert, W. and C. Negev, Moav, Edom, Sinai; Saharo-Arabian element.

Growth rings distinct. Vessels diffuse, 180–300/mm^2, in clusters often tangentially very wide, forming a dendritic pattern together with paratracheal parenchyma and vascular tracheids; angular to rounded in cross-section, tangential diameter 10–40 μm, radial diameter up to 60 μm, walls ca. 2 μm thick. Vessel member length 250(140–360) μm. Perforations simple in oblique end walls. Intervessel pits opposite to alternate, rounded, 6–8 μm in diameter, with slit-like, often crossed apertures. Vessel-parenchyma and vessel-ray pits similar but half-bordered. Many vessels and vascular tracheids with faint spiral thickenings. Many vessels with gummy contents.
Fibres 630(450–790) μm long, thick- to very thick-walled, with simple pits mainly in radial walls.
Parenchyma very scanty paratracheal, in 2–4-celled strands.
Rays 8–12/mm, 1–4-seriate, up to 22 cells high; weakly heterocellular, with marginal cells less procumbent than central ones or square, sometimes upright. Many cells crystalliferous.
Crystals solitary, prismatic, sometimes as druses, in ray cells.

Rhamnus palaestinus Boiss.
Plate 57B and C; Fig. 8C

Shrub; deciduous; maquis and garigue, also semi-steppes, on sandy and calcareous ground, often among rocks; Coastal Galilee, Acco Plain, Sharon Plain, Philistean Plain, Upper and Lower Galilee, Mt. Carmel, Mt. Gilboa, Samaria, Judean Mts., Golan, Gilead, Ammon, Moav, Edom; East Mediterranean element.

Growth rings distinct. Vessels diffuse, 170–265/mm^2, solitary and in short multiples and small clusters, forming a dendritic pattern together with paratracheal parenchyma and vascular tracheids; angular to rounded in cross-section, tangential diameter 15–40 μm, radial diameter up to 55 μm, walls ca. 2 μm thick. Vessel member length 250(180–360) μm. Perforations simple in oblique end walls. Inter

vessel pits opposite to alternate, rounded, 4–6 μm in diameter, with slit-like, often crossed apertures. Vessel-parenchyma and vessel-ray pits similar but half-bordered. Vessels and vascular tracheids with spiral thickenings.
Fibres 670(380–1,000) μm long, thick- to very thick-walled, with small simple pits more numerous in radial than in tangential walls; sometimes gelatinous; some crystalliferous.
Parenchyma very scanty paratracheal, in 2–4-celled strands.
Rays 7–13/mm, 1–4-seriate, up to 20 cells high; weakly heterocellular to homocellular (Kribs' heterogeneous type III), composed of weakly upright to procumbent marginal cells and procumbent central cells; many cells crystalliferous.
Crystals usually solitary, prismatic, in ordinary ray cells and chambered fibres; as druses in ordinary or enlarged ray cells.
Trabeculae in single and double files found in rays, external to traumatic tissue.

Rhamnus punctatus Boiss.
Plate 57D

Shrub; semi-deciduous; maquis, growing under relatively moist conditions; Upper and Lower Galilee, Mt. Carmel, Judean Mts., Upper Jordan Valley, Ammon; East Mediterranean element.

Growth rings distinct. Wood diffuse- to semi-ring-porous. Vessels 170–220/mm^2, solitary, in multiples of 2–3, rarely in clusters, forming, together with paratracheal parenchyma and vascular tracheids, a dendritic pattern and often, at the beginning of growth rings, a more or less continuous band, 1 or more vessels wide; rounded to angular in cross-section, diameter 10–40 μm, walls 1–2 μm thick. Vessel member length 330(160–370) μm. Perforations simple in oblique end walls. Inter-vessel pits alternate, ca. 7 μm in diameter, with slit-like apertures. Vessel-parenchyma and vessel-ray pits similar but half-bordered. Many vessels and vascular tracheids with spiral thickenings.
Fibres 510(360–620) μm long, thick-walled, with simple pits mainly in radial walls.
Parenchyma very scanty paratracheal, in 2–4-celled strands.
Rays 10–15/mm, 1–3(4)-seriate, up to 17(22) cells high; weakly heterocellular (Kribs' heterogeneous type III), composed of weakly procumbent or square marginal cells and procumbent central cells; many cells crystalliferous.
Crystals usually solitary, prismatic, sometimes as druses, in ray cells.

Characteristics of the Genus Rhamnus

Growth rings distinct. Vessels forming a dendritic pattern together with paratracheal parenchyma and vascular tracheids; angular to rounded. Perforations simple. Vessels and vascular tracheids with spiral thickenings. Fibres with simple pits. Parenchyma very scanty paratracheal. Rays 1–4-seriate, many cells crystalliferous. Crystals prismatic, sometimes as druses, in ray cells.

Separation of Rhamnus Species

The four species are very similar in their wood anatomy and, on the basis of the material studied, cannot be unambiguously separated.

Ziziphus lotus (L.) Lam.
Plate 58A–C

Shrub; deciduous; stony slopes and alluvial plains, mainly on basalt ground, up to 500 m above sea level; Acco Plain, Sharon Plain, Upper and Lower Galilee, Mt. Carmel, Esdraelon Plain, Mt. Gilboa, Samaria, Dan Valley, Hula Plain, Upper Jordan Valley, Beit Shean Valley, Lower Jordan Valley, Golan, Gilead; Mediterranean and Sudanian element.

Growth rings distinct. Vessels diffuse, 20–35/mm^2, solitary (28–40%) or in radial multiples of 2–4(5), rarely in small radial clusters; rounded in cross-section, tangential diameter 30–120 μm, radial diameter up to 160 μm, walls 3–8 μm thick. Vessel member length 290(210–380) μm. Perforations simple in oblique end walls. Inter-vessel pits alternate, rounded to polygonal, 7–8 μm in diameter, with slit-like, sometimes coalescent apertures. Vessel-parenchyma and vessel-ray pits similar but half-bordered.
Fibres 570(410–750) μm long, medium thick- to thick-walled, with simple pits in radial and tangential walls.
Parenchyma vasicentric and in 1–2-seriate marginal bands; in 2–6-celled strands.
Rays 11–18/mm, 1(2)-seriate, 2–45 cells high; weakly heterocellular, composed of square, upright and weakly procumbent cells; some cells crystalliferous.
Crystals solitary, prismatic, sometimes slightly elongate, in some of the ray cells.

Archaeological records: Beit Shean Valley; Chalcolithic Period (N. Liphschitz & Y. Waisel).

Ziziphus spina-christi (L.) Desf.
Plate 58D

Tree; evergreen; oases in hot deserts, wadi beds, coastal foothills, fields of alluvial soils, usually from 350 m below to 500 m above sea level; Acco Plain, Sharon Plain, Philistean Plain, Upper and Lower Galilee, Esdraelon Plain, Samaria, Shefela, Judean Mts., Negev, Dan Valley, Hula Plain, Upper Jordan Valley, Beit Shean Valley, Lower Jordan Valley, Dead Sea area, Arava Valley, Golan, valleys in Moav and Edom; Sudanian element.

Growth rings distinct. Vessels diffuse, 14–24/mm^2, solitary (25–30%), in radial multiples of 2–3, occasionally in small clusters; rounded in cross-section, tangential diameter 30–150 μm, radial diameter up to 190 μm, walls 5–12 μm thick. Vessel member length 430(270–650) μm. Perforations simple in oblique end walls. Inter-vessel pits alternate, rounded to polygonal, 6–10 μm in diameter, with slit-like, sometimes coalescent apertures. Vessel-parenchyma and vessel-ray pits similar but half-bordered. Many vessels with gummy contents.

Wood Anatomical Descriptions

Fibres 810(590–1,130) μm long, medium thick- to very thick-walled, with simple pits mainly in radial walls.
Parenchyma scanty paratracheal to vasicentric and in 1(3)-seriate marginal bands; in 2–6-celled strands; infrequently crystalliferous.
Rays 8–11/mm, 1(2)-seriate, up to 20(24) cells high; weakly heterocellular, composed of square, upright and infrequent weakly procumbent cells; some cells crystalliferous.
Crystals solitary, prismatic, in some of the ray cells and very infrequent in the parenchyma cells.

Archaeological records: 5 locations in the coastal plain and the Negev; 9th century B. C. to the 3rd century A. D. (N. Liphschitz & Y. Waisel).

Characteristics of the Genus Ziziphus

Growth rings distinct. Vessels diffuse, less than 50/mm^2; solitary or in radial multiples and small clusters; rounded in cross-section. Perforations simple in oblique end walls. Fibres with simple pits. Parenchyma mainly vasicentric and in marginal bands. Rays 1(2)-seriate.

Separation of Ziziphus Species

In the material studied the two species differ mainly in some quantitative characters, which probably cannot be relied on for an unambiguous separation based on wood anatomy.

ROSACEAE

Most woody Rosaceae, native or cultivated in Israel, are characterized by a high proportion of solitary vessels, fibres with distinctly bordered pits (fibre-tracheids) and scanty paratracheal plus diffuse apotracheal parenchyma. High proportions of vessels in multiples occur in *Prunus, Amygdalus* and *Sarcopoterium*. The latter genus moreover stands out because it has libriform fibres and lacks apotracheal parenchyma. In *Armeniaca* the fibres are of an intermediate type, with the bordered pits mainly confined to the radial walls.
The representatives of the Maloideae (= Pomoideae)—*Cotoneaster, Crataegus, Cydonia, Eriolobus* and *Pyrus*—constitute a particularly homogeneous group with narrow rays and prismatic crystals in most species. The Prunoideae — *Amygdalus, Armeniaca, Cerasus* and *Prunus* — are all treated under *Prunus* by some authors, while Dahlgren (1980) goes to the other extreme and recognizes the separate family of the Amygdalaceae. The species of Israel are fairly diverse in their wood anatomy and show considerable differences in ray width and proportion of solitary vessels. The shared presence of druses in the ray cells in *Amygdalus, Armeniaca* and *Prunus* is, however, indicative of a close phylogenetic relationship. The Rosoideae, *Rosa* and *Sarcopoterium*, differ considerably

from each other in fibre type and degree of vessel grouping, but share wide and tall rays.

Amygdalus communis L.
Plate 59A

Tree; deciduous; maquis; Upper and Lower Galilee, Mt. Carmel, Judean Mts.; East Mediterranean and West Irano-Turanian element.

Growth rings distinct. Wood semi-ring- to ring-porous. Vessels 150–270/mm², solitary (ca. 50%), in mainly radial multiples of 2–5 and in loose clusters; rounded to angular in cross-section, tangential diameter 30–120 μm, radial diameter up to 130 μm, walls 3–4 μm thick. Vessel member length 260(140–330) μm. Perforations simple in oblique end walls. Inter-vessel pits alternate, rounded to polygonal, 5–8 μm in diameter, with slit-like apertures. Vessel-parenchyma and vessel-ray pits similar but half-bordered. Vessel walls with fine spiral thickenings. Many vessels with tyloses and gummy contents.
Fibres 840(400–1,120) μm long, thick-walled, with distinctly bordered pits in radial and tangential walls.
Parenchyma very sparse, occasionally scanty paratracheal, and apotracheal diffuse; in 2–5-celled strands.
Rays 6–13/mm, tending to be of two distinct sizes: 1(2)-seriate and 3–8-seriate, the latter up to ca. 40 cells high; weakly heterocellular, composed mainly of procumbent cells, with weakly procumbent to square marginal cells (Kribs' heterogeneous type III); some cells crystalliferous.
Crystals infrequently present as clusters or druses in ray cells.

Amygdalus korschinskii (Hand.-Mazzetti) Bornm.
Plate 59B

Tree or high shrub; deciduous; impoverished maquis and maquis-steppes; Upper Galilee, Mt. Gilboa, Samaria, Judean Mts., Judean Desert, Central Negev, Upper Jordan Valley, Gilead, Moav; West Irano-Turanian element.

Growth rings distinct. Wood ring-porous. Vessels 74–150/mm², mostly solitary, sometimes in pairs, rarely in multiples of 3–5 or small clusters; rounded to angular in cross-section, diameter 12–100 μm, walls 3–5 μm thick. Vessel member length 210(120–290) μm. Perforations simple in oblique end walls. Inter-vessel pits alternate, rounded to polygonal, 5–7 μm in diameter, with oval to slit-like apertures. Vessel-parenchyma and vessel-ray pits similar but half-bordered. Vessels with fine spiral thickenings. Some vessels with gummy contents.
Fibres 690(390–870) μm long, thick- to very thick-walled, with distinctly bordered pits in radial and tangential walls.
Parenchyma very sparse, scanty paratracheal and apotracheal diffuse; in 2–4-celled strands.
Rays 7–13/mm, of two distinct sizes: 1(2)-seriate and 3–8-seriate, the latter up to

44 cells high; rarely compound; heterocellular, composed mainly of procumbent cells and weakly procumbent to upright marginal cells; some cells crystalliferous. Crystals infrequently present as clusters in ray cells.

Characteristics of the Genus Amygdalus

Growth rings distinct. Vessels rounded to angular in cross-section. Perforations simple in oblique end walls. Inter-vessel pits alternate, rounded to polygonal. Vessels with fine spiral thickenings. Some vessels with gummy contents. Fibres with distinctly bordered pits in the radial and the tangential walls. Parenchyma sparse, scanty paratracheal and diffuse apotracheal. Rays uni- and multiseriate, up to 8 cells wide.

Key to Amygdalus Species

Most vessels solitary, widely spaced in latewood. **A. korschinskii**
Vessels often in short multiples, densely spaced in latewood. **A. communis**

Armeniaca vulgaris Lam.
Plate 59C and D; Fig. 9C

Cultivated tree; deciduous; native in China.

Growth rings distinct. Wood ring- to semi-ring-porous. Vessels 55–155/mm^2, many solitary, some in multiples of 2–3(4), rarely in clusters; earlywood vessels embedded in ground tissue of tracheids; rounded to angular in cross-section, tangential diameter 15–100 μm, radial diameter up to 150 μm, walls ca. 2 μm thick. Vessel member length 250(150–360) μm. Perforations simple in oblique end walls. Inter-vessel pits alternate, rounded to polygonal, 4–7 μm in diameter, with short slit-like apertures. Vessel-parenchyma and vessel-ray pits rounded, half-bordered. Vessels and some vasicentric tracheids with very fine, loose spiral thickenings.
Fibres 650(340–1,120) μm long, thick-walled, with minutely but distinctly bordered pits mainly in radial walls.
Parenchyma scanty paratracheal and diffuse apotracheal; in 2–6-celled strands.
Rays 4–5/mm, 1–2(3)- and (4)5–12-seriate, the wider ones up to 38 cells high; weakly heterocellular, composed of procumbent central cells and square to slightly upright marginal cells; often tangentially distended at the boundary between growth rings.
Crystals infrequently present as druses in ray cells.
Traumatic gum ducts often present.

Cerasus prostrata (Labill.) Ser.
Plate 60A and B

Shrub; deciduous; rocks; Mt. Hermon; Mediterranean element.

Growth rings distinct. Wood ring-porous. Vessels ca. 200–250/mm^2, mostly soli-

tary, occasionally in variously directed multiples of 2–3; angular to round in cross-section, tangential diameter 10–40 μm, radial diameter up to 45 μm, walls ca. 1 μm thick. Vessel member length 250(170–380) μm. Perforations simple in oblique end walls. Inter-vessel pits alternate, rounded, ca. 5 μm in diameter, with elliptic to slit-like apertures. Vessel-parenchyma and vessel-ray pits similar but half-bordered. Vessel walls with prominent spiral thickenings.
Fibres 510(320–620) μm long, very thick-walled, with distinctly bordered pits in radial and tangential walls; occasionally with spiral thickenings.
Parenchyma very sparse, scanty paratracheal and diffuse apotracheal; in few-celled strands.
Rays 12–18/mm, 1–3-seriate, up to 0.8 mm, rarely up to 2 mm or more in height; heterocellular, composed of square, weakly procumbent and upright cells.
Crystals not observed.

Cotoneaster nummularia Fisch. et Mey.
Plate 60C and D

Shrub; deciduous; on rocks; Mt. Hermon; West Irano-Turanian and Mediterranean element.

Growth rings distinct. Wood semi-ring-porous. Vessels 300–400/mm^2, mostly solitary, few in tangential or radial multiples of 2–3 or very small clusters; angular in cross-section, diameter 10–45 μm, walls 1–2 μm thick. Vessel member length 470(210–560) μm. Perforations simple in oblique end walls. Inter-vessel pits not observed. Vessel-parenchyma and vessel-ray pits alternate, rounded, half-bordered, with slit-like to gash-like apertures. Vessel members with fine spiral thickenings.
Fibres 690(450–920) μm long, very thick-walled, with distinctly bordered pits in radial and tangential walls; with fine spiral thickenings.
Parenchyma scanty paratracheal and diffuse apotracheal, or diffuse-in-aggregates; in 3–6-celled strands; many chambered crystalliferous.
Rays 8–11/mm, 1–2(3)-seriate, up to 21 cells high; heterocellular, composed of procumbent central cells and one to several rows of square to upright marginal cells.
Crystals solitary, prismatic, in enlarged chambered parenchyma cells.

Crataegus aronia (L.) Bosc.

Tree or shrub; deciduous; maquis and maquis-steppes; Sharon Plain, Philistean Plain, Upper and Lower Galilee, Mt. Carmel, Esdraelon Plain, Mt. Gilboa, Samaria, Shefela, Judean Mts., Judean Desert, Gilead, Ammon, Moav, Edom; East Mediterranean and West Irano-Turanian element.

Growth rings distinct. Vessels diffuse, 180–250/mm^2, mostly solitary, sometimes in variously directed multiples of 2; angular to rounded in cross-section, diameter 20–50 μm, walls ca. 1–2 μm thick. Vessel member length 380(190–590) μm.

Perforations simple in oblique end walls. Inter-vessel pits alternate, rounded, ca. 4 μm in diameter, with slit-like apertures. Vessel-parenchyma and vessel-ray pits alternate, rounded, the former simple, the latter half-bordered.
Fibres 770(550–1,120) μm long, thick-walled, with distinctly bordered pits in radial and tangential walls.
Parenchyma scanty paratracheal and diffuse apotracheal, sometimes diffuse-in-aggregates; in 2–4-celled strands; some chambered crystalliferous.
Rays 7–13/mm, 1–3(4)-seriate, up to 25 cells high; mainly homocellular, composed of procumbent cells; marginal cells less strongly procumbent than central cells, sometimes almost square.
Crystals solitary, prismatic, in enlarged chambered parenchyma cells.

Crataegus azarolus L.
Plate 61A and B

Tree; deciduous; in more humid maquis; Upper Galilee (Mt. Meiron), Mt. Carmel, Judean Mts.; East Mediterranean element.

Growth rings distinct. Vessels diffuse, ca. 250/mm^2, mostly solitary, occasionally in variously directed multiples of 2(3), or very small clusters; angular to round in cross-section, diameter 20–70 μm, walls ca. 1.5 μm thick. Vessel member length 440(250–640) μm. Perforations simple in oblique end walls. Inter-vessel pits diffuse, rounded, 5–9 μm in diameter, with slit-like, sometimes coalescent apertures. Vessel-parenchyma and vessel-ray pits similar but half-bordered. Some vessels with very faint spiral thickenings.
Fibres 740(530–960) μm long, thick-walled, with distinctly bordered pits in radial and tangential walls; some with fine spiral thickenings.
Parenchyma scanty paratracheal and diffuse apotracheal, sometimes diffuse-in-aggregates; in 2–4-celled strands; occasionally crystalliferous.
Rays 11–15/mm, 1–3(4)-seriate, up to 25(35) cells high; homocellular, composed of procumbent cells; marginal cells less strongly procumbent than the central cells.
Crystals sometimes in clusters or prismatic, in enlarged chambered parenchyma cells.

Crataegus monogyna Jack.

Shrub or small tree; deciduous; oak maquis, in shady valleys, terra rossa, about 1,000 m; Upper Galilee; Mediterranean and Euro-Siberian element.

Growth rings distinct. Vessels diffuse, 300–350/mm^2; mostly solitary, occasionally in multiples of 2(3); angular in cross-section, diameter 10–42 μm, walls 1–2 μm thick. Vessel member length 320(180–470) μm. Perforations simple in oblique end walls. Vessel-parenchyma and vessel-ray pits rounded, small, alternate, simple or half-bordered. Very faint spiral thickenings in some of the vessels.

Wood Anatomy

Fibres 570(410–750) μm long, thick-walled, with distinctly bordered pits in radial and tangential walls; many with distinct spiral thickenings.
Parenchyma scanty paratracheal and diffuse apotracheal, occasionally diffuse-in-aggregates; in 2–4-celled strands; often crystalliferous.
Rays 11–15/mm, 1–3(4)-seriate, some up to 1.5 mm high; heterocellular, composed of procumbent central cells and square to upright marginal cells.
Crystals rare, solitary, prismatic, in chambered parenchyma cells.

Crataegus sinaica Boiss.
Plate 61C and D

Shrub; deciduous; Sinai; West Irano-Turanian element.

Growth rings distinct. Vessels diffuse, 165–265/mm², mostly solitary, occasionally in 2–3(4)-celled multiples or clusters; angular in cross-section, diameter 16–52 μm, walls ca. 2 μm thick. Vessel member length 580(310–830) μm. Perforations simple in oblique end walls. Inter-vessel pits diffuse, rounded, 5–7 μm in diameter, with slit-like, sometimes coalescent apertures. Vessel-parenchyma and vessel-ray pits similar but half-bordered. Very faint spiral thickenings in some of the vessels.
Fibres 1,000(540–1,460) μm long, thick-walled, with distinctly bordered pits in radial and tangential walls; some with faint spiral thickenings.
Parenchyma scanty paratracheal, and apotracheal diffuse and diffuse-in-aggregates; in (2)4-celled strands; many-chambered crystalliferous.
Rays 5–10/mm, 1–3(4)-seriate, up to 25(35) cells high; weakly heterocellular, mainly composed of procumbent cells; marginal cells square or weakly upright or procumbent (Kribs' heterogeneous type III to homogeneous).
Crystals solitary, prismatic, in enlarged chambered parenchyma cells.

Characteristics of the Genus Crataegus

Growth rings distinct; vessels diffuse, densely arranged, mostly solitary, mostly angular in cross-section, wall thickness 1–2 μm. Perforations simple in oblique end walls. Vessel-parenchyma and vessel-ray pits alternate, small, rounded. Fibres thick-walled with distinctly bordered pits in the radial and tangential walls. Parenchyma scanty paratracheal and diffuse apotracheal, sometimes diffuse-in-aggregates; in 2–4-celled strands; often crystalliferous. Rays 1–3 (or –4)-seriate. Crystals in chambered axial parenchyma.

Separation of Crataegus Species

The species are very similar in their wood anatomy and cannot be unambiguously separated.

Archaeological records: Tel Afek (coastal plain) and Tel Arad; ranging from 8th century B. C. to 8th century A. D. (N. Liphschitz & Y. Waisel).

Cydonia oblonga Mill. (*C. vulgaris* Pers.)
Plate 62A and B

Cultivated tree; deciduous; native in Central and East Asia.

Growth rings distinct. Vessels diffuse, ca. 150/mm², almost exclusively solitary; angular in cross-section, diameter 25–50 μm, walls 1–1.5 μm thick. Vessel member length 440(360–620) μm. Perforations simple, very rarely reticulate, in oblique end walls. Vessel-fibre pits rounded, 4–7 μm in diameter, with slit-like apertures. Vessel-parenchyma and vessel-ray pits alternate, round to elliptic, small, half-bordered. Some vessels with spiral thickenings.
Fibres 660(470–850) μm long, thick- to very thick-walled, with distinctly bordered pits in radial and tangential walls; some with fine spiral thickenings.
Parenchyma scanty paratracheal, and apotracheal diffuse and diffuse-in-aggregates; in 4–8-celled strands.
Rays 15–19/mm; (1)2(3)-seriate, up to ca. 55 cells high; predominantly homocellular, composed of procumbent cells occasionally with square marginal cells.
Crystals not observed.

Eriolobus trilobatus (Labill. ex Poir.) M. Roem.
Plate 62C and D

Tree or shrub; deciduous; shady maquis; Upper Galilee (Mt. Meiron); East Mediterranean element.

Growth rings distinct. Wood diffuse- to semi-ring-porous. Vessels ca. 100–120/mm², mostly solitary, some in multiples of 2–3(4), rarely in small clusters; angular in cross-section, diameter 20–70 μm, walls ca. 3 μm thick. Vessel member length 640(240–970) μm. Perforations simple in oblique end walls. Intervessel pits alternate to diffuse, rounded, ca. 5–7 μm in diameter, with slit-like apertures. Vessel-parenchyma and vessel-ray pits similar but half-bordered. Many vessels with faint spiral thickenings.
Fibres 1,140(470–1,500) μm, thick-walled, with distinctly bordered pits in radial and tangential walls; occasionally with faint spiral thickenings.
Parenchyma scanty paratracheal, and apotracheal diffuse to diffuse-in-aggregates; in (2)4–7-celled strands; occasionally crystalliferous.
Rays 7–10/mm, 1–3-seriate, up to ca. 15(25) cells high; heterocellular (Kribs' heterogeneous types II and III), composed of procumbent central cells and one to several rows of weakly procumbent to upright marginal cells.
Crystals solitary, prismatic, infrequent in chambered parenchyma cells.
Trabeculae observed in a radial file in vessels and parenchyma cells.

Prunus ursina Ky.
Plate 63A and B

Small tree; deciduous; maquis; Upper Galilee; East Mediterranean element.

Growth rings distinct. Wood semi-ring-porous. Vessels ca. 100–200/mm²; solitary (ca. 25%), in radial to oblique multiples of 2–4(5), occasionally in small clusters; rounded to angular in cross-section, tangential diameter 17–55 μm, radial diameter up to 70 μm, walls ca. 1 μm thick. Vessel member length 250 (100–360) μm. Perforations simple in oblique end walls. Inter-vessel pits alternate, polygonal to rounded, ca. 5 μm in diameter, with elliptic to slit-like apertures. Vessel-parenchyma and vessel-ray pits similar but half-bordered or with reduced borders. Vessels with widely spaced distinct spiral thickenings. Some vessels with gummy contents.

Fibres 810(390–1,170) μm long, thick- to very thick-walled, with small but distinctly bordered pits in radial and tangential walls; many with spiral thickenings.

Parenchyma scanty paratracheal and diffuse apotracheal; fusiform or in 2–4-celled strands; occasionally chambered crystalliferous.

Rays 10–14/mm, 1–4(5)-seriate, up to ca. 40(100) cells high; heterocellular, composed of square to upright marginal, sometimes sheath cells, and procumbent to square central cells; rarely crystalliferous.

Crystals infrequently present as druses in ray and chambered parenchyma cells.

Pyrus syriaca Boiss.

Plate 63C and D

Tree; deciduous; maquis, forests and devastated woodland; Upper Galilee, Mt. Carmel, Samaria, Judean Mts., Judean Desert, Gilead, Ammon, Edom; East Mediterranean and West Irano-Turanian element.

Growth rings distinct. Vessels diffuse, ca. 230/mm^2, almost exclusively solitary, rarely in pairs or small clusters; angular in cross-section, diameter 20–45 μm, walls 1–2 μm thick. Vessel member length 470(320–570) μm. Perforations simple in oblique end walls. Inter-vessel pits diffuse, round, ca. 6 μm in diameter. Vessel-parenchyma pits similar but half-bordered. Vessel-ray pits rounded, small, half-bordered. Many vessels with gummy contents.

Fibres 710(530-900) μm long, medium thick-walled, with distinctly bordered pits in radial and tangential walls.

Parenchyma scanty paratracheal, and apotracheal diffuse to diffuse-in-aggregates; in 4–7-celled strands; rarely crystalliferous.

Rays 11–17/mm, 1–2(3)-seriate, up to 43(55) cells high; predominantly homocellular, composed of procumbent cells, occasionally with square or weakly upright marginal cells.

Crystals solitary, prismatic, very infrequent in chambered parenchyma cells.

Rosa arabica Crep.

Shrub; rocks; Sinai; East Saharo-Arabian element.

Growth rings distinct. Wood semi-ring- to ring-porous. Vessels ca. 150/mm^2, mostly solitary, some in mostly tangential multiples of 2–3; rounded to angular in cross-section, diameter 16–65 μm, walls ca. 1–2 μm thick. Vessel member length 390(250–510) μm. Perforations simple in oblique end walls. Inter-vessel pits alternate, round, 5–6 μm in diameter, with oval to slit-like apertures. Vessel-parenchyma and vessel-ray pits similar but half-bordered. Many vessels with very fine spiral thickenings.

Fibres 680(400–850) μm long, thick-walled, with distinctly bordered pits in radial and tangential walls; many with spiral thickenings.

Parenchyma scanty paratracheal and sparse diffuse apotracheal; in 2–3-celled strands.

Rays 11–15/mm, 1–6-seriate; uni- and biseriate rays low, some multiseriates more than 7 mm high; uniseriates homocellular, composed of upright cells; multiseriates heterocellular, composed of square, upright and weakly procumbent cells mixed together; sheath cells occasionally present; some rays compound; some cells crystalliferous.

Crystals solitary, prismatic, in some of the ray cells.

Rosa canina L.

Shrub; mountain slopes; Upper Galilee (Mt. Meiron); Euro-Siberian, Mediterranean and West Irano-Turanian element.

Growth rings distinct. Wood ring- to semi-ring-porous. Vessels numerous, mostly solitary, some in tangential multiples of 2–4; angular in cross-section, diameter 13–55 μm, walls ca. 2 μm thick. Vessel member length 400(320–560) μm. Perforations simple in oblique end walls. Inter-vessel pits alternate to opposite, polygonal, ca. 5 μm in diameter, with slit-like apertures. Vessel-parenchyma and vessel-ray pits rounded, half-bordered. Vessels, except for the widest, with fine spiral thickenings.

Fibres 670(390–870) μm long, thick- to very thick-walled, with distinctly bordered pits in radial and tangential walls; some with spiral thickenings.

Parenchyma scanty paratracheal and diffuse apotracheal (difficult to distinguish from uniseriate rays).

Rays 8–12/mm, 1–5(6)-seriate, uniseriates low, multiseriates more than 7 mm high; uniseriates predominantly homocellular, composed mainly of upright cells; multiseriates heterocellular, composed of upright, square and procumbent cells mixed together. Some cells crystalliferous.

Crystals solitary, prismatic, in some of the ray cells.

Rosa phoenicia Boiss.

Shrub; river banks and swamp edges; Acco Plain, Sharon Plain, Philistean Plain, Upper Galilee, Mt. Carmel, Samaria, Upper Jordan Valley, Gilead; East Mediterranean element.

Growth rings distinct. Wood ring- to semi-ring-porous. Vessels 100–160/mm^2, mostly solitary, some, mainly at the beginning of growth ring, in mostly tangential multiples of 2–3; angular in cross-section, diameter 20–60 μm, walls 1.5–2 μm thick. Vessel member length 380(160–530) μm. Perforations simple in

oblique end walls. Inter-vessel pits alternate to diffuse, round to polygonal, ca. 5 μm in diameter, with short slit-like apertures. Vessel-parenchyma and vessel-ray pits similar but half-bordered. Vessels with spiral thickenings.
Fibres 780(380–1,010) μm long, very thick-walled, with distinctly bordered pits in radial and tangential walls; some with fine spiral thickenings.
Parenchyma very scanty paratracheal and diffuse apotracheal; in 2–4-celled strands.
Rays 6–14/mm, 1–7(10)-seriate, of various heights, up to more than 3 mm; uniseriates predominantly homocellular, composed mainly of upright cells; multiseriates heterocellular, composed of procumbent (sometimes square and upright) central cells and square to upright marginall cells; many compound; some cells crystalliferous.
Crystals solitary, prismatic, in some of the ray cells.

Rosa pulverulenta Bieb. (*R. glutinosa* Sm.)
Plate 64A and B

Shrub; rocks; Mt. Hermon; Mediterranean element.

Growth rings distinct. Wood ring-porous. Vessels ca. 100–130/mm², mostly solitary, some in mainly tangential multiples of 2(4); round to angular in cross-section, diameter 16–60 μm, walls ca. 2 μm thick. Vessel member length 400 (280–530) μm. Perforations simple in oblique end walls. Inter-vessel pits alternate, round or polygonal, 5–6 μm in diameter, with slit-like apertures. Vessel-parenchyma and vessel-ray pits similar but half-bordered. Most vessels with fine spiral thickenings.
Fibres 830(450–1,070) μm, thick- to very thick-walled, with distinctly bordered pits in radial and tangential walls; with spiral thickenings.
Parenchyma scanty paratracheal and diffuse apotracheal; mainly in 2-celled strands.
Rays 11–15/mm, 1–7-seriate, of arious heights, up to ca. 5 mm; uniseriates predominantly homocellular, composed mainly of upright cells; multiseriates heterocellular, composed of square, slightly procumbent and upright cells mixed together; with many sheath cells; many cells crystalliferous.
Crystals variously shaped, varying in numbers, in ray cells.

Characteristics of the Genus Rosa

Most vessels solitary, not exceeding the diameter of 70 μm; walls very thin (not thicker than 2 μm), most of them with fine spiral thickenings. Fibres with distinctly bordered pits; some with spiral thickenings. Parenchyma scanty paratracheal, and diffuse apotracheal. Rays uniseriate, and multiseriate heterocellular, up to more than 3 mm high; some cells with prismatic crystals.

Separation of Rosa Species

Differences in vessel distribution and ray width and height might be due to young age of the samples, so that we must refrain from attempting to key out the species on the basis of our limited material.

Sarcopoterium spinosum (L.) Sp.
Plate 64C and D

Shrub; batha; Acco Plain, Sharon Plain, Philistean Plain, Upper and Lower Galilee, Mt. Carmel, Esdraelon Plain, Mt. Gilboa, Samaria, Shefela, Judean Mts., Judean Desert, N. Negev, Dan Valley, Hula Plain, Upper Jordan Valley, Gilead, Ammon, Moav, Edom; East Mediterranean element.

Growth rings distinct. Wood diffuse- to semi-ring porous. Vessels 150–300/mm^2, solitary (ca. 12%), in clusters and in variously directed multiples of 2–5; angular to rounded in cross-section, tangential diameter 12–70 μm, radial diameter up to 80 μm, walls ca. 2 μm thick; wider earlywood vessels embedded in ground tissue of very narrow vessels intergrading with vascular tracheids. Vessel member length 250(140–360) μm. Perforations simple in oblique end walls. Inter-vessel pits alternate, rounded to polygonal, ca. 5 μm in diameter, with slit-like apertures. Vessel-parenchyma and vessel-ray pits similar but half-bordered.
Fibres 340(240–440) μm long, medium thick-walled, with minutely bordered to simple pits in radial and tangential walls; partly septate.
Parenchyma very scanty paratracheal, mainly in 2-celled strands.
Rays 5–8/mm, uniseriate and (2)6–12-seriate; uniseriates 1–5 cells high, multiseriates up to ca. 2 mm high; not always well delimited; heterocellular, composed of square and upright and weakly procumbent cells; many cells crystalliferous.
Crystals solitary, prismatic, in ordinary or chambered ray cells.

RUTACEAE

The two woody genera, *Ruta* and *Citrus,* representing the Rutaceae in Israel differ widely in their wood anatomy, especially in vessel and parenchyma distribution and frequency, and in occurrence of crystals (see the descriptions). This is at least partly due to their widely different habit and ecology, and probably also to their different alliance in the Rutaceae: *Ruta* is a member of the Rutoideae; *Citrus* of the Aurantioideae.

Citrus aurantium L.
Plate 65 A–C; Fig. 8B

Cultivated fruit tree; evergreen; native in Asia.

Growth rings very faint. Vessels diffuse, 10–25/mm^2, solitary (15–40%), in predominantly radial multiples of 2–4 and small clusters; rounded in cross-section, tangential diameter 30–110 μm, radial diameter up to 120 μm, walls ca. 3 μm thick. Vessel member length 210(150–280) μm. Perforations simple in mainly

oblique end walls. Inter-vessel pits alternate to opposite, round to polygonal, minute, ca. 2–3 μm in diameter, with slit-like, occasionally coalescent apertures. Vessel-parenchyma and vessel-ray pits similar but half-bordered. Many vessels with gummy contents.
Fibres 600(450–840) μm long, medium thick-walled, with simple to minutely bordered pits mainly in radial walls.
Parenchyma partly vasicentric, mostly confluent, forming tangential bands 2–7 cells wide in 2–4-celled strands; many cells chambered crystalliferous and often enlarged.
Rays 7–11/mm, 1–4-seriate, up to ca. 23(30) cells high; homocellular to weakly heterocellular, composed of procumbent central cells and procumbent to square marginal cells.
Crystals solitary, prismatic, in chambered, often enlarged, parenchyma cells.

Ruta chalepensis L.
Plate 65D

Shrub; batha and garigue; sandy and calcareous soils; Acco Plain, Sharon Plain, Philistean Plain, Upper and Lower Galilee, Mt. Carmel, Esdraelon Plain, Samaria, Judean Mts., Upper Jordan Valley, Golan, Gilead; Mediterranean element.

Growth rings distinct. Vessels diffuse, 120–200/mm^2, in mainly radial multiples of 2–6, sometimes in clusters or solitary, tending to form a dendritic pattern; rounded to angular in cross-section, tangential diameter 15–40 μm, radial diameter up to 65 μm, walls 2–3 μm thick. Vessel member length 180(90–210) μm. Perforations simple in mainly oblique end walls. Inter-vessel pits alternate, rounded, 7–8 μm in diameter, with slit-like apertures. Vessel-ray pits similar but half-bordered.
Fibres 360(300–430) μm long, thin- to medium thick-walled, with sparse, simple to minutely bordered pits mainly in radial walls.
Parenchyma very sparse, scanty paratracheal; fusiform or in 2-celled strands.
Rays 6–8/mm, 1–4(5)-seriate, up to ca. 35 cells high; heterocellular, composed of procumbent, square and sometimes upright cells.
Crystals not observed.

SALICACEAE

The genera *Populus* and *Salix* are very similar in their wood anatomy (also in species from outside Israel, cf. Metcalfe & Chalk, 1950). Vessels with large polygonal inter-vessel pits and simple vessel-ray pits in only part of the ray cells (mostly the marginal ones), libriform fibres, scanty paratracheal and sparse marginal parenchyma, together with very narrow rays characterize the family. The genera can be differentiated only on the basis of the ray structure, which is typically homocellular in *Populus* and heterocellular in *Salix*.

Populus euphratica Oliv.
Plate 66A–C

Tree; deciduous; river banks and springs; Central Negev, Upper and Lower Jordan Valley, Dead Sea area, Moav; forms riverine forests, especially on the banks of the lower course of the Jordan River; Irano-Turanian and Saharo-Arabian element.

Growth rings distinct. Wood diffuse- to weakly semi-ring-porous. Vessels ca. 90–120/mm², solitary (17–32%), in radial multiples of 2–4(5) and occasionally in small clusters; angular to rounded in cross-section, tangential diameter 20–90 μm, radial diameter up to 110 μm, walls 2–3 μm thick. Vessel member length 380(180–560) μm. Perforations simple in oblique end walls. Inter-vessel pits polygonal to rounded, alternate, ca. 9–10 μm in diameter, with elliptic to slit-like apertures. Vessel-ray pits mostly confined to 1–3 rows of marginal ray cells, simple, rectangular or rounded, sometimes giving the cross-fields a reticulate appearance, diameter 8–10 μm.
Fibres 700(290–1,090) μm long, thin- to medium thick-walled, with sparse, minute simple pits in radial walls.
Parenchyma very sparse, scanty paratracheal and in narrow discontinuous marginal bands; in 4–6-celled strands.
Rays 10–15/mm, 1(2)-seriate, up to 40 cells high, typically homocellular, sometimes weakly heterocellular with marginal cells somewhat less procumbent than central cells or square.
Crystals not observed.

Archaeological records: Arava Valley, Negev and Sinai; ranging from Middle Bronze Period to 7th century A. D. (N. Liphschitz & Y. Waisel).

Salix acmophylla Boiss.

Tree or shrub; deciduous; near water; Coastal Galilee, Acco Plain, Sharon Plain, Philistean Plain, Upper and Lower Galilee, Esdraelon Plain, Samaria, Judean Mts., Judean Desert, Dan Valley, Hula Plain, Upper and Lower Jordan Valley; Dead Sea area, Gilead, Ammon, Moav; East Mediterranean and Irano-Turanian element.

Growth rings distinct. Vessels diffuse, ca. 90–120/mm², solitary (ca. 25–50%), in radial multiples of 2–3(4), occasionally in small clusters; angular in cross-section, tangential diameter 30–90 μm, radial diameter up to 100 μm, walls 1–1.5 μm thick. Vessel member length 430(260–530) μm. Perforations simple in oblique end walls. Inter-vessel pits alternate, polygonal, 8–9 μm in diameter, with elliptic or slit-like apertures. Vessel-parenchyma and vessel-ray pits mostly simple, usually polygonal, sometimes rectangular, rarely elongate in various directions, sometimes giving the cross-fields a reticulate appearance, 6–8 μm in diameter. Vessel-ray pits mostly confined to upright marginal cells. Tyloses sometimes present.

Fibres 720(450–870) μm long, thin- to medium thick-walled, with minute simple pits mainly in radial walls.
Parenchyma sparse, scanty paratracheal and apotracheal in narrow discontinuous marginal bands; in 2–5-celled strands.
Rays 7–13/mm, uniseriate, very rarely biseriate in the centre; up to 15 cells high; mostly heterocellular, composed of 1–3 rows of upright marginal cells and procumbent or sometimes square and even upright central cells.
Crystals not observed.

Salix alba L.

Tree; deciduous; near water. Var. *alba*: Philistean Plain, Esdraelon Plain, Dan Valley, Hula Plain, Upper Jordan Valley, Ammon. Var. *micans*: Acco Plain, Sharon Plain, Philistean Plain, Upper Galilee, Samaria, Hula Plain, Ammon. Mediterranean, Euro-Siberian and Irano-Turanian element.

Growth rings distinct. Vessels diffuse, ca. 65–95/mm², solitary (ca. 50%), in radial multiples of 2–3(4), occasionally in small clusters; angular in cross-section, tangential diameter 30–90 μm, radial diameter up to 100 μm, walls 1–1.5 μm thick. Vessel member length 380(210–510) μm. Perforations simple in oblique end walls. Inter-vessel pits alternate, polygonal or rounded, 5–7 μm in diameter, with elliptic apertures. Vessel-parenchyma and vessel-ray pits mostly simple, usually polygonal, sometimes rectangular, rarely elongate in various directions, sometimes giving the cross-fields a reticulate appearance, 5–8 μm in diameter. Vessel-ray pits mostly confined to upright marginal cells. Tyloses sometimes present.
Fibres 720(450–990) μm long, thin- to medium thick-walled, with minute simple pits mainly in radial walls; sometimes chambered crystalliferous close to a false(?) growth ring boundary.
Parenchyma sparse, scanty paratracheal and diffuse apotracheal and in narrow discontinuous marginal bands; in 2–5-celled strands.
Rays 12–15/mm, uniseriate, rarely biseriate in the centre; up to 15(20) cells high; mostly heterocellular, composed of 1–3 rows of upright, sometimes irregularly shaped marginal cells and procumbent or sometimes square and even upright central cells.
Crystals prismatic, solitary, in chambered fibres and chambered marginal parenchyma (at false growth ring boundary?).

Salix babylonica L.
Plate 66D and E; Fig. 5A

Tree; deciduous; escaped from cultivated humid sites; Sharon Plain, Philistean Plain, Upper and Lower Galilee, Dan Valley; origin unclear (temperate China or Iran?).

Growth rings distinct. Vessels diffuse, 130–160/mm², solitary (ca. 50%), in mul-

tiples of 2–3(6) and small clusters; angular in cross-section, tangential diameter 20–80 μm, radial diameter up to 100 μm, walls 1–1.5 μm thick. Vessel member length 500(250–650) μm. Perforations simple in oblique end walls. Inter-vessel pits alternate, polygonal, 6–8 μm in diameter, with elliptic apertures. Vessel-parenchyma and vessel-ray pits mostly simple, polygonal, rounded or rectangular. Vessel-ray pits mostly confined to upright marginal cells.
Fibres 770(490–1,160) μm long, thin- to medium thick-walled, with minute simple pits mainly in radial walls.
Parenchyma sparse, scanty paratracheal and apotracheal in narrow discontinuous marginal bands; in 2–4-celled strands.
Rays 7–13/mm, uniseriate, very rarely biseriate in the centre, up to 24 cells high; mostly heterocellular, composed of 1–3 rows of marginal upright cells and procumbent or sometimes square and even upright central cells.
Crystals not observed.

Characteristics of the Genus Salix

Growth rings distinct. Vessels diffuse, solitary (up to 50%) and in mainly short multiples, occasionally in small clusters; maximal vessel diameter ca. 100 μm, walls 1–1.5 μm thick. Perforations simple. Inter-vessel pits alternate, mainly polygonal. Vessel-parenchyma and vessel-ray pits mostly simple, the latter mainly confined to upright marginal cells. Fibres thin- to medium thick-walled, with minute simple pits mainly in the radial walls. Parenchyma sparse, scanty paratracheal and apotracheal in narrow discontinuous marginal bands. Rays uniseriate, very rarely biseriate in the centre; mostly heterocellular, composed of 1–3 rows of marginal upright cells and procumbent or sometimes square and even upright central cells.

Separation of Salix Species

The species are very similar in their wood anatomy and can not be separated.

SALVADORACEAE

Salvadora persica L.
Plate 67A and B

Tree or shrub; evergreen; hot desert oases; mostly on damp soils, sometimes also on salines; Judean Desert, Lower Jordan Valley, Dead Sea area, Arava Valley, Edom; Sudanian element.

Wood with included phloem of the foraminate or foraminate to concentric type. Growth rings faint. Vessels diffuse, 14–40/mm^2, in radial multiples of up to 10, or clusters, occasionally solitary; rounded in cross-section; of two size classes:

wider vessels (80–130 μm in tangential diameter, radial diameter up to 150 μm, walls ca. 6 μm thick) mixed together with narrow ones (diameter 13–35 μm). Vessel member length 160(120–250) μm. Perforations simple in horizontal end walls. Inter-vessel pits alternate, rounded, 3–4 μm in diameter, with slit-like, often coalescent apertures. Vessel-parenchyma and vessel-ray pits similar but half-bordered. Many vessels with gummy contents. Vessel elements storied together with the parenchyma and included phloem elements.
Included phloem groups partly on the outer side of vessel groups, partly not associated with the latter; round to oval in cross-section, single or coalescent, forming tangential bands of varying lengths together with lignified conjunctive parenchyma.
Fibres 590(340–790) μm long, thick-walled, with minutely bordered pits mainly in radial walls.
Parenchyma scanty paratracheal and conjunctive forming 5–20-seriate concentric bands of varying lengths together with phloem groups; fusiform and in 2-celled strands; storied together with vessel members and included phloem elements.
Rays 7–13/mm, 1–3(4)-seriate, up to 16(26) cells high; weakly heterocellular, composed of weakly procumbent and square cells; some cells crystalliferous.
Crystals solitary, prismatic, in some of the ray cells, or absent.

Archaeological records: Arava Valley; Iron Age (N. Liphschitz & Y. Waisel).

SOLANACEAE

The woody Solanaceae of Israel, *Lycium, Nicotiana* and *Withania*, can easily be separated wood anatomically: *Lycium* stands out because of its dendritic vessel pattern and the occurrence of crystal sand in most of its species; *Nicotiana* and *Withania* lack these attributes but share a high proportion of radial vessel multiples, and can be separated according to ray width (see the descriptions).

Lycium depressum Stocks

Shrub; hedges and wadis; Judean Desert, Central Negev, Upper and Lower Jordan Valley, Dead Sea area, Edom; possibly introduced near Jerusalem; mainly Irano-Turanian element.

Growth rings distinct. Wood diffuse- to semi-ring-porous. Vessels 18–40/mm² (excluding the narrowest), mostly in clusters forming a dendritic pattern, occasionally in multiples of up to 6, or solitary; rounded in cross-section; of two size classes: wider vessels (tangential diameter 40–130 μm, radial diameter up to 150 μm, walls ca. 2–3 μm thick) in the early- and intermediate-wood clustered with very narrow ones, intergrading with vascular tracheids (diameter 15–42 μm); only narrow elements in latewood. Vessel member length 240(170–290) μm. Perforations simple in horizontal to oblique end walls. Inter-vessel pits alternate, round to polygonal, 5–7 μm in diameter, with short oval to slit-like apertures.

Vessel-parenchyma and vessel-ray pits similar but half-bordered or simple. Vessels and vascular tracheids with prominent spiral thickenings. Some vessels with tyloses.
Fibres 550(340–650) μm long, medium thick- to thick-walled, with small minutely bordered pits more numerous in radial than in tangential walls.
Parenchyma very scanty paratracheal, and apotracheal diffuse to diffuse-in-aggregates; in 2–4-celled strands.
Rays 5–9/mm, 1–2(3)-seriate, up to ca. 50 cells high; heterocellular, composed of mostly procumbent and partly square or upright cells.
Crystals not observed.

Lycium europaeum L.
Plate 67C and D

Shrub; along roads, fields or wadis; coast of Carmel, Sharon Plain, Philistean Plain, Upper and Lower Galilee, Esdraelon Plain, Mt. Gilboa, Samaria, Shefela, Judean Mts., N. Negev, Upper Jordan Valley, Beit Shean Valley, Dead Sea area, Golan, Gilead, Ammon, Moav; Mediterranean element.

Growth rings distinct. Wood ring-porous. Vessels 50–65/mm², mostly solitary and in tangential pairs in the earlywood, in clusters and forming a dendritic pattern in the intermediate- and latewood; rounded to angular in cross-section; of two size classes: wider vessels (tangential diameter 40–140 μm, radial diameter up to 200 μm, walls 3–5 μm thick) clustered with very narrow ones intergrading with vascular tracheids (diameter ca. 15–25 μm). Vessel member length 320(210–400) μm. Perforations simple in mainly oblique end walls. Inter-vessel pits alternate, round to polygonal, ca. 6 μm in diameter, with short, oval to slit-like apertures. Vessel-parenchyma and vessel-ray pits similar but half-bordered. Vessels and vascular tracheids with prominent spiral thickenings.
Fibres 650(450–860) μm long, medium thick-walled, with simple pits more numerous in radial than in tangential walls.
Parenchyma very scanty paratracheal and sparsely diffuse apotracheal; in 2–5-celled strands; some cells crystalliferous.
Rays 8–13/mm, predominantly uniseriate, some 2(3)-seriate, up to ca. 25(40) cells high; weakly heterocellular, composed of square, upright and weakly procumbent cells.
Crystal sand present in some apotracheal parenchyma cells.

Lycium schweinfurthii Dammer
Plate 68A and B

Shrub. Var. *schweinfurthii*: sandy loam and calcareous sandstone; Sharon Plain, Philistean Plain, W. and N. Negev. Var. *aschersonii*: sandy soils; Philistean Plain. South Mediterranean element.

Growth rings distinct. Wood semi-ring-porous. Vessels 35–100/mm², mostly in clusters, often forming a dendritic pattern, rarely solitary; rounded to angular in

cross-section; of two size classes: wider vessels (diameter 30–120 μm, walls 4–6 μm thick) clustered with very narrow ones intergrading with vascular tracheids (diameter ca. 20–30 μm). Vessel member length 340(170–550) μm. Perforations simple in oblique to horizontal end walls. Inter-vessel pits alternate, mainly rounded, ca. 6 μm in diameter, with short slit-like apertures. Vessel-parenchyma and vessel-ray pits similar but half-bordered. Vessels with fine spiral thickenings.
Fibres 530(260–660) μm long, medium thick- to thick-walled, with minutely bordered pits more numerous in radial than in tangential walls.
Parenchyma very scanty paratracheal, and sparse apotracheal diffuse and diffuse-in-aggregates and in 1–2-seriate marginal bands; in 2–4-celled strands; some cells with crystal sand.
Rays 9–11/mm; 1(2)-seriate, up to 12 cells high; heterocellular, composed of upright, square and some weakly procumbent cells.
Crystal sand in some apotracheal parenchyma cells.

Lycium shawii Roem. & Schult.
Plate 68C and D; Fig. 8D

Shrub; hammada or beds of wadis in desert; Judean Desert, N., C. and S. Negev, Lower Jordan Valley, Dead Sea area, Arava Valley, Moav, Edom; Saharo-Arabian and East Sudanian element.

Growth rings distinct. Vessels diffuse, 30–40/mm^2, mostly in clusters, some solitary, forming a dendritic pattern; rounded to angular in cross-section; of two size classes: the wider vessels (diameter 30–130 μm, walls ca. 4 μm thick) clustered with very narrow ones intergrading with vascular tracheids (diameter ca. 15–30 μm). Vessel member length 270(200–330) μm. Perforations simple in oblique end walls. Inter-vessel pits alternate to diffuse, mostly round, sometimes polygonal, ca. 6–8 μm in diameter, with slit-like to elliptic apertures. Vessel-parenchyma and vessel-ray pits similar but half-bordered. Some vessels with very faint spiral thickenings.
Fibres 580(490–690) μm long, medium thick- to thick-walled, with minutely bordered pits more numerous in radial than in tangential walls.
Parenchyma scanty paratracheal, and apotracheal diffuse to diffuse-in-aggregates and in 1–2-seriate marginal bands; in 2–4-celled strands; many cells crystalliferous.
Rays 7–14/mm, uniseriate, occasionally biseriate, 1–16 cells high; many homocellular, composed of upright cells, some heterocellular and including also square and weakly procumbent cells.
Crystal sand present in many apotracheal parenchyma cells.

Characteristics of the Genus Lycium

Vessels forming a dendritic pattern; of two size classes: the wider vessels clustered with very narrow ones intergrading with vascular tracheids. Perforations simple. Fibres with simple or minutely bordered pits more numerous in radial than in tangential walls. Rays mostly uniseriate. Crystal sand often present in part of the apotracheal parenchyma cells.

Separation of Lycium Species

Differences between the species, as reflected in the descriptions, might be due to the young age of some of the samples and therefore cannot be relied upon for identification.

Nicotiana glauca R. C. Graham

Plate 69A and B

Shrub; evergreen; rocks, walls and waste places in cities; naturalized from South America.

Growth rings faint to distinct. Vessels diffuse, 45–100/mm², in radial multiples of 2–8(16) and in clusters, rarely solitary; angular in cross-section, diameter 20–100 μm, walls 3–5 μm thick. Vessel member length 340(210–460) μm. Perforations simple in oblique end walls. Inter-vessel pits alternate, rounded to polygonal, 7–10 μm in diameter, with elliptic to slit-like, sometimes coalescent apertures. Vessel-parenchyma and vessel-ray pits alternate, 6–8 μm in diameter, half-bordered or with reduced borders to simple.
Fibres 670(430–900) μm long, thin-walled, with distinctly bordered pits only in radial walls.
Parenchyma scanty paratracheal and diffuse apotracheal; in 2(3)-celled strands.
Rays 7–9/mm, 1–5-seriate, up to ca. 40 cells high; weakly heterocellular, composed of upright, square and very weakly procumbent cells.
Crystals not observed.
Trabeculae observed in a radial file in vessels and an isolated one in a parenchyma cell.

Withania somnifera (L.) Dunal

Plate 69C and D

Shrub; waste places; coast of Carmel, Sharon Plain, Philistean Plain, Upper and Lower Galilee, Mt. Carmel, Samaria, Shefela, Judean Mts., Judean Desert, N. Negev, Upper and Lower Jordan Valley, Dead Sea area, Gilead, Ammon, Moav; mainly Mediterranean and West Irano-Turanian element, also tropical.

Growth rings absent or very faint. Vessels diffuse, 20–80/mm², in multiples of 2–5 and in small clusters, occasionally solitary; angular to rounded in cross-section, diameter 25–130 μm, walls ca. 3 μm thick. Vessel member length 320 (180–420) μm. Perforations simple in oblique end walls. Inter-vessel pits alternate, sometimes opposite, rounded, occasionally polygonal, 5–7 μm in diameter, with slit-like apertures. Vessel-parenchyma and vessel-ray pits similar or horizontally elliptic, half-bordered to simple.

Fibres 580(400–740) μm long, thin-walled, with minutely bordered to simple pits in radial walls.
Parenchyma sparse, scanty paratracheal; in 2-celled strands.
Rays 6–9/mm, 1–5(6)-seriate, up to 26 cells high; heterocellular, composed of upright, square and weakly procumbent cells.
Crystals not observed.

STYRACACEAE

Styrax officinalis L.
Plate 70A–C

Shrub or small tree; deciduous; maquis and forests; Sharon Plain, Upper and Lower Galilee, Mt. Carmel, Mt. Gilboa, Samaria, Judean Mts., Golan, Gilead, Ammon; East Mediterranean element.

Growth rings distinct. Wood semi-ring-porous. Vessels 40–50/mm^2, solitary (12–25%) and in radial multiples of 2–5(9), rarely in clusters; rounded to angular in cross-section, tangential diameter 30–70 μm, radial diameter up to 110 μm, walls ca. 3 μm thick. Vessel member length 600(430–870) μm. Perforations scalariform with (2)3–8 coarse, widely spaced bars (rarely reticulate) in very oblique end walls. Inter-vessel pits alternate, rounded, 4–5 μm in diameter, with slit-like apertures. Vessel-parenchyma and vessel-ray pits similar or horizontally elongate, half-bordered. Many vessels with gummy contents.
Fibres 1,200(830–1,720) μm long, medium thick-walled, with simple to minutely bordered pits more numerous in radial than in tangential walls.
Parenchyma scanty paratracheal and diffuse apotracheal; in long strands of 4–8(10) cells; sometimes chambered, crystalliferous.
Rays 9–11/mm, uniseriate and (2)3–5-seriate, multiseriates up to ca. 40 cells high; heterocellular (Kribs' heterogeneous types I–II).
Crystals solitary, prismatic, in chambered parenchyma cells.

TAMARICACEAE

The two genera of Tamaricaceae in Israel, *Reaumuria* and *Tamarix*, share a number of general wood anatomical characters, such as short vessel members with small bordered pits, storied structure (but only weakly so in *Reaumuria*) and heterocellular rays. They differ markedly, however, in degree of vessel grouping (mostly solitary in *Tamarix*, in multiples in *Reaumuria*) and ray size (narrow in *Reaumuria*, wide in *Tamarix*).

Reaumuria hirtella Jaub. et Sp.
Plate 70D and E

Dwarf shrub. Var. *hirtella*: deserts; wadi beds, rocky ground; N. Sinai and probably within the Negev, Dead Sea area, Edom. Var. *palaestina*: steppes and deserts; stony or marly ground; Judean Mts., Judean Desert, W. Negev, Lower Jordan Valley, Dead Sea area, Arava Valley, Moav. Var. *brachylepis*: steppes and deserts; stony or marly ground; Judean Desert, W. Negev. Saharo-Arabian and West Irano-Turanian element.

Growth rings distinct. Vessels diffuse, 145–210/mm^2, solitary (ca. 16%), in radial multiples of up to 10, or in clusters; rounded in cross-section, tangential diameter 20–65 μm, radial diameter up to 70 μm, walls 2–5 μm thick. Vessel member length 140(100–180) μm. Perforations simple in horizontal to oblique end walls. Inter-vessel pits alternate, rounded, 3–4 μm in diameter, with slit-like apertures. Vessel-parenchyma and vessel-ray pits similar but half-bordered. Vessel members weakly storied together with parenchyma cells.
Fibres 310(250–420) μm long, medium thick-walled, with simple pits more numerous in radial than in tangential walls.
Parenchyma scanty paratracheal and very sparsely diffuse apotracheal and in 1–2-seriate marginal bands; fusiform and in 2-celled strands; weakly storied together with vessel members.
Rays 2–4/mm, 1–3-seriate, up to 12 cells high; heterocellular, composed of procumbent, upright and square cells.
Crystals not observed.

Tamarix amplexicaulis Ehrenb.
Plate 71A and B

Shrub; evergreen; saline soil; Dead Sea area, Arava Valley; Saharo-Arabian and East Sudanian element.

Growth rings distinct; rays hardly or not distended tangentially at the boundary between growth rings. Wood diffuse-porous. Vessels 65–70/mm^2, mostly solitary, sometimes in multiples of 2(4) and clusters of 3–4; rounded to angular in cross-section, diameter 25–95 μm, walls 4–8 μm thick. Vessel member length 110(90–130) μm. Perforations simple in horizontal end walls. Inter-vessel pits alternate, rounded, 3–4 μm in diameter, with slit-like, often coalescent apertures. Vessel-parenchyma and vessel-ray pits similar but half-bordered. Vessel members storied together with parenchyma cells. Many vessels with gummy contents.
Fibres 320(220–540) μm long, medium thick- to thick-walled, with numerous simple pits mainly in radial walls.
Parenchyma vasicentric; fusiform; storied together with vessel members.
Rays 3–4/mm, 3–14-seriate, up to 4 mm or more in height; heterocellular, composed of weakly procumbent, square and upright cells mixed together; many crystalliferous.
Crystals usually solitary, prismatic, in ray cells.

Tamarix aphylla (L.) Karst.

Plate 71C and D

Tree or high shrub; evergreen; sandy plains and dunes, salty deserts and wadi beds; Negev, Dead Sea area, Arava Valley, Sinai (cultivated); Sudanian element.

Growth rings faint to distinct; rays usually distended tangentially at the boundary between growth rings. Wood diffuse- to semi-ring-porous. Vessels 6–15(50)/mm², mostly solitary, sometimes in multiples of 2(4) and clusters of 3–4; rounded in cross-section, diameter 25–280 μm, walls ca. 7 μm thick. Vessel member length 120(90–140) μm. Perforations simple in horizontal end walls. Inter-vessel pits alternate, rounded, 3–4 μm in diameter, with slit-like, often coalescent apertures. Vessel-parenchyma and vessel-ray pits similar but half-bordered. Vessel members storied together with parenchyma cells.
Fibres 500(300–780) μm long, thin- to medium thick-walled, with numerous simple pits mainly in radial walls.
Parenchyma vasicentric; mostly fusiform, sometimes in 2-celled strands; storied together with the vessel members.
Rays 2–3/mm, 5–23-seriate, up to 1.5(2.5) mm high; heterocellular, composed of procumbent central cells and 1–4 rows of square marginal cells; sometimes compound; peripheral cells often crystalliferous.
Crystals solitary, prismatic, in peripheral ray cells.

Tamarix chinensis Lour.

Plate 72A

Shrub or tree; evergreen; grown as an ornamental; sometimes escaped from cultivation; Sino-Japanese element.

Growth rings distinct; rays tangentially distended at the boundary between growth rings. Wood ring- to semi-ring-porous. Vessels ca. 50/mm², mostly solitary, sometimes in multiples of 2(4) and clusters of 3–4; rounded in cross-section, diameter 16–130 μm, walls ca. 5 μm thick. Vessel member length 100(70–130) μm. Perforations simple in horizontal end walls. Inter-vessel pits alternate, rounded, 3–4 μm in diameter, with slit-like, often coalescent apertures. Vessel-parenchyma and vessel-ray pits similar but half-bordered. Vessel members storied together with parenchyma cells.
Fibres 440(250–640) μm long, thin- to medium thick-walled, with simple pits mainly in radial walls.
Parenchyma vasicentric; mostly fusiform, sometimes in 2-celled strands; storied together with vessel members.
Rays 4–5/mm, 4–10-seriate, up to 2 mm or more in height; very weakly heterocellular, most cells weakly procumbent to square, some cells upright; sometimes compound.
Crystals not observed.

Tamarix gennessarensis Zoh.
Plate 72B

Tree; near water; evergreen; Upper Jordan Valley (shores of Sea of Galilee); East Mediterranean element.

Growth rings distinct; rays tangentially distended at the boundary between growth rings. Wood semi-ring-porous. Vessels 20–40/mm^2, mostly solitary, some in multiples of 2–4 and in clusters; rounded in cross-section, diameter 30–150 μm, walls 4–8 μm thick. Vessel member length 120(100–140) μm. Perforations simple in horizontal end walls. Inter-vessel pits alternate, rounded, 3–4 μm in diameter, with slit-like, often coalescent apertures. Vessel-parenchyma and vessel-ray pits similar but half-bordered. Vessel members storied together with parenchyma cells.
Fibres 500(300–660) μm long, thin- to medium thick-walled, with simple pits mainly in radial walls.
Parenchyma vasicentric; fusiform, sometimes in 2-celled strands; storied together with vessel members.
Rays 3–5/mm, (1)3–8-seriate, up to ca. 1 mm high; heterocellular, composed of variously procumbent and square cells as well as some upright cells as sheath cells; sometimes compound.
Crystals not observed.

Tamarix nilotica (Ehrenb.) Bge.
Plate 72C and D

Tree or shrub; evergreen; 9 varieties; sandy soil of coastal plain, stony and sandy dry wadi beds, marshes and near saline springs, saline soils, sand dunes; Coastal Galilee, Acco Plain, Sharon Plain, Philistean Plain, Shefela, Judean Mts., Negev, Upper and Lower Jordan Valley, Dead Sea area, Arava Valley, Sinai, Edom; Saharo-Arabian element.

Growth rings distinct; rays tangentially distended at the boundary between growth rings. Wood ring- to semi-ring-porous. Vessels 11–26/mm^2, mostly solitary, sometimes in multiples of 2(4) and clusters of 3–4; rounded in cross-section, diameter 20–160 μm, walls ca. 5 μm thick. Vessel member length 100(80–130) μm. Perforations simple in horizontal end walls. Inter-vessel pits alternate, rounded, small, 3–4 μm in diameter, with slit-like, often coalescent apertures. Vessel-parenchyma and vessel-ray pits similar but half-bordered. Vessel members storied together with parenchyma cells.
Fibres 630(350–750) μm long, thin- to medium thick-walled, with numerous simple pits mainly in radial walls.
Parenchyma vasicentric; fusiform, sometimes in 2-celled strands; storied together with vessel members.
Rays 2–4/mm, (3)6–10-seriate, up to 1 mm or more in height; weakly heterocel-

lular, composed of procumbent central cells and marginal square cells; sometimes compound; cells occasionally crystalliferous.
Crystals prismatic, occasionally present in ray cells.

Tamarix palaestina Betrol.
Plate 73A

Tree or shrub; evergreen; near water containing a high percentage of chlorides; Upper and Lower Jordan Valley, Dead Sea area; endemic; Saharo-Arabian element.

Growth rings distinct; rays hardly distended tangentially at the boundary between growth rings. Wood ring- to semi-ring-porous. Vessels 20–30/mm^2, solitary, sometimes in multiples of 2–3; rounded in cross-section, diameter 25–160 μm, walls 4–9 μm thick. Vessel member length 120(100–140) μm. Perforations simple in horizontal end walls. Inter-vessel pits alternate, rounded, 3–4 μm in diameter, with slit-like, often coalescent apertures. Vessel-parenchyma and vessel-ray pits similar but half-bordered. Vessel members storied together with parenchyma cells. Many vessels with gummy contents.
Fibres 580(300–760) μm long, medium thick-walled, with simple pits mainly in radial walls.
Parenchyma scanty paratracheal to vasicentric; fusiform, sometimes in 2-celled strands; storied together with vessel members.
Rays 3–6/mm, 2–9-seriate, up to ca. 2.5 mm high; weakly heterocellular, composed of procumbent central cells and square, procumbent and some weakly upright marginal cells; sometimes compound.
Crystals not observed.

Tamarix parviflora DC.
Plate 73B and C

Low tree or shrub; evergreen. Var. *parviflora*: near river banks; Sharon Plain (cultivated), Hula Plain; East Mediterranean element. Var. *sodomensis*: salines, Dead Sea area; Saharo-Arabian element.

Growth rings distinct; rays tangentially distended at the boundary between growth rings. Wood diffuse- to semi-ring-porous. Vessels ca. 60/mm^2, mostly solitary, sometimes in multiples of 2–3(4) and clusters of 3–4; rounded in cross-section, diameter 16–110 μm, walls ca. 4 μm thick. Vessel member length 110 (80–130) μm. Perforations simple in horizontal end walls. Inter-vessel pits alternate, rounded, very small, ca. 2 μm in diameter, with slit-like, often coalescent apertures. Vessel-parenchyma and vessel-ray pits similar but half-bordered. Vessel members storied together with parenchyma cells.
Fibres 490(300–780) μm long, thin- to medium thick-walled, with simple pits in radial and tangential walls.
Parenchyma vasicentric; fusiform, sometimes in 2-celled strands; storied together with vessel members.

Table 4.5
Differences between *Tamarix* species.*

Species	Vessels per mm²	Vessel diameter in μm	Vessel wall thickness in μm	Ray width in cells	Ray height in mm	Ray sheath cells	Crys- tals	Porosity
amplexicaulis	65–70	25–95	4–8	3–14	4	—	++	D
aphylla	6–15(50)	25–280	7	5–23	1.5(2.5)	—	++	D–SR
chinensis	50	16–130	5	4–10	>2	—	—	R–SR
gennessarensis	20–40	30–150	4–8	(1)3–8	1	+	—	SR
nilotica	11–26	20–160	5	(3)6–10	>1	—	+	R–SR
palaestina	20–30	25–160	4–9	2–9	2.5	—	—	R–SR
parviflora	60	16–110	4	2–10	1.5	+	—	D–SR
passerinoides	50–60	22–130	4–8	8–20(22)	1.5(2)	—	+	R–SR
tetragyna	6–46	20–145	6	5–12	>2	—	—	R–SR

D— diffuse-porous; R — ring-porous; SR — semi-ring-porous.

* We did not include all species listed in *Flora Palaestina* in our survey. This is because of difficulties in obtaining reliably identified specimens of the species. The anatomical differences between the species, mainly in quantitative features listed here, should be used with caution for identification purposes. The overlap between the species is probably too large to allow reliable wood anatomical identification.

Rays ca. 4–5/mm, 2–10-seriate, up to 1.5 mm high; weakly heterocellular, composed mostly of procumbent and some square central cells and some upright marginal and sheath cells; sometimes compound.
Crystals not observed.

Tamarix passerinoides Del.
Plates 73D and 74A

Shrub; evergreen; salines in warm deserts; Dead Sea area, Arava Valley, deserts east of Ammon; South Saharo-Arabian and Sudanian element.

Growth rings fairly distinct; rays not distended tangentially at the boundary between growth rings. Wood ring- to semi-ring-porous. Vessels 50–60/mm^2, mainly solitary, some in multiples of 2(3) and a few in very small clusters; rounded in cross-section, diameter 22–130 μm (tending to be of two size classes), walls 4–8 μm thick. Vessel member length 100(70–120) μm. Perforations simple in horizontal end walls. Inter-vessel pits alternate, rounded, 3–4 μm in diameter, with slit-like, often coalescent apertures. Vessel-parenchyma and vessel-ray pits similar but half-bordered. Vessel members storied together with parenchyma cells. Many vessels with gummy contents.
Fibres 490(270–620) μm long, thick- to very thick-walled, with simple pits mainly in radial walls.
Parenchyma vasicentric; fusiform, sometimes in 2-celled strands; storied together with vessel members.
Rays 3–4/mm; 8–20(22)-seriate, up to ca. 1.5(2) mm high; heterocellular, composed of procumbent central cells and square, rarely upright, mainly marginal cells; sometimes compound; some cells crystalliferous.
Crystals solitary, prismatic, in some of the ray cells.

Tamarix tetragyna Ehrenb.
Plate 74B

Small tree or shrub; evergreen; saline and brackish swamps. Var. *tetragyna*: Central Negev, Lower Jordan Valley, Dead Sea area. Var. *deserti*: Negev, Dead Sea area. Var. *meyeri*: Acco Plain, Sharon Plain, N. Negev, Upper Jordan Valley. East Mediterranean and Saharo-Arabian element.

Growth rings distinct; rays slightly distended tangentially at the boundary between growth rings. Wood ring- to semi-ring-porous. Vessels 6–46/mm^2, mostly solitary, sometimes in pairs; rounded in cross-section, diameter 20–145 μm, walls ca. 6 μm thick. Vessel member length 120(100–140) μm. Perforations simple in horizontal end walls. Inter-vessel pits alternate, rounded, 3–4 μm in diameter, with slit-like, often coalescent apertures. Vessel-parenchyma and vessel-ray pits similar but half-bordered. Vessel members storied together with parenchyma cells.

Wood Anatomical Descriptions

Fibres 530(210–740) μm long, thin- to medium thick-walled, with numerous simple pits mainly in radial walls.
Parenchyma vasicentric; fusiform, sometimes in 2-celled strands; storied together with vessel members.
Rays ca. 5/mm, 5–12-seriate, up to 2 mm or more in height; weakly heterocellular, composed of square, weakly procumbent and weakly upright cells; sometimes compound.
Crystals not observed.

Characteristics of the Genus Tamarix

Vessels mostly solitary. Perforations simple in horizontal end walls. Inter-vessel pits alternate, rounded, with slit-like, often coalescent apertures, not exceeding 4 μm in diameter. Vessel-parenchyma and vessel-ray pits similar but half-bordered. Vessel members storied together with parenchyma cells. Fibres with simple pits mainly in the radial walls. Parenchyma mainly vasicentric; mainly fusiform; storied together with vessel members. Rays wide and up to 1 mm or more in height; sometimes compound.

Archaeological records: 12 locations, mainly in the Negev and Sinai; since very early periods (N. Liphschitz & Y. Waisel; E. Werker; A. Fahn; A. Fahn & E. Werker; A. Fahn & E. Zamski).

THYMELAEACEAE

Thymelaea hirsuta (L.) Endl.
Plate 74C and D

Shrub; steppes and semi-steppe batha, rarely Mediterranean batha; sandy, loamy and stony ground; Coastal Galilee, Acco Plain, Sharon Plain, Philistean Plain, Mt. Carmel, S. Judean Mts., W., N. and C. Negev, Hula Plain, Sinai, Moav, Edom; Mediterranean and Saharo-Arabian element.

Growth rings faint to fairly distinct. Wood diffuse- to semi-ring-porous. Vessels 100–180/mm^2, in many-celled clusters, forming a dendritic pattern together with vascular tracheids, sometimes in tangential multiples in earlywood; angular to rounded in cross-section, tangential diameter 12–65 μm, radial diameter up to 100 μm, walls 2–4 μm thick. Vessel member length 130(100–170) μm. Perforations simple in oblique end walls. Inter-vessel pits alternate, vestured, round to polygonal, 6–7 μm in diameter, with oval apertures. Vessel-parenchyma and vessel-ray pits similar but half-bordered. Faint spiral thickenings in some vessels. Some vessels with gummy contents.
Fibres 790(530–1,010) μm long, medium thick-walled, with distinctly bordered pits more numerous in radial than in tangential walls.

Wood Anatomy

Parenchyma scanty paratracheal, and apotracheal diffuse to diffuse-in-aggregates; fusiform or in 2-celled strands.
Rays 12–18/mm, 1–2-seriate, up to ca. 30 cells high; heterocellular, composed of procumbent central cells and square to weakly upright marginal cells.
Crystals not observed.

TILIACEAE

Grewia villosa Willd.
Plate 75A and B

Shrub or small tree; evergreen; hot valleys, stony ground; Lower Jordan Valley, Dead Sea area, Arava Valley; mainly Sudanian element but also occurring in other tropical regions of the Old World.

Growth rings distinct. Vessels diffuse, 45–145/mm², solitary (ca. 30%), in mostly radial multiples of 2–6 and in clusters; rounded in cross-section, diameter 20–90 μm, walls 5–8 μm thick. Vessel member length 200(140–260) μm. Perforations simple in horizontal to oblique end walls. Inter-vessel pits alternate, rounded, ca. 6 μm in diameter, with slit-like apertures. Vessel-parenchyma and vessel-ray pits similar but half-bordered. Tyloses sometimes present.
Fibres 710(420–880) μm long, thick- to very thick-walled, with simple pits mainly in radial walls.
Parenchyma scanty paratracheal to vasicentric, sometimes aliform, and diffuse apotracheal and marginal in 1–3-seriate bands; in 2–4-celled strands; many cells chambered crystalliferous.
Rays 9–12/mm, 1–3(4)-seriate, of various heights up to ca. 45 cells, sometimes fused; heterocellular, composed of procumbent central cells and square to strongly upright marginal cells (Kribs' heterogeneous type II); many cells vertically and horizontally chambered crystalliferous.
Crystals mostly solitary, prismatic, in chambered parenchyma and ray cells.

ULMACEAE

Celtis australis and *Ulmus canescens*, the two ulmaceous tree species in Israel, are very similar in their wood anatomy, despite the fact that they are treated in different subfamilies (Cronquist, 1981). They share ring-porosity, latewood vessels in clusters forming a more or less tangential pattern, spiral thickenings in the narrow vessels, libriform fibres, paratracheal parenchyma and fairly wide rays. They can be separated according to ray composition (markedly heterocellular in *Celtis*; typically homocellular in *Ulmus*) and crystal distribution (in ray cells in *Celtis*; in axial parenchyma in *Ulmus*).

Celtis australis L.
Plates 75C and D, 76A–C

Tree; deciduous; planted, rarely subspontaneous in Upper Galilee, Samaria, Judean Mts., Gilead; native in the Mediterranean countries and other countries in the Middle East up to Iran.

Growth rings distinct. Wood ring-porous. Earlywood vessels 15–40/mm², solitary and in multiples of 2(3), often associated with very narrow vessels; latewood vessels ca. 100/mm², in many-celled clusters forming an oblique to tangential pattern together with vascular tracheids and paratracheal parenchyma; rounded to angular in cross-section, tangential diameter 20–230 μm, radial diameter up to 280 μm, walls 1.5–2 μm thick. Vessel member length 240(140–370) μm. Perforations simple in oblique end walls. Inter-vessel pits alternate, rounded, 7–9 μm in diameter, with slit-like apertures. Vessel-parenchyma and vessel-ray pits similar but half-bordered or simple. Walls of smaller vessels and vascular tracheids with spiral thickenings. Tyloses sometimes present.
Fibres 1,060(530–1,420) μm long, medium thick- to thick-walled, with minutely bordered pits in radial walls; sometimes gelatinous.
Parenchyma paratracheal, difficult to distinguish from narrow vessels, forming together with them an oblique to tangential pattern in the latewood, and sparsely diffuse apotracheal; in 2–4(5)-celled strands.
Rays 4–9/mm, 1–8(10) cells wide, up to 50 cells high; heterocellular (Kribs' heterogeneous type II), composed of procumbent central cells and 1–3 rows of square to upright marginal cells, sometimes with sheath cells; some cells crystalliferous.
Crystals solitary, prismatic, in ray cells.

Ulmus canescens Melv.
Plate 76D and E

Tree; deciduous; shady places near water; Lower Galilee, Mt. Carmel, Samaria; North and East Mediterranean element.

Growth rings distinct. Wood ring-porous. Earlywood vessels ca. 10/mm², solitary or in pairs, often associated with narrow vessels; latewood vessels 150–200/mm², clustered in tangential to oblique bands together with parenchyma and vascular tracheids; round to angular in cross-section, tangential diameter 25–270 μm, radial diameter up to 300 μm, walls 2 μm thick. Vessel member length 230(140–300) μm. Perforations simple in horizontal to oblique end walls. Inter-vessel pits alternate, rounded to polygonal, 8–10 μm in diameter, with slit-like to elliptic apertures. Vessel-parenchyma and vessel-ray pits similar but half-bordered or simple. Walls of smaller vessels and vascular tracheids with fine spiral thickenings. Tyloses sometimes present.
Fibres 1,240(810–1,760) μm long, medium thick- to thick-walled, with simple pits mainly in radial walls; sometimes gelatinous.

Wood Anatomy

Parenchyma vasicentric around large vessels and around the tangential to oblique vessel bands; in 2–4-celled strands; sometimes crystalliferous.
Rays 4–6/mm, 1–6-seriate, up to 100 cells (1 mm) high; typically homocellular and composed of strongly procumbent cells, sometimes weakly heterocellular with weakly procumbent or square marginal cells.
Crystals solitary, prismatic, in chains of short parenchyma cells.

VERBENACEAE

The genus *Avicennia*, included in the Verbenaceae by many taxonomists (e.g., Melchior, 1964; Feinbrun-Dothan, 1978; Cronquist, 1981), is treated by us in a family of its own. See under Avicenniaceae for wood anatomical arguments.

Vitex agnus-castus L.
Plate 77A and B

Shrub; banks of streams and in dry wadis; Coastal Galilee, coast of Carmel, Sharon Plain, Upper and Lower Galilee, Samaria, Shefela, Judean Mts., Judean Desert, Hula Plain, Upper Jordan Valley, Beit Shean Valley, Lower Jordan Valley, Dead Sea area, Golan, Gilead, Ammon, Moav; Mediterranean element.

Growth rings distinct. Wood ring-porous. Vessels 25–35/mm^2, solitary (35–50%), in radial multiples of 2–3(5), and occasionally in very small clusters; rounded in cross-section, tangential diameter 20–130 μm, radial diameter up to 140 μm, walls 3–6 μm thick. Vessel member length 230(160–360) μm. Perforations simple in oblique to horizontal end walls. Inter-vessel pits alternate, round to polygonal, ca. 5 μm in diameter, with slit-like apertures. Vessel-parenchyma and vessel-ray pits similar but half-bordered to simple.
Fibres 820(560–1,090) μm long, thick-walled, with minute simple pits mainly in radial walls.
Parenchyma scanty paratracheal to vasicentric; in 4(6)-celled strands.
Rays 5–8/mm, 1–4-seriate, up to ca. 65 cells (1.5 mm) high; heterocellular, composed of mainly procumbent central cells and square to upright marginal cells (Kribs' heterogeneous types II–III).
Crystals not observed.

Vitex pseudo-negundo (Hausskn. ex Bornm.) Hand.-Mazz.
Plate 77C and D

Shrub; near water; shore of the Sea of Galilee; West Irano-Turanian element.

Growth rings distinct. Wood ring- to semi-ring-porous. Vessels 30–40/mm^2, mostly solitary, some in mainly radial multiples of 2–3; rounded to angular in cross-section, tangential diameter 20–90 μm, radial diameter up to 115 μm, walls 3–5 μm thick. Vessel member length 190(120–270) μm. Perforations simple in oblique to horizontal end walls. Inter-vessel pits alternate, rounded to polygonal,

3–5 μm in diameter, with slit-like apertures. Vessel-parenchyma and vessel-ray pits similar but half-bordered to simple.
Fibres 580(450–850) μm, medium thick-walled, with minute simple pits mainly in radial walls.
Parenchyma scanty paratracheal to vasicentric; in 2–4-celled strands.
Rays 9–10/mm; 1–3-seriate, up to ca. 40 cells high; heterocellular, composed of mainly procumbent central cells and square or weakly procumbent, occasionally upright, marginal or central cells.
Crystals not observed.

Characteristics of the Genus Vitex

Growth rings distinct. Wood ring- or semi-ring-porous. Perforations simple. Inter-vessel pits alternate. Vessel-parenchyma and vessel-ray pits similar but half-bordered to simple. Fibres with minute simple pits mainly in the radial walls. Parenchyma scanty paratracheal to vasicentric. Rays heterocellular.

Separation of Vitex Species

There are some differences in quantitative characters between the two *Vitex* species in the material seen by us (especially in ray size), which need further testing before they can be used reliably to separate *V. agnus-castus* from *V. pseudo-negundo*.

VITACEAE

Vitis vinifera L.
Plate 78A and B; Fig. 5B

Climber; deciduous; cultivated and escaped from cultivation; native in South-east Europe up to West India.

Growth rings distinct. Wood ring- to semi-ring-porous. Vessels 60–100/mm^2, mostly in radial multiples of 2–15 and in clusters, rarely solitary; rounded in cross-section; of two size classes: wider vessels (tangential diameter 60–220 μm, radial diameter up to 300 μm) often associated with very narrow ones (diameter 15–50 μm) intergrading with vascular tracheids; walls 3–7 μm thick. Vessel member length 590(240–750) μm. Perforations simple in mostly oblique end walls, scalariform in some of the narrow vessels. Inter-vessel pits scalariform, in larger vessels in several rows. Vessel-parenchyma and vessel-ray pits similar but half-bordered. Narrow vessel members occasionally with irregular spiral thickenings. Tyloses sometimes present. Vascular tracheids with irregular, very fine spiral thickenings.
Fibres 870(580–1,250) μm long, medium thick-walled, septate, with simple pits in radial and tangential walls.

Wood Anatomy

Parenchyma scanty paratracheal, in up to ca. 7–13-celled strands.
Rays 2–4/mm, (3)7–13 cells wide, very high, of indeterminate height (over 20 mm) or dissected into smaller units (of 1–12 mm); heterocellular, composed mostly of procumbent cells and some square or upright cells; some enlarged cells crystalliferous.
Crystals present as raphides, in some of the ray cells.

ZYGOPHYLLACEAE

In Israel there are four woody genera of Zygophyllaceae: *Balanites, Fagonia, Nitraria* and *Zygophyllum*. Some authors prefer to treat *Balanites* in a family of its own, the Balanitaceae. Parameswaran & Conrad (1982) have discussed the wood anatomical similarities and differences between *Balanites* and Zygophyllaceae, and provided some support for the former's family status on account of its very wide rays. Although our study confirms the differential and shared characters listed by Parameswaran & Conrad, we consider the similarities to be so marked that we support the treatment of *Balanites* as a subfamily within the Zygophyllaceae (e.g., Melchior, 1964; Cronquist, 1981). For *Nitraria* a family status has also been advocated, and within the Zygophyllaceae of Israel it does stand out on account of its vessel clusters, libriform fibres and aliform parenchyma. However, Metcalfe & Chalk (1950) also recorded these features for some other genera of the Zygophyllaceae of which the taxonomic status within the family is undisputed. *Fagonia* and *Zygophyllum* have many wood anatomical features in common and can only be separated on the basis of characters of minor diagnostic value (see the descriptions).

Balanites aegyptiaca (L.) Del.
Plates 78C and D; 79A

Shrub or tree; evergreen; hot deserts, oases and wadis; East Judean Desert, Beit Shean Valley, Lower Jordan Valley, Dead Sea area, N. Arava Valley; Sudanian element.

Growth rings indistinct, although ray distention may mark growth ring boundaries. Vessels diffuse, 2–15/mm^2, solitary (ca. 25%) and, together with vasicentric tracheids, in clusters and variously directed multiples of 2–3; rounded in cross-section, diameter (50)120–180 μm, walls 5-10 μm thick. Vessel member length 150(100–190) μm. Perforations simple in horizontal to slightly oblique end walls. Inter-vessel pits alternate, vestured, round, ca. 3 μm in diameter, with slit-like, extended, sometimes coalescent apertures. Vessel-ray pits similar but half-bordered. Vasicentric tracheids storied together with vessel members and parenchyma cells.
Fibres 1,100(610–1,300) μm long, thick-walled, with minutely but distinctly bordered pits in radial and tangential walls.

Parenchyma scanty paratracheal mixed with vasicentric tracheids, and apotracheal diffuse and abundantly diffuse-in-aggregates; fusiform or sometimes in 2-celled strands; some parenchyma cells near rays chambered crystalliferous; storied.
Rays 2/mm, 9–20(35)-seriate (uniseriates very infrequent), up to 2.5 mm or more high; heterocellular, composed mostly of procumbent cells and some square marginal and sheath cells; partly chambered crystalliferous.
Crystals prismatic (solitary or grouped), in horizontally and longitudinally chambered peripheral ray cells, occasionally in chambered axial parenchyma cells.
Secretory ducts normally absent, but axial traumatic ducts may occasionally be present (Parameswaran & Conrad, 1982).

Fagonia mollis Del.
Plate 79B–D

Dwarf shrub; 3 varieties; deserts, on chalks and soil rich in gypsum, granite, also in sandy or gravel wadi beds; Judean Desert, Lower Jordan Valley, Dead Sea area, Arava Valley, Negev, Sinai, Moav, Edom; East Saharo-Arabian.

Growth rings distinct. Vessels diffuse, ca. 200/mm^2, mostly solitary, sometimes in pairs; rounded in cross-section, tangential diameter 15–50 μm, radial diameter up to 65 μm, walls ca. 3 μm thick. Vessel member length 110(80–140) μm. Perforations simple in horizontal to oblique end walls. Inter-vessel pits alternate, rounded, ca. 3 μm in diameter, with slit-like apertures. Vessel-parenchyma and vessel-ray pits similar but half-bordered. Vasicentric tracheids present, but not very distinct from fibre-tracheids.
Fibres 380(190–500) μm long, medium thick- to thick-walled, with numerous distinctly bordered pits in radial and tangential walls.
Parenchyma scanty paratracheal, and apotracheal diffuse and diffuse-in-aggregates and in narrow marginal bands; fusiform and in 2-celled strands.
Rays 6–9/mm, 1–4-seriate, up to ca. 30 cells high; heterocellular, composed of procumbent, square and weakly upright cells mixed together; not well defined and cells elliptic and rather thick-walled in tangential section.
Crystals not observed.

Nitraria retusa (Forssk.) Aschers.
Plate 80A and B; Fig. 5E

Shrub; evergreen; saline deserts; E. Judean Desert, Negev, Lower Jordan Valley, Dead Sea area, Arava Valley, Sinai, deserts of Ammon, Moav and Edom; Saharo-Arabian element with extensions into Sudanian territories.

Growth rings indistinct. Vessels diffuse, ca. 80/mm^2, of two size classes: wider vessels solitary and in multiples of 2–3(4) (diameter 40–100 μm, walls ca. 4 μm thick), and narrow vessels [solitary, in multiples of 2–4(6) and clusters, diameter

10-30 μm], both forming a pattern of clusters together with vascular tracheids; rounded in cross-section. Vessel member length 150(120–180) μm. Perforations simple in mostly transverse end walls. Inter-vessel pits alternate, rounded, 2–3 μm in diameter, with slit-like apertures. Vessel-parenchyma and vessel-ray pits similar but half-bordered. Many vessels with gummy contents, some with spiral thickenings. Vessel members storied together with parenchyma cells.
Fibres 490(320–680) μm long, thick- to very thick-walled; pits simple to minutely bordered mainly in radial walls; some chambered crystalliferous.
Parenchyma paratracheal, mostly aliform to confluent in 2–5-seriate, more or less tangential bands; fusiform; storied together with vessel members; some cells chambered crystalliferous.
Rays 3/mm, 1–3-seriate, up to 25 cells high; heterocellular, composed of procumbent, square and upright cells; some cells crystalliferous.
Crystals prismatic, of various sizes, in chambered parenchyma cells and in fibres and ray cells.

Zygophyllum coccineum L.
Plates 80C and D; 81A

Perennial; hot deserts, gravelly or sandy, often saline soils; C. and S. Negev, Lower Jordan Valley, Dead Sea area, Arava Valley, Edom; Saharo-Arabian element.

Growth rings absent or faint. Vessels diffuse, 65–80/mm², mostly solitary, some in mostly tangential multiples of 2–3, rarely in small clusters, often forming a tangential pattern; rounded in cross-section, tangential diameter 12–65 μm, radial diameter up to 85 μm, walls 5–9 μm thick. Vessel member length 150 (90–210) μm. Perforations simple in horizontal to oblique end walls. Inter-vessel pits alternate, rounded, ca. 5 μm in diameter, with slit-like apertures. Vessel-parenchyma and vessel-ray pits similar but half-bordered.
Fibres 450(320–620) μm long, thick-walled, with distinctly bordered pits in radial and tangential walls.
Parenchyma very scanty paratracheal and sparsely diffuse apotracheal; fusiform or in 2-celled strands.
Rays 8–11/mm, 1–3(4)-seriate, up to ca. 25 cells high; heterocellular, composed of square, procumbent and weakly upright cells.
Crystals not observed.

Zygophyllum dumosum Boiss.
Plate 81B–D

Dwarf shrub; deserts, stony or rocky ground on hillsides and plateaus; Judean Desert, Negev, Lower Jordan Valley, Dead Sea area, Arava Valley, Sinai, Moav, Edom; East Saharo-Arabian element.

Growth rings distinct. Wood semi-ring-porous. Vessels 80–110/mm², mostly solitary, sometimes in mostly tangential multiples of 2–6 or more rarely in clusters;

rounded in cross-section, diameter 15–80 μm, walls ca. 5 μm thick. Vessel member length 140(100–160) μm. Perforations simple in mostly horizontal end walls. Inter-vessel pits alternate, rounded, 2–3 μm in diameter, with slit-like apertures. Vessel-parenchyma and vessel-ray pits similar but half-bordered. Vessel members storied together with parenchyma cells.
Fibres 420(270–680) μm long, very thick-walled, with small but distinctly bordered pits in radial and tangential walls.
Parenchyma scanty paratracheal, and apotracheal diffuse and diffuse-in-aggregates, and marginal in 1–2-seriate bands; fusiform; storied together with vessel members; some cells crystalliferous, enlarged, solitary or in chains.
Rays 3–8/mm, 1–3-seriate, up to 12 cells high; homocellular, composed of procumbent cells of varying width.
Crystals prismatic in enlarged, solitary or 2–3-celled chains of parenchyma cells.

Archaeological records: Negev; 8th century B. C. to 10th century A. D. (N. Liphschitz & Y. Waisel).

Characteristics of the Genus Zygophyllum

Vessels mostly solitary, sometimes in tangential multiples; rounded in cross-section. Perforations simple. Inter-vessel pits alternate. Vessel-parenchyma and vessel-ray pits similar but half-bordered. Fibres with distinctly bordered pits. Parenchyma scanty paratracheal and diffuse apotracheal. Rays mainly 1–3 cells wide.

Key to Zygopyllum Species

Vessels diffuse; marginal parenchyma absent or indistinct; apotracheal diffuse parenchyma sparse. **Z. coccineum**
Wood semi-ring-porous; marginal parenchyma distinct; apotracheal parenchyma diffuse-in-aggregates, common. **Z. dumosum**

PALMAE

The only woody representatives of the monocotyledons in Israel, the palms *Hyphaene* and *Phoenix* can be recognized immediately by their stem structure with scattered primary vascular bundles. The type of vessel perforation offers a reliable means of differentiating the genera: they are scalariform in the stem of *Phoenix* (simple in the root, however, according to Tomlinson, 1961) and simple in *Hyphaene*. The following generalized descriptions are based on small stem segments, and we refrain therefore from great detail, because, depending on the position in the stem, characters like fibre wall thickness, distance between bundles, etc. can vary greatly.

Hyphaene thebaica (Del.) Mart.
Plate 82A

Tree; desert, sandy soil; Arava Valley, Sinai; Sudano-Decanian and Irano-Turanian element.

Stem with primary vascular bundles embedded in ground parenchyma. Vascular bundles mostly large (600–2,100 μm in diameter). Each vascular bundle contains a very thick cap consisting of fibres on the phloem side. Most vascular bundles include 2 wide metaxylem vessels. Vessel perforations simple in horizontal to slightly oblique end walls. Vessel pits opposite to almost alternate, mostly elongate with slit-like apertures.

Phoenix dactylifera L.
Plate 82B

Tree; cultivated and escaped from cultivation; native in North Africa and Southwest Asia.

Stem with primary vascular bundles embedded in ground parenchyma. Vascular bundles of various sizes (240–1,500 μm in diameter) and at the periphery of the stem small fibre strands scattered in ground parenchyma. Each vascular bundle with a thick cap of fibres on the phloem side and a very narrow strip of thick-walled parenchyma cells on the xylem side. Each vascular bundle with (1)2–3(4) relatively wide metaxylem vessels. Vessel perforations scalariform with few bars, sometimes partly reticulate, in oblique end walls. Vessel pits elongate to various degrees, opposite to scalariform, in several rows.

5. ECOLOGICAL CONSIDERATIONS

5.1. *Ecological Trends within the Woods from the Region*

The woody flora of Israel and the adjacent regions occupies a great diversity of ecological niches (cf. Zohary, 1962, 1966, 1972, 1973), differing in climatic and edaphic conditions. This makes it attractive to use the wood anatomical information on the trees and shrubs from the region for an analysis of possible ecological trends. Such trends have been established in recent years in comparisons of mesic and xeric vegetations or taxa and/or of cool temperate and warm tropical ones (i.a. Carlquist, 1966, 1975, 1977, 1980; Baas, 1973, 1976, 1982, 1985), and were also reported for vessel characters of woods from Israel in a precursory study of the same material as described in the present book (Baas et al., 1983).
Ecological trends can be studied by comparing entire woody floras from different regions with salient differences in, for instance, climatic conditions, within closely related groups of species belonging to the same genus or family but occupying a diversity of ecological niches, or within a single species growing in different localities. In this chapter the emphasis will be on comparing the anatomical trends within ecologically defined vegetation types or florulas. In the discussion of factors underlying the floristic trends, attention will also be paid to ecological trends existing within genera of a wide ecological amplitude.
The trends may be indicative of the adaptive value of a given wood anatomical feature for specific conditions and may shed light on its functional significance in the survival strategies of the species. The functions of different wood elements and their structural variations can be established only by careful physiological experimentation (Zimmermann, 1983), but ecological wood anatomy can provide useful hypotheses in this respect. At the same time ecological wood anatomy contributes to our understanding of the factors influencing xylem evolution (cf. Bailey & Tupper, 1918; Carlquist, 1975; Baas, 1976, 1985).
In our previous analysis (Baas et al., 1983), four rough ecological categories were recognized for the region of study, viz.: arid, Mediterranean, hydrophyllic and synanthropic. In the arid flora a distinction was made between the tropical elements surviving in wadis and near oases, and the more xeric remainder of the desert flora. In the present analysis a more refined subdivision will be used, especially of the large arid category, primarily based on different soil conditions related to water retention capacity, salinity and/or the availability of subterranean water. A brief characterization of the different ecological groups is given

below, together with the listing of species typical for each group. Tropical elements are marked by (t).

A. *Desert species.* These grow in habitats with a very low annual rainfall of 25 to 250 mm and very prolonged dry periods.

1. Species growing in crevices of smooth-faced rock outcrops. Most rainfall passes into the crevices and is protected against evaporation, so that water conditions are relatively favourable for this category.
Anacardiaceae: *Pistacia atlantica, Pistacia khinjuk, Rhus tripartita*; Asclepiadaceae: *Periploca aphylla* (t); Leguminosae: *Colutea istria*; Rhamnaceae: *Rhamnus dispermus*; Rosaceae: *Crataegus sinaica, Rosa arabica*; Tiliaceae: *Grewia villosa* (t).

2. Species from wadis and oases, with access to various amounts of subterranean water. Many of these species are tropical elements.
Asclepiadaceae: *Calotropis procera* (t); Boraginaceae: *Cordia sinensis* (t); Capparidaceae: *Capparis cartilaginea, Capparis decidua* (t), *Cleome droserifolia* (t), *Maerua crassifolia* (t); Chenopodiaceae: *Atriplex halimus*; Cruciferae: *Zilla spinosa*; Leguminosae: *Acacia gerrardii* (t), *Acacia raddiana* (t), *Acacia tortilis* (t), *Retama raetam*; Malvaceae: *Abutilon fruticosum* (t), *Abutilon pannosum* (t), *Hibiscus micranthus* (t); Menispermaceae: *Cocculus pendulus* (t); Moraceae: *Ficus pseudo-sycomorus* (t), *Ficus sycomorus* (t); Moringaceae: *Moringa peregrina* (t); Nyctaginaceae: *Commicarpus africanus* (t); Resedaceae: *Ochradenus baccatus* (t); Rhamnaceae: *Ziziphus spina-christi* (t); Salvadoraceae: *Salvadora persica* (t); Solanaceae: *Lycium depressum, Lycium shawii*; Tamaricaceae: *Tamarix aphylla* (t), *Tamarix nilotica, Tamarix tetragyna*; Thymelaeaceae: *Thymelaea hirsuta*; Zygophyllaceae: *Balanites aegyptiaca* (t).

3. Species from stony soils made up of rock or conglomerate hamadas, with a low water retention capacity.
Caryophyllaceae: *Gymnocarpos decandrum*; Chenopodiaceae: *Anabasis articulata, Noaea mucronata*; Zygophyllaceae: *Fagonia mollis, Zygophyllum dumosum.*

4. Species from saline, chalky soils: xerohalophytes. The high salinity adds physiological drought to the physical drought of the environment. This category can be considered to be the most water-stressed of the entire flora of Israel and the adjacent regions, and is almost exclusively composed of members of the Chenopodiaceae.
Chenopodiaceae: *Aellenia lancifolia, Anabasis setifera, Anabasis syriaca, Atriplex halimus, Chenolea arabica, Halogeton alopecuroides, Hammada negevensis, Hammada scoparia, Salsola baryosma, Salsola tetrandra, Salsola vermiculata, Suaeda asphaltica, Suaeda palaestina*; Tamaricaceae: *Reaumuria hirtella.*

5. Species from salt marshes: hydrohalophytes. Although these species grow on a wet substratum, a high salinity is responsible for considerable physiological drought.
Chenopodiaceae: *Arthrocnemum macrostachyum, Arthrocnemum perenne,*

Seidlitzia rosmarinus, Suaeda fruticosa, Suaeda monoica; Compositae: *Inula crithmoides*; Tamaricaceae: *Tamarix amplexicaulis, Tamarix palaestina, Tamarix parviflora, Tamarix passerinoides*; Zygophyllaceae: *Nitraria retusa, Zygophyllum coccineum.*

6. Species from desert sands. Although subjected to an extremely arid climate with an annual precipitation of 50–100 mm, the desert sands have a comparatively rich vegetation, thanks to water condensation during the cool nights. Chenopodiaceae: *Haloxylon persicum, Hammada salicornica*; Compositae: *Artemisia monosperma*; Leguminosae: *Retama raetam*; Polygonaceae: *Calligonum comosum*; Solanaceae: *Lycium schweinfurthii*; Thymelaeaceae: *Thymelaea hirsuta.*

B. *Mediterranean species.* Trees and shrubs from the coastal plain and mountain regions in the northern and central parts of Israel. All of these species are subjected to a long dry period, but enjoy an annual rainfall of 400–1,000 mm.

1. Species from the maquis and batha (=dwarf shrub associations), occupying a diversity of habitats at fairly low elevation.
Aceraceae: *Acer obtusifolium*; Anacardiaceae: *Pistacia lentiscus, Pistacia palaestina, Pistacia × saportae, Rhus coriaria, Rhus pentaphylla*; Araliaceae: *Hedera helix*; Caprifoliaceae: *Lonicera etrusca, Viburnum tinus*; Cistaceae: *Cistus creticus, Cistus salvifolius*; Ericaceae: *Arbutus andrachne*; Euphorbiaceae: *Euphorbia hierosolymitana*; Fagaceae: *Quercus boissieri, Quercus calliprinos, Quercus ithaburensis*; Labiatae: *Coridothymus capitatus, Prasium majus, Salvia fruticosa, Teucrium creticum*; Lauraceae: *Laurus nobilis*; Leguminosae: *Anagyris foetida, Calycotome villosa, Ceratonia siliqua, Cercis siliquastrum, Colutea cilicica, Genista fasselata, Gonocytisus pterocladus, Ononis natrix, Spartium junceum*; Myrtaceae: *Myrtus communis*; Oleaceae: *Jasminum fruticans, Olea europaea, Phillyrea latifolia*; Ranunculaceae: *Clematis cirrhosa*; Rhamnaceae: *Paliurus spina-christi, Rhamnus alaternus, Rhamnus palaestinus, Rhamnus punctatus*; Rosaceae: *Amygdalus communis, Amygdalus korschinskii, Crataegus aronia, Crataegus azarolus, Crataegus monogyna, Eriolobus trilobatus, Prunus ursina, Pyrus syriaca, Rosa canina, Sarçopoterium spinosum*; Rutaceae: *Ruta chalepensis*; Styracaceae: *Styrax officinalis.*

2. Species from the pseudo-savanna: savanna species mixed with Mediterranean elements. Ecological conditions comparable with B1.
Leguminosae: *Acacia albida* (t), *Prosopis farcta* (t); Rhamnaceae: *Ziziphus lotus.*

3. Species from high mountains or "oro-mediterranean elements", mainly from Mt. Hermon at elevations exceeding 1,000 m. These species are subjected to less xeric conditions than those of categories B1 and B2 because of a lower average temperature and somewhat higher precipitation and relative humidity.
Aceraceae: *Acer hermoneum*; Fagaceae: *Quercus libani*; Rosaceae: *Cerasus prostrata, Cotoneaster nummularia, Rosa pulverulenta.*

C. *Synanthropic species*: shrubs growing near dwellings, mostly on fertilized soil and profiting from various amounts of irrigation water. Often these species are fast-growing, ruderal or weedy shrubs.

Capparidaceae: *Capparis ovata, Capparis spinosa*; Compositae: *Artemisia arborescens, Inula viscosa*; Euphorbiaceae: *Ricinus communis*; Solanaceae: *Lycium europaeum, Nicotiana glauca, Withania somnifera.*

D. *Hydrophyllic species*: trees and shrubs growing along permanent or ephemeral streams and on the shores of lakes.

Apocynaceae: *Nerium oleander*; Compositae: *Pluchea dioscoridis*; Oleaceae: *Fraxinus syriaca* ; Platanaceae: *Platanus orientalis*; Rosaceae: *Rosa phoenicea*; Salicaceae: *Populus euphratica, Salix acmophylla, Salix alba*; Tamaricaceae: *Tamarix gennessarensis*; Ulmaceae: *Ulmus canescens*; Verbenaceae: *Vitex agnus-castus, Vitex pseudo-negundo.*

In the above classification gymnosperms (conifers and *Ephedra*) and palms have been omitted, as well as cultivated species which occasionally have escaped from cultivation (e.g., *Elaeagnus angustifolia, Rosmarinus officinalis, Salix babylonica*) and grow wild in one of the above habitat categories. A number of the listed species are climbers (species of *Cocculus, Clematis, Hedera, Lonicera, Prasium, Vitis*); these will be discussed separately from the erect shrubs and trees.

Plant habit is known to be significantly related to certain wood anatomical features such as vessel member and fibre length, vessel diameter and frequency, proportion of upright ray cells, etc. This also applies to the woody species of Israel, as illustrated in Fig. 10A–C. Therefore some characters have been analyzed for ecological trends in trees and shrubs separately. This is necessary because the proportion of trees and shrubs varies greatly in the different ecological categories, and would result in false wood anatomical trends if all habit categories were combined.

Vessel Characters

Vessel perforations. Almost all species of the region have simple perforations. Scalariform perforations are restricted to *Arbutus, Lonicera, Platanus, Styrax* and *Viburnum.* In the first three genera scalariform perforations occur sporadically and most of the perforations are of the simple type. Four of these genera have one species each in the Mediterranean maquis; *Platanus orientalis* is a hydrophyllic species. None of the desert species have scalariform perforations.

Vessel distribution and grouping. In Table 5.1 the incidence of ring-porosity is given for the different ecological categories. It should be stressed that such percentages are not very meaningful for categories containing only a small number of species (this also applies to the discussion of other characters in Tables 5.2–6). Yet it will be obvious that ring-porosity is quite common in the Mediterranean flora (except in the small category of pseudo-savanna species) and in the hydrophyllic species, while in the various categories of the desert flora and in the synanthropic species it is rare. In Table 5.1 all species with even a weak ten-

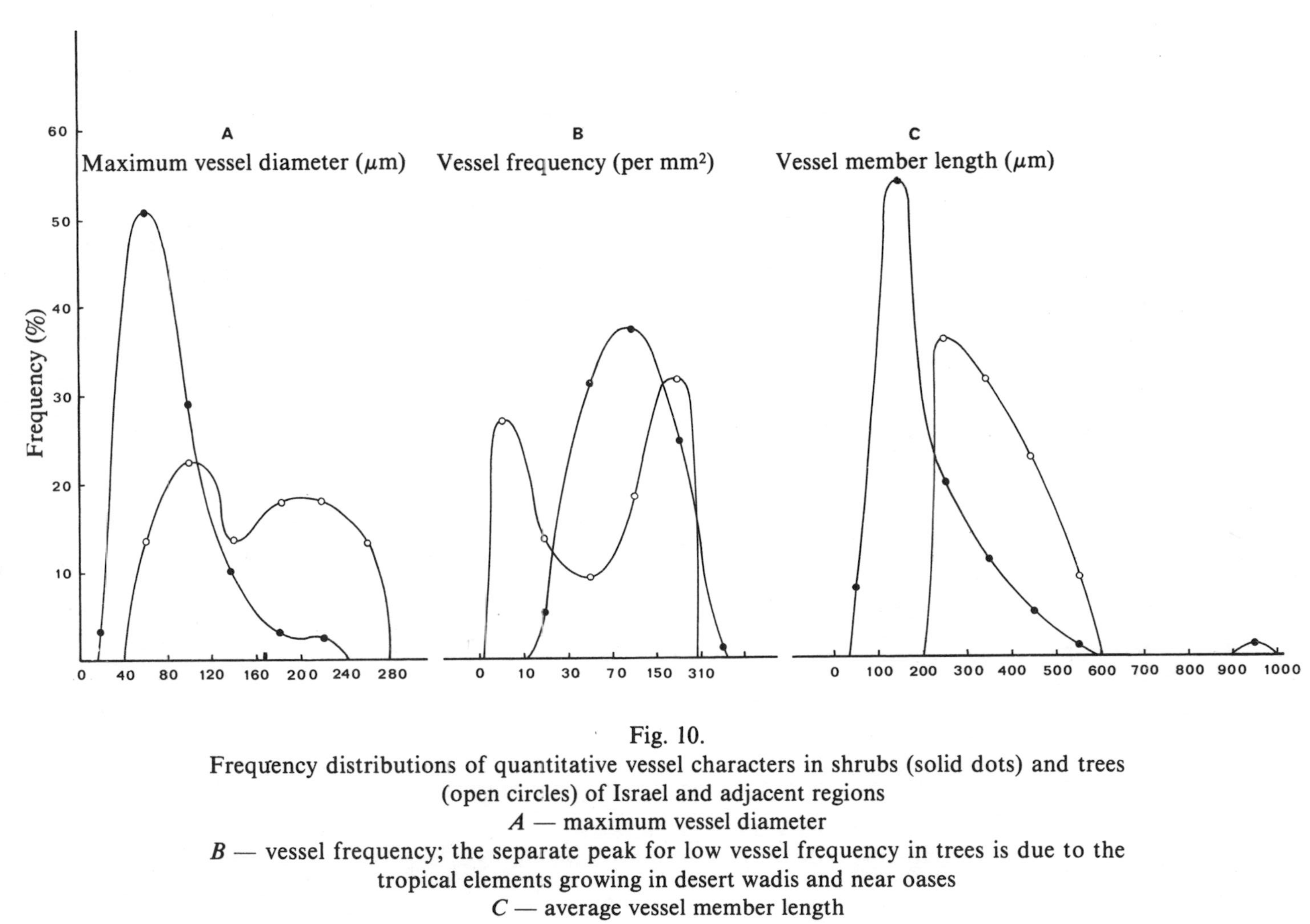

Fig. 10.
Frequency distributions of quantitative vessel characters in shrubs (solid dots) and trees (open circles) of Israel and adjacent regions
A — maximum vessel diameter
B — vessel frequency; the separate peak for low vessel frequency in trees is due to the tropical elements growing in desert wadis and near oases
C — average vessel member length

Table 5.1. Incidence (expressed as % of number of species) of ring-porosity and vessel grouping in the woody flora of Israel and adjacent regions.

Character	Ecological category (number of species)												
	A. DESERTS (77)	A. 1. Crevices (9)	A. 2. Wadis and oases (29)	A. 3. Stony soils (5)	A. 4. Xerohalophytes (14)	A. 5. Hydrohalophytes (13)	A. 6. Desert sands (7)	B. MEDITERRANEAN (55)	B. 1. Maquis and batha (47)	B. 2. Pseudosavanna (3)	B. 3. High mountains (5)	C. SYNANTHROPIC (8)	D. HYDROPHYLLIC (12)
Ring-porous tendency	**17**	44	10	20	0	23	43	**61**	63	0	80	**25**	**58**
Over 80% of vessels in multiples	**50**	50	33	40	68	38	100	**28**	33	0	0	**38**	**8**
Over 80% of vessels solitary	**6**	9	3	40	0	8	0	**27**	26	0	60	**0**	**16**

dency for ring-porosity have been scored as ring-porous (see the descriptions for further details). We did not find great differences in the incidence of ring-porosity between the evergreen and deciduous components of the various ecological categories.

Vessel grouping is a common phenomenon in most woody species. The majority of species have both solitary vessels and vessel multiples in a ratio varying between 1:5 and 5:1. Species with over 80% of the vessels in multiples are most common in the arid desert flora (Table 5.1). The tropical elements of the desert in wadis and oases, and the Mediterranean and synanthropic flora show a somewhat intermediate frequency of species with a high degree of vessel grouping, while the latter feature is rare in the hydrophyllic flora and in the small categories of the pseudo-savanna (mostly tropical elements!) and the high mountains. The other extreme, i.e., all or most vessels (over 80%) solitary, is of less common occurrence throughout the woody flora of Israel and adjacent regions. The highest percentages are found in the Mediterranean flora, mainly due to the relatively large number of species belonging to the Rosaceae, a family with solitary vessels in many of its genera. The relatively high percentages in the small categories A. 3 (stony desert soils) and B. 3 (high mountains) are probably not meaningful in view of the small number of species involved.

Vessel diameter and different size classes. Vessel diameter shows very wide ranges within species and individual wood samples. Average values for maximum vessel diameter are different for trees and shrubs, as will be evident from Table 5.2 and Fig. 10A, from which the intermediate category of large shrubs to small trees has been omitted. The numerical data in Table 5.2 should be interpreted with great caution: in many ecological categories the average is based on very few species only, especially for trees which are poorly represented in the desert (even totally absent in categories A. 3–6). Yet there is a distinct trend for tropical elements (predominant in wadis, oases and the pseudo-savanna) to show the highest values, surpassed only by the average for species with a climbing habit. The shrubs belonging to the extremely xeric categories of the stony soils and the xerohalophytes have the lowest value, only parallelled in the high mountain shrubs of the Mediterranean flora. Considered as a whole, the desert flora still shows higher values for average maximum vessel diameter than the Mediterranean flora, and is more or less similar in this respect to the most contrasting ecological category, viz. the hydrophyllic species growing along streams and lakes.

In Baas et al. (1983), minimum vessel diameter has also been analyzed for the different major ecological categories. The extratropical desert species and the Mediterranean species had more or less identical average values of 15 and 16 μm respectively, while the tropical elements of the desert and the hydrophyllic and synanthropic categories showed values ranging from 23 to 25 μm.

Vessel diameter ranges do not take into account the relative frequency of vessels with a narrow or wide diameter. In diffuse-porous woods, both the vessels with approximately maximum diameter values and those with minimum diameters

Table 5.2. Maximum vessel diameter (μm) in shrubs and trees and incidence of species with two different size classes of vessels in the woody flora of Israel and adjacent regions.

Character / Ecological category	A. DESERTS	A. 1. Crevices	A. 2. Wadis and oases	A. 3. Stony soils	A. 4. Xerohalophytes	A. 5. Hydrohalophytes	A. 6. Desert sands	B. MEDITERRANEAN	B. 1. Maquis and batha	B. 2. Pseudosavanna	B. 3. High mountains	C. SYNANTHROPIC	D. HYDROPHYLLIC	CLIMBERS
Average maximum tangential vessel diameter in shrubs	**82**	74	119	49	52	71	100	**73**	71	130	48	**109**	**88**	127
Ibid. in trees	**182**	145	198	–	–	–	–	**124**	109	200	170	–	**160**	–
Percentage of species with two vessel size classes (all species)	**64**	61	41	50	97	88	65	**36**	38	0	40	**28**	**29**	50

are usually quite infrequent, and the majority of vessels show intergrading values. In ring-porous woods, but also in a number of woods with other vessel distribution patterns, both relatively wide and very narrow vessels are quite frequent (the latter often intergrading with vascular tracheids), e.g., in many Chenopodiaceae, Papilionoideae of the Leguminosae, Anacardiaceae and Capparidaceae. The narrow vessels are then often arranged together with multiples or clusters of wider vessels. The syndrome of two vessel size classes, irrespective of whether it is caused by the latter pattern or by ring-porosity, has been recorded in Table 5.2. Note that it is a very common phenomenon in the desert categories; the relatively low percentage for wadis and oases is due to the high percentage of tropical elements in this category, which rarely show two vessel size classes. The very high percentages for xero- and hydrohalophytes are mainly due to the abundance of Chenopodiaceae in these ecological categories. The figures in Table 5.2 are slightly biassed due to the predominance of shrubs in the desert flora. In shrubs the occurrence of two vessel size classes is somewhat more frequent (57%) than in small to big trees (36–38%) of the region.

Vessel frequency and relative conductivity. The number of vessels per square millimetre varies markedly within individual species and within the ecological categories. The data in Table 5. 3 should therefore not be taken as absolutely characteristic for each category; the averages usually have very high standard deviations and this, together with the small number of trees or shrubs in some of the categories, renders most of the differences between categories statistically insignificant. However, some trends are apparent: 1. vessel frequency in shrubs is usually much higher than that in trees (see also Fig. 10B); only the small categories of trees from rock crevices in the desert and hydrophyllic trees do not follow this trend; 2. high vessel frequencies are common in the most xeric desert categories (A.3 and A.5), the Mediterranean flora, and in species with a climbing habit; 3. the lowest values are found in the wadis and oases in the desert and in the pseudo-savanna. This is due to the high percentage of tropical elements in these ecological categories, which are almost invariably characterized by infrequent, wide vessels. The frequency distribution diagrams for vessel diameter and vessel frequency of trees (Fig. 10A and B) show two peaks on account of the deviating behaviour of these tropical elements. Since there are relatively few shrubs among the tropical elements, the distribution diagrams for shrubs do not show this effect.

Both vessel frequency and vessel diameter are relevant to the conductivity of a unit of functioning sapwood. Conductivity is proportional to the 4th power of the vessel radius, in accordance with the Hagen–Poiseuille equation (Zimmermann, 1983; Van den Oever et al., 1981), and to the vessel frequency. In Table 5. 3 proportional values have been derived using this formula and assuming that average vessel diameter is more or less proportional to maximum vessel diameter, and that the frequency of the latter is in turn proportional to total vessel frequency. It is quite obvious that these assumptions do not hold true for ring-porous woods or other types of wood with two vessel size classes. Yet the calculations have been made in order to give some impression of the compensatory

Table 5.3. Vessel frequency and relative conductivity (= average maximum vessel radius to the fourth power × average vessel frequency divided by this value for xerohalophytes) for trees and shrubs from Israel and adjacent regions.

Character	Ecological category: A. DESERTS	A. 1. Crevices	A. 2. Wadis and oases	A. 3. Stony soils	A. 4. Xerohalophytes	A. 5. Hydrohalophytes	A. 6. Desert sands	B. MEDITERRANEAN	B. 1. Maquis and batha	B. 2. Pseudosavanna	B. 3. High mountains	C. SYNANTHROPIC	D. HYDROPHYLLIC	CLIMBERS
Vessel frequency in shrubs	**90**	120	53	146	104	80	79	**158**	159	31	240	**69**	**65**	152
Vessel frequency in trees	**52**	162	9	–	–	–	–	**111**	135	4	26	–	**92**	–
Average relative conductivity in shrubs	**5.4**	4.8	14.2	1.1	1.0	2.7	10.5	**5.9**	5.3	11.8	1.7	**12.9**	**5.2**	52.7
Average relative conductivity in trees (see text)	**76.2**	96.0	18.5	–	–	–	–	**35.0**	25.4	85.4	49.1	–	**80.5**	–

role of vessel diameter in conductive efficiency, balancing or often overriding the tremendous differences in vessel frequency. Relative conductivity, in our arbitrary calculations, appears to be much higher in trees than in shrubs. Within shrubs it is highest in species with a climbing habit, followed by the largely tropical elements of the wadis, oases and pseudo-savanna, and the species from desert sands and the synanthropic elements. Extremely low values occur in the desert categories A. 3–5 and in high mountain shrubs. The behaviour of the hydrophyllic species seems paradoxical: the species of shrub habit show more or less medium to low relative conductivity values, while the trees have the highest conductivity of all categories. The latter is probably due to the two ring-porous tree species (*Fraxinus* and *Ulmus*) which bias the average maximum vessel diameter for this ecological category.

Vessel member length. In Table 5.4 the data on vessel member length are summarized. Vessel members tend to be much shorter in shrubs than in trees (Fig. 10C). Within shrubs the highest value is found in climbers. The shortest vessel members are found in the driest habitats (A. 3–6). The shrubs from rock crevices in the desert are similar in vessel member length to those from the Mediterranean maquis. The synanthropic and hygrophyllic shrubs have relatively long vessel members. Thus there is general agreement between our data and the trend noted by Carlquist (i. a., 1975) that shorter elements are associated with drier conditions. This trend is roughly parallelled in the tree species, although chance fluctuations due to small numbers of species obscure it in some categories. The number of high mountain species is too low to permit a meaningful analysis of altitudinal trends as proposed by Baas (1973) and Van der Graaff & Baas (1974). In our material there is certainly no evidence for element-length shortening with increasing elevation. This is not surprising in view of the overriding influence of the less severe xeric conditions of the montane biotopes, favouring longer elements. Altitudinal (and latitudinal) trends can be expected only within floras with similar water availability (Van den Oever et al., 1981).

Vessel wall characters. Vessel wall thickness has been discussed in our previous analysis (Baas et al., 1983). The arid desert flora appeared to show the thickest vessel walls (4.6 μm on average), while the Mediterranean, synanthropic and hydrophyllic categories were — on average — more or less similar in this respect (2.7–3.6 μm).

Spiral or helical thickenings on the vessel walls are most common in the Mediterranean flora and in species from rock crevices in the desert (Table 5.5). Percentages are given for the trees and shrubs jointly, because there is no difference in the incidence of spiral thickenings between these two habit categories. Spiral thickenings are also fairly common in climbers. They are quite rare in most desert categories as well as in the hydrophyllic and synanthropic flora.

Fibre Characters

Fibre length. Average data for fibre length in shrubs are summarized in Table 5.4. Fibre length is dependent on cambial initial length as well as on subsequent

Table 5. 4. Average vessel member and fibre length (μm) in woods from Israel and adjacent regions. F/V = average fibre length divided by average vessel member length.

Character	A. DESERTS	A. 1. Crevices	A. 2. Wadis and oases	A. 3. Stony soils	A. 4. Xerohalophytes	A. 5. Hydrohalophytes	A. 6. Desert sands	B. MEDITERRANEAN	B. 1. Maquis and batha	B. 2. Pseudosavanna	B. 3. High mountains	C. SYNANTHROPIC	D. HYDROPHYLLIC	CLIMBERS
Vessel member length in shrubs	**150**	264	170	142	110	133	162	**275**	265	240	373	**299**	**308**	405
Vessel member length in trees (including small trees and large shrubs)	**225**	315	223	–	–	140	–	**341**	351	230	285	–	**338**	–
Fibre length in shrubs	**454**	636	552	396	307	374	497	**602**	593	595	676	**557**	**748**	633
Average F/V ratio in shrubs	**3.0**	2.4	3.2	2.8	2.8	2.8	3.1	**2.2**	2.2	2.5	1.8	**1.9**	**2.4**	1.6

intrusive growth. There is a parallel between the trends for fibres and vessel members (elements which show little or no intrusive growth) to be shortest in the most xeric category. For fibres the ecological trend is much weaker, however, due to the fact that, on average, intrusive growth in the desert flora (except for the category from rock crevices) is higher (×2.8–3.2) than in the other ecological categories (×1.8–2.4).

Fibre-tracheids and septate fibres. In the descriptions, fibres are mostly reported as having minutely bordered to simple pits mainly confined to the radial walls. In the wood anatomical literature, such fibres are commonly referred to as libriform fibres. The other extreme, i.e., fibres with distinctly bordered pits common in both the radial and tangential walls, can be assigned to the fibre-tracheid type. We refrained from applying these terms in the descriptions because in some species intermediates between libriform fibres and fibre-tracheids are the most common fibre type. In Table 5.5 the incidence of fibre-tracheids is given; species with intermediate fibre type have been scored as 50% fibre-tracheid and 50% libriform. Generally, fibre-tracheids are of rare occurrence in the desert flora (with the exception of the small category on stony soils). Their more common occurrence in the Mediterranean maquis and batha is mainly due to the high number of Rosaceae restricted to this ecological category.

Septate fibres are uncommon throughout the woody flora of Israel and the adjacent regions (Table 5.5). It may be assumed that septate fibres have living protoplasts. In addition to septate fibres, non-septate fibres may retain living protoplasts in many species from Israel (Fahn & Arnon, 1963; Fahn & Leshem, 1963), notably in the woods of *Tamarix* and various Chenopodiaceae. Our research material (dried wood samples) was unfortunately unsuitable for determination of the presence of living fibres in different ecological categories. It may tentatively be anticipated that living fibres will be found to be especially common in the various desert habitats.

Fibre wall thickness. Data on fibre wall thickness, in accordance with Chattaway's terminology of 1932, are summarized in Table 5.5. Thick- to very thick-walled fibres are common in both the desert floras and the Mediterranean flora. They are uncommon in the hydrophyllic and synanthropic flora, which is not unexpected for these categories of generally fast-growing trees or shrubs. Thin-walled fibres are rare throughout the flora of Israel, and most common in the synanthropic category. Woods of medium fibre wall thickness are common in all ecological categories (the complement of the percentages given in Table 5.5 for thick- and thin-walled fibres to 100).

Parenchyma Characters

Parenchyma strand length. In Table 5.5 the incidence of the two extreme character states, viz. fusiform parenchyma and parenchyma commonly in strands of more than 4 cells, is summarized. The species with fusiform parenchyma often also have some parenchyma strands consisting of two cells. The complement of both percentages is for woods with parenchyma strands of 2–4 cells. There is a

Table 5.5. Incidence of miscellaneous characters (% of species) in trees and shrubs of Israel and adjacent regions.

Character / Ecological category	A. DESERTS	A. 1. Crevices	A. 2. Wadis and oases	A. 3. Stony soils	A. 4. Xerohalophytes	A. 5. Hydrohalophytes	A. 6. Desert sands	B. MEDITERRANEAN	B. 1. Maquis and batha	B. 2. Pseudosavanna	B. 3. High mountains	C. SYNANTHROPIC	D. HYDROPHYLLIC	CLIMBERS
Incidence of spiral thickenings on vessel walls	**17**	50	13	10	7	8	36	**53**	53	0	80	**14**	**17**	75
Incidence of fibre-tracheids	**11**	33	3	60	0	16	7	**32**	32	0	60	**0**	**17**	33
Incidence of septate fibres	**3**	11	3	0	0	0	0	**14**	12	33	0	**0**	**0**	33
Incidence of thick-walled fibres	**44**	56	32	60	57	38	43	**58**	64	0	60	**16**	**16**	0
Incidence of thin-walled fibres	**3**	0	7	0	0	0	0	**0**	0	0	0	**43**	**8**	0
Incidence of fusiform parenchyma	**72**	22	62	80	93	100	86	**24**	26	33	0	**29**	**8**	33
Incidence of parenchyma strands of 4 or more cells	**6**	11	10	0	0	0	0	**33**	33	33	44	**29**	**50**	33
Incidence of abundant parenchyma	**51**	11	45	40	93	54	43	**15**	13	66	0	**0**	**0**	17
Incidence of scarce parenchyma	**33**	67	31	20	7	46	43	**57**	56	33	80	**100**	**83**	83
Incidence of storied structure	**57**	11	45	60	93	77	57	**16**	16	33	0	**14**	**7**	17
Occurrence of crystals	**68**	89	66	40	86	62	57	**68**	67	100	60	**14**	**25**	50

striking tendency for the desert flora (except for the species from rock crevices) to show a high incidence of species with fusiform parenchyma, while parenchyma strands with more than 4 cells are most common in the Mediterranean, synanthropic and hydrophyllic floras.

Parenchyma abundance. In our studies we have not made a quantitative analysis of the percentage surface area or volume of axial parenchyma. Therefore qualifications in our descriptions, such as "infrequent", "sparse" or "abundant", tend to be subjective. Yet the opposite character states, "parenchyma sparse" and "parenchyma abundant", have been analyzed and are recorded in Table 5.5. The qualification "abundant parenchyma" encompasses not only those woods with conspicuous broad bands, ample aliform to confluent parenchyma, or profusely diffuse apotracheal parenchyma, but also those species with included phloem and consequently large amounts of conjunctive parenchyma. It appears that, again with the exception of the species from the rock crevices, the various desert florulas show the highest incidence of species with abundant parenchyma, paralleled only in the small Mediterranean category of the pseudo-savanna. In all other ecological categories the predominant feature is parenchyma sparse.

Ray Characters

Ray width and frequency. Average ray width shows little variation among the ecological categories (Table 5.6). Within each ecological category there is a great deal of variation related to the taxonomic affinity of individual species. Thus the high average value for the hydrohalophytes in Table 5.6 is due to the fact that 3 out of 9 species have extremely wide rays (*Inula crithmoides, Tamarix amplexicaulis* and *T. passerinoides*) and 4 species are Chenopodiaceae with vaguely delimited rays intergrading with broad radial parenchyma strips. Similarly, the low value for shrubs from stony desert soils (5 species only) is due to the fact that two of the species have exclusively uniseriate rays. Perhaps the high average value for ray width in climbers reflects a truly general tendency for this habit category to be characterized by broad rays (Pfeiffer, 1926), although 2 out of 6 species included in our survey show medium values (4 cells in *Lonicera* and *Prasium*).

Ray frequency tends to be inversely proportional to average ray width. In Table 5.6 it appears that this relationship also exists, more or less, between ray frequency and *maximum* ray width. Ray volume distribution is probably more or less similar in the different ecological categories. Obviously, this does not apply to the rayless woods with anomalous structure, fairly common in the extremely xeric desert categories, which have been omitted from Table 5.6.

Ray composition is strongly influenced by distance from the pith (Barghoorn, 1940, 1941; De Bruyne, 1952), and shrubs with narrow stem diameters always have a higher proportion of erect cells than mature wood of big trees. This is also evident from figures for trees and shrubs of Israel and the adjacent regions: in normal trees of all ecological categories, 50% of the species have heterocellular

Table 5.6. Quantitative ray characters and incidence (% of species) of homocellular rays.

Ray character	Ecological category: A. DESERTS	A. 1. Crevices	A. 2. Wadis and oases	A. 3. Stony soils	A. 4. Xerohalophytes	A. 5. Hydrohalophytes	A. 6. Desert sands	B. MEDITERRANEAN	B. 1. Maquis and batha	B. 2. Pseudosavanna	B. 3. High mountains	C. SYNANTHROPIC	D. HYDROPHYLLIC	CLIMBERS
Average maximum ray width in shrubs (number of cells)	**6.0**	4.4	4.5	2.5	6.4	11.4	5.7	**4.1**	4.2	3.0	4.0	**5.3**	**4.4**	12.2
Average ray frequency in shrubs	**7**	11	8	12	6	4	10	**9**	9	12	13	**8**	**10**	6
Incidence of homocellular rays of procumbent cells only in trees and shrubs	**8**	0	12	20	0	8	14	**21**	21	33	10	**0**	**17**	3

rays (or heterogeneous rays according to Kribs' terminology); in small trees to large shrubs, the corresponding figure is 77% (including some species with exclusively erect cells); and for shrubs it is 95% (including many species with exclusively erect cells). The percentages of species with homogeneous rays are given for all three habit categories jointly in Table 5.6. These numerical data suggest that the slightly higher proportion of species with this feature in the Mediterranean and hydrophyllic flora than in the other ecological categories is thus mainly due to a higher proportion of tree species in the former habitats, and not to an ecological trend for ray composition within groups of the same habit.

Miscellaneous Features

Storied structure. Storied cambia are a feature of highly specialized families and are usually related to a whole suit of specialized features (Chalk, 1983), such as short cambial initials, paratracheal parenchyma, homocellular rays, libriform fibres, simple vessel perforations, etc. The incidence of storied structure, in the mature xylem usually only retained in the parenchyma strands and vessel members, is summarized in Table 5.5. With the exception of the woody vegetation from rock crevices, all desert categories show a high incidence of storied structure. This is not due solely to the high incidence of Chenopodiaceae in the desert flora, but also to the relatively large number of families with storied structure (12). In the large category of the Mediterranean maquis and batha, species of only one family, i.e., the Leguminosae, account for the 16% of species (8) with storied structure. Storied structure is often associated with fusiform parenchyma, and in Table 5.5 a striking parallel between the incidence of both features in the different ecological categories is apparent.

Crystals. It is well known that the frequency and presence or absence of crystals can vary greatly within individual species or from one plant part to another. Abundance of crystals has not been taken into account in Table 5.5, where incidence of "crystals present" is summarized. It is striking that within the desert and Mediterranean categories the incidence of species with crystals is virtually the same, while it is much lower in the synanthropic and hydrophyllic categories. Whether it can be concluded from these figures that relatively mesic conditions are associated with lack of crystals remains questionable, in view of the small number of species in the latter categories and the absence of any difference between the desert and the Mediterranean flora.

Included phloem. Included phloem is a feature of restricted occurrence. In the flora of Israel and the adjacent regions it occurs in all Chenopodiaceae, the aberrant genus *Cleome* of the Capparidaceae, the Menispermaceae, Nyctaginaceae, and *Avicennia* of the Avicenniaceae. All these representatives, except *Avicennia*, are restricted to desert habitats. Chenopodiaceae occupy the most xeric subcategories (A. 3–6); the other taxa with included phloem occupy less extreme niches in wadis and near oases or in the mangroves (*Avicennia*). The distribution of species with included phloem seems to confirm that it is an adaptive device for protection of phloem sap flow under extremely exposed and water-stressed conditions (Fahn & Shchori, 1968). The fairly common occurrence of included phloem in tropical vines in rainforests calls for another interpretation: mechanical protection of the phloem against twisting has been invoked in this case (cf. Pfeiffer, 1926).

5.2 *Comparison with Woody Floras from More Temperate or Tropical Regions*

In order to present our results on the ecological trends within the woody flora of Israel and adjacent regions in a broader context, the incidence of some wood anatomical features will be compared with the situation in tropical rainforest floras and mesic temperate to arctic florulas. Detailed wood anatomical surveys exist for only few areas of the world, but in recent years sufficient data have been accumulated to allow some tentative generalizations.

Vessel characters have been discussed in great detail in our precursory study (Baas et al., 1983), and here only a summary of the conclusions is given.

Scalariform *vessel perforations* are of common occurrence in cool tropical high-mountain floras, mesic temperate floras and arctic vegetations. Their absence from the desert flora and scarcity in the other ecological categories in Israel is shared by tropical lowland forests, especially the canopy species of the rainforest and all size classes in the seasonal forest, and by all xeric floras analyzed to date (Baas, 1976).

Vessel distribution and grouping have not been analyzed in such great detail in other floras. The ring-porous tendencies in the Mediterranean woody flora (61% of the species) seem quite exceptionally strong when compared with the percentage in the European temperate flora (40%) and tropical forests (0% in the rainforest, 10% in the monsoon forests). High degrees of vessel grouping in the desert flora also remain unequalled in more mesic temperate or tropical floras, where this feature was noted in only 10% and 2% of the species respectively (Baas et al., 1983). The relatively high incidence (27%) of species with (almost) exclusively solitary vessels in the Mediterranean flora is parallelled in the temperate European flora (28% of the species).

Maximum *Vessel diameter* in shrubs and trees from Israel tends to be high in comparison with shrubs and trees from mesic temperate zones and in most cases also when compared with tropical rain- and monsoon forest species. Only the shrubs from stony desert soils and from high elevations, and the xerohalophytes have very narrow vessels, comparable with those from arctic regions. Within the desert the tropical elements, especially those from wadis and oases, stand out by having an average maximum vessel diameter which is higher than that in related species from more mesic tropial habitats.

Different *vessel size classes* have been found in considerably higher percentages in the desert flora of Israel and adjacent regions than in other floras. In tropical forests the phenomenon is quite rare and mainly confined to seasonally dry monsoon forests (15%). In the temperate European flora it is about as frequent (23%) as in the synanthropic and hydrophyllic categories in Israel.

Data on *vessel frequency* are not available on the same level of comprehensiveness for other floras. The lowest frequency levels in the tropical elements from wadis and oases and also from the pseudo-savanna fit in with the well-known trend that vessel frequency in tropical lowland species tends to be much lower than in temperate or tropical high-mountain species.

Vessel member length is very low for all ecological categories from Israel when compared with either temperate or tropical floras. This is also borne out in Fig. 11 in which the frequency distribution of different vessel element length classes is given for Israel, as well as a random sample of the world flora (1,800 species; data from Chalk, 1983). The very low values for the Mediterranean and desert floras are parallelled in other xeric floras which have been studied in this respect [the desert and sclerophyllous scrub of California (Webber, 1936), and several arid florulas in West Australia (Carlquist, 1977), as well as arctic shrubs (Miller, 1975)].

Spiral thickenings are about equally common in various subtropical and temperate floras, regardless of whether they have a predominantly mesic or xeric ecology, as in the Mediterranean maquis and batha of Israel. The low incidence of spiral thickenings we found in the desert flora is in contrast with reports in the literature on desert plants in other regions of the world. Webber (1936) reported spirals in 52% of the Californian desert flora, and Carlquist (1975) also suggested that spirals are a very common feature in the desert flora in general.

Fibre type has been roughly compared in floras from different latitudinal zones by Baas (1982, 1985). In the tropical flora of Java, fibre-tracheids tend to be infrequent (18%) and largely restricted to mountain species; in the temperate floras of Japan, Europe and New Zealand, the percentages are comparable to those for the Mediterranean flora of Israel (31–42%). The percentage in the desert flora (11%) remains exceptionally low in this comparison.

Data on *parenchyma strand length* are not available for other floristic regions, but it can be anticipated that the high incidence of fusiform parenchyma in the desert flora represents an extreme case. *Parenchyma abundance* in the different ecological categories of the flora of Israel and the adjacent regions can be compared with data on Europe, New Zealand, Formosa, South East Asia, Tropical West Africa and Java, summarized by Baas (1982). The incidence of species with abundant parenchyma in the desert florulas is comparable to that in tropical regions (44–57%). In the Mediterranean flora of Israel abundant parenchyma (15%) is only marginally less common than in temperate floras (19–32%).

Ray width and frequency have not been analyzed for other floras. *Ray composition* did not show a consistent ecological trend in comparisons of temperate and tropical floras by Baas (1982, 1985). The percentages for the desert flora, the

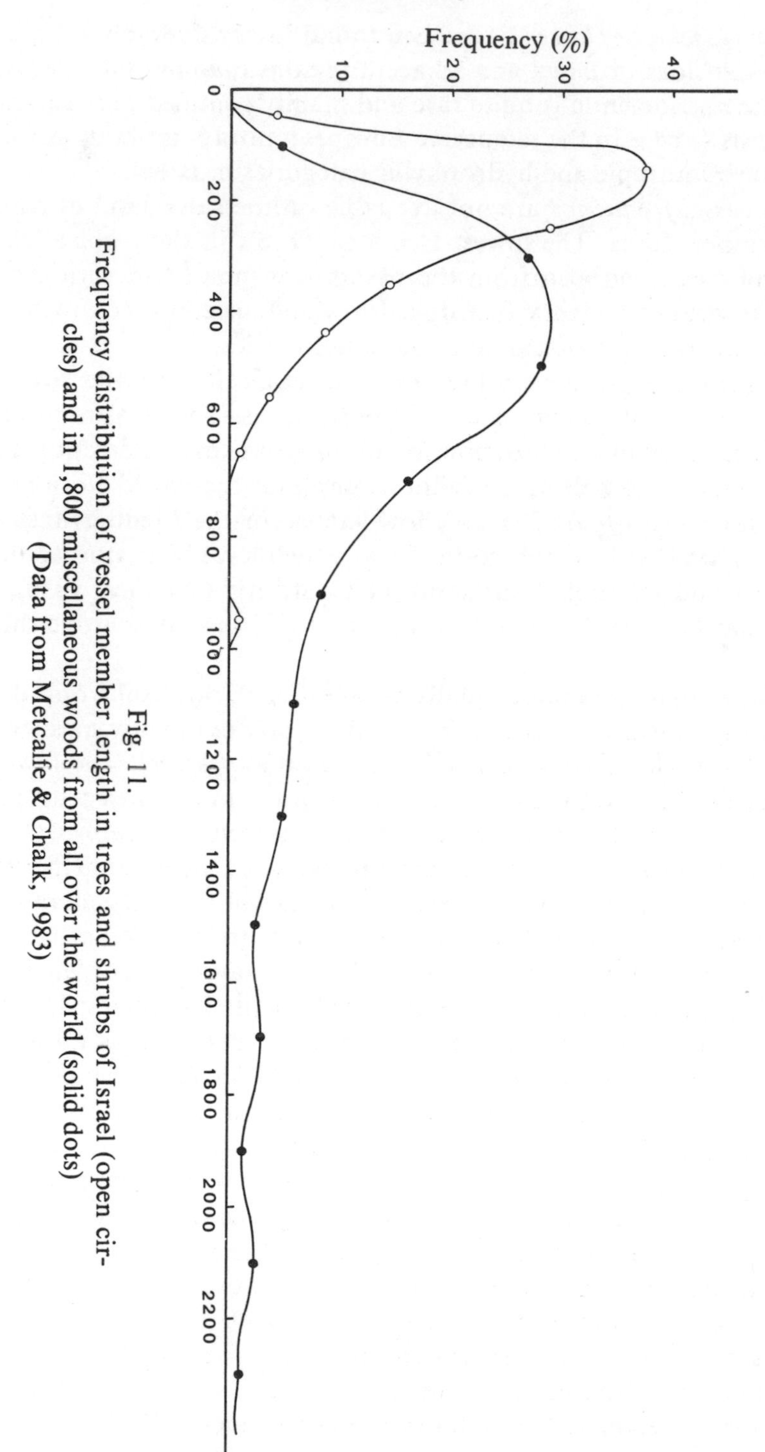

Fig. 11.
Frequency distribution of vessel member length in trees and shrubs of Israel (open circles) and in 1,800 miscellaneous woods from all over the world (solid dots)
(Data from Metcalfe & Chalk, 1983)

Mediterranean flora as well as the hydrophyllic group of species (Table 5.6) are all within ranges reported for the various tropical as well as temperate floras.
Storied structure occurs in 14% of 1,800 woods from all over the world (Metcalfe & Chalk, 1950). This figure is almost identical to that for the Mediterranean and synanthropic flora in Israel (Table 5.5). Thus the percentages for the desert florulas are exceptionally high.
The occurrence of *crystals* in the flora of Israel can be compared with detailed wood anatomical accounts on temperate and tropical floras by Grosser (1977) and Moll & Janssonius (1906–1936), respectively. In temperate Europe 26% of the species have been reported as featuring crystals, a figure comparable with that for the hydrophyllic and synanthropic components in Israel (14–25%). For Java the incidence of crystals is closer to that for the Mediterranean and desert flora, i.e., 59%.

5.3 *Trends within Genera and Biological Causes and Significance of the Ecological Trends*

Before discussing the possible causes and functional and evolutionary significance of the ecological trends surveyed in the previous pages, it should be stressed that within almost all ecological categories there is an enormous range for each quantitative as well as qualitative wood anatomical feature. The trends expressed in averages of quantitative characters or as percentages of species in a certain habitat category do not always reflect statistically significant differences between the individual categories, especially if the latter contain only a small number of woody species. Thus, the hydrophyllic flora, despite its tendency for relatively long vessel members, also contains species with very short vessel members (*Tamarix gennessarensis* and *Vitex* spp.), and the flora of wadis and oases has species with narrow vessels in addition to the majority of species with wide vessels.
We have asked the question whether the trends observed are merely due to the different floristic composition of the ecological categories and are based on the restricted anatomical ranges within the constituting genera and families, in combination with the incidence of species belonging to these genera and families. This would be the case if no trace of ecological trends is apparent within families or genera covering more than one ecological category. In Table 5.7 data on vessel member length and maximum vessel diameter are summarized for some taxa with a wide ecological amplitude. In only few instances (6 out of 24) the trends based on the floristic comparisons are confirmed by variation within genera or families. In more instances (9) the trends are contradicted or even reversed. In the remaining instances the trends are only weakly supported (5) or the data are neutral with respect to the trends (4). Thus it seems likely that for

Table 5.7. Vessel member length and maximum values (averages of very few, often only one, species) within mutually related taxa in different ecological categories.

+ = general ecological trend for entire flora is confirmed; ± = general ecological trend is only very weakly or partly confirmed; 0 = data confirm nor contradict general trends; – = general trend contradicted or reversed within taxon concerned.

Character and taxon	Ecological category	A. 1. Crevices	A. 2. Wadis and oases	A. 3. Stony soils	A. 4. Xerohalophytes	A. 5. Hydrohalophytes	A. 6. Desert sands	B. 1. Maquis and batha	B. 2. Pseudosavanna	B. 3. High mountains	C. Synanthropic	D. Hydrophyllic	Agreement with general trends
VESSEL MEMBER LENGTH													
Anacardiaceae													
shrubs		250						280					±
trees		240						330					±
Capparidaceae													
Capparis (shrubs)			165								165		–
Chenopodiaceae			160	130	110	140	110						+
Compositae													
Artemisia							180				210		+
Inula						120					190		+
Leguminosae													
Papilionoideae (shrubs)		120	130				130	130	190				–
Rhamnaceae													
Rhamnus (shrubs)		250						290					±
Rosaceae													
shrubs		390						330		370		380	–
small trees		580						410					–
Solanaceae (shrubs)			260				340				330		–
Tamaricaceae													
small trees			110			115						120	–
MAXIMUM TANGENTIAL VESSEL DIAMETER													
Anacardiaceae													
shrubs		100						105					0
trees		145						100					0
Capparidaceae													
Capparis (shrubs)			110								110		0
Chenopodiaceae			100	40	55	55	75						+
Compositae													
Artemisia							70				60		–
Inula						50					50		0
Leguminosae													
Papilionoideae (shrubs)		90	100				100	85	140				+
Rhamnaceae													
Rhamnus (shrubs)		40						40					+
Rosaceae													
shrubs		60						60		50		60	±
small trees		50						65					–
Solanaceae (shrubs)			105				120				120		–
Tamaricaceae													
small trees			195			130						130	±

vessel member length and maximum vessel diameter floristic composition is a more important factor than ecological trends within genera or families. On the other hand, when related species from outside the Middle East were included in the comparison, it was demonstrated (Baas et al., 1983) that for several genera the vessel member length in the desert and Mediterranean flora is located at the extreme end of the trend of a mesic to xeric series (from long to short elements), and quite substantial differences between tropical lowland species and the species from Israel were revealed. We therefore suggest that at least the trends for vessel member length are due to the reinforced effects of different floristic composition and to trends existing within closely related groups of species. However, the generally arid nature of both the desert flora and the Mediterranean flora of Israel obscures any trends within genera. This interpretation may equally well apply to some other features analyzed, for which the variation within genera is briefly summarized below.

Ring-porosity is variable within some genera (e.g., *Acer, Quercus, Rosa*), but the variation does not show a parallel to the ecological trend that ring-porosity is most common in the Mediterranean maquis and batha. Vessel grouping usually varies within certain limits, but in genera covering more than one ecological category no major differences parallelling the general trend that vessel grouping is most prominent in the desert were found. Only in *Rhus* does the most xeric species also show the highest percentage of vessels in multiples (*R. tripartita*). In *Artemisia* it is the other way round. However, the desert species *A. monosperma* has interxylary cork layers, and each growth layer is thus protected on its inner and outer side. In this case vessel grouping as an adaptation to drought is not "needed". Vessels of two size classes usually characterize all species of a genus in the flora of Israel. Only in *Ficus* the variation parallels the general trend: the most xeric species, *F. pseudo-sycomorus*, tends to have two vessel size classes, while vessel size class distribution is more even in the other two species. Presence or absence of spiral thickenings is usually constant for the species of a genus within the geographical restriction of Israel and the adjacent regions. In those genera in which variation occurs (*Acacia, Crataegus, Rhamnus*) there is no relation with the ecological category. Vessel wall thickness is also mostly uniform within each genus, and only in *Rhus* and *Ziziphus* is there some tendency towards thicker vessel walls in the more xeric species.

If the infrageneric comparisons were extended to species from outside the Middle East, the general lack of ecological trends within genera reported here would be partly eliminated. For instance tropical rainforest species of *Acacia* and *Rhus* all lack spiral thickenings; the degree of vessel grouping in species belonging to genera extending into the tropical rainforest would in most cases be less than that in the species from Israel; and vessels of two size classes would hardly occur in the tropics but be more typical of cool temperate relatives (in *Rhamnus* for instance) (cf. Baas, 1973; Van der Graaff & Baas, 1974).

With regard to qualitative and semi-quantitative features, such as type of fibre tissue, storied structure, parenchyma strand length and abundance, the ecological trends observed in the flora of Israel are certainly due to differences in floristic composition of the individual ecological categories. Thus the high number of Chenopodiaceae in the desert categories largely accounts for the high percentage of species with storied structure and fusiform parenchyma, and the Rosaceae in the Mediterranean flora are largely responsible for the high percentage there of species with solitary vessels and fibre-tracheids. Other examples could readily be cited, but the reader is referred to the generic and specific descriptions for further details.

In addition to the ecological trends involving genetically fixed characters, and thus reflecting the results of long-term evolutionary processes, part of the trends, especially of quantitative features, may be due to phenotypic variation induced by various climatic and edaphic conditions. That such variation does exist is well known, and Bissing (1982), for instance, has shown that the percentage of solitary vessels and the degree of ring-porosity change in transplants from dry to more mesic habitats and *vice versa*, partly in parallel with the trends reported here for different floristic groups. Baas et al. (1984) have demonstrated that dwarf growth is associated with shorter and narrower vessel elements, and the hypothesis has been put forward that any ecological stress in nature might induce this as a phenotypic effect. This could well be true for a number of shrubs growing under almost permanently water-stressed conditions in Israel. However, the data in Table 5.7, although based on comparisons of different genotypes, do not suggest any major phenotypic response, parallel to the general trends.

Irrespective of whether adaptive changes within genera or families, or the relative success of certain genera and families in specific biotopes, are at the basis of the trends observed, we should enquire into their biological significance. In our previous paper (Baas et al., 1983) we have interpreted the trends for vessel diameter, frequency and two vessel size classes in terms of safety and efficiency of xylem sap transport. The high proportion of species with two vessel size classes in the arid categories (Table 5.2) provides for safe (the very narrow vessels) as well as efficient (the wide vessels) transport. The overall low values for vessel diameter in the most xeric categories (plants of stony soils, xerohalophytes and hydrohalophytes) indicate further emphasis on safety, while the wide vessels in many species from wadis and oases, desert sands, the pseudo-savanna and the synanthropic flora represent options for efficient rather than safe water translocation. On the whole, in comparison with other, more mesic floras, there is a surprising lack of extra provisions for safe water transport (i.e., reduction of the risks through embolisms) in major parts of the xeric woody flora of Israel: rather than featuring vessels narrower than comparable components of the tropical rainforest or the mild temperate forests, most categories have, on average, wider vessels. This is also evident when one compares the least xeric categories in the flora of Israel, i.e., the hydrophyllic species and the Mediterranean high-mountain species, with the various desert categories: their hydraulic architecture is as safe as or safer than that in some desert florulas (cf. Table 5.2). Conductive efficiency, which is consistently higher in trees and climbers than in erect shrubs, is lowest in the extreme desert categories from stony soils and in the xero- and hydrohalophytes, but is equally low in shrubs from the high, less xeric mountains (Table 5.3). On the other hand, the high levels of relative conductivity in shrubs from desert sands, wadis and oases, and the pseudo-savanna is quite remarkable. Baas et al. (1983) have discussed other examples of xeric conditions associated with increased efficiency rather than safety of the hydraulic system. If rootwood were to be considered in addition to the stemwood, it is likely that desert species with their extensive root systems would show even stronger options for conductive efficiency. Fahn (1964) demonstrated that in *Retama raetam* the average vessel diameter and the proportional surface area of vessels increase in the very long horizontal roots from their place of attachment outwards. This phenomenon is not restricted to desert species, but has also been demonstrated for rootwood in temperate regions. However, in the desert flora it is especially significant because the root system often represents a much greater bulk than the above-ground stems and branches.

Both safety (two vessel size classes) and efficiency (relatively wide vessels) can be considered of adaptive value in desert species which show high transpiration rates during active growth in the short periods when water is available, and need features insuring the safety of the vessels during the long dry periods.

The salient trends in characters such as vessel member length, spiral thickenings and degree of vessel grouping, existing in the flora of Israel and the adjacent regions, and also part of more general, well-established trends, have been discussed at length previously (Baas et al., 1983; Baas, 1985). Their functional interpretation remains elusive and speculative in the absence of experimental data on their precise role in the physiology of trees and shrubs.

The trends for fibre length and incidence of fibre-tracheids also seem functionally elusive. Densely pitted fibres with large pit membranes could contribute to sap transport by the xylem, but it is difficult to grasp why this would be more useful in the Mediterranean flora than in the desert flora (Table 5.5); on a worldwide basis such a functional explanation would seem even more pointless, because a relatively high percentage of species with fibre-tracheids is also typical of cool mesic temperate and tropical mountain floras. Baas (1982, 1985) has invoked the correlative constraints existing within the evolutionary trends in xylem specialization, e.g., with regard to fibre type and type of vessel perforation, as established by I. W. Bailey and his students, to account for some general ecological trends. In this interpretation the more rigid elimination of the scalariform perforation from xeric floras as well as from the tropical lowland forests would also account for the scarcity of species with fibre-tracheids in these floras. Although primitive woods with scalariform perforations are extremely rare throughout the flora of Israel and the adjacent regions, a similar interpretation might also apply, because floristic composition, or in other words the correlative constraints imposed by the limited wood anatomical ranges within genera or families, is entirely responsible for the ecological trend in fibre types. Similar correlative constraints may be the basis of the ecological trends for parenchyma strand length, intrusive fibre tip growth and storied structure. In any event, it is impossible at this stage to offer a plausible functional explanation of these trends.

Parenchyma abundance, likewise a character which usually is under constraints imposed by taxonomic affinity, might be of relevance to water storage in addition to its well-documented function in the storage and mobilization of carbohydrates. The abundant parenchyma in many desert species would then be a sensible device for extra storage in the xylem, more effective than alternative storage in living fibres where a higher proportion of the volume is usually taken up by cell wall material. However, the above interpretation is highly speculative and could not be used to explain the high incidence of species with abundant parenchyma in tropical rainforests.

Occurrence of crystals in the wood is probably not of primary functional significance. Apart from being strongly correlated with taxonomic grouping, it is likely to be influenced by edaphic and climatic factors. The high percentages in both the tropical flora and in the desert and Mediterranean floras of Israel, and their lower incidence in synanthropic and hydrophyllic species from Israel and in the temperate flora, invite an analysis of common and differing factors in the calcium and oxalate metabolisms in these floras, which is far beyond the scope of this book.

5.4 *The Ecological Categories in Retrospect*

The ecological categories delimited at the beginning of this chapter can be regrouped on the basis of our wood anatomical analysis. Such a regrouping does not imply any "correction" of the classification of vegetation types as determined by phytogeographical and abiotic factors. However, it may reveal more clearly some aspects of functional adaptation, and at the same time indicate remaining enigmas in the functional interpretation of ecological trends in wood anatomy.

Within the various categories of the desert, the species from rock crevices stand out in many respects, and are on the whole more similar in their wood anatomy to the Mediterranean maquis and batha species than to the flora of the desert. This seems to make sense if one considers that, despite the lower annual rainfall, species in this category can utilize water (protected from rapid evaporation in the crevices) more effectively than species in the other desert categories. Moreover, the majority of species from rock crevices have close relatives in the Mediterranean flora (*Pistacia, Rhus, Colutea, Rhamnus, Crataegus, Rosa*), which implies that constraints imposed by taxonomic affinity also contribute to the similarity in many wood anatomical features between the two habitat types.

The vegetation from wadis and oases is also very atypical within the desert flora. Its species are mainly tropical elements showing many wood anatomical similarities with species growing at present in the tropical lowland, either in rainforests or in seasonally dry monsoon forests. Only their vessel members are much shorter than those of their relatives from tropical latitudes (Baas et al., 1983). The availability of subterranean water or surface water may account for the fact that these trees and shrubs are successful in the deserts of the Middle East, despite the lack of any provisions for safety in their hydraulic architecture. The small Mediterranean category from the pseudo-savanna also comprises a number of tropical elements and is similar in wood anatomy to the flora from wadis and oases. The non-tropical elements in these habitat categories resemble species from other desert habitats or from the maquis or batha.

In the remaining desert categories, wood anatomical adaptations tend to increase in the series: plants of desert sands, hydrohalophytes, plants of stony soils, and xerohalophytes. Vessel member length, maximum vessel diameter and various qualitative wood anatomical features have been used as indicators to construct this series, without attributing any functional significance to the term adaptation in this respect.

Within the Mediterranean flora, the small group of species from the high mountains shows some similarity in wood structure to the North temperate flora (e.g., in vessel member length, vessel diameter and frequency, incidence of spiral thickenings, fibre-tracheids, etc.).

Both the synanthropic and hydrophyllic categories show, quite expectedly, the least signs of adaptations to dry conditions. Their vessel member length values are among the highest for the flora as a whole, and their vessel diameters allow for efficient rather than safe water transport; yet the tropical elements in the drier habitats have even wider vessels. The scarcity of species with thick- to very thick-walled fibres in both categories may be related to relatively high growth rates in their favourable habitats.

The hydraulic architecture of the climbers in our study reflects an option for high conductive efficiency, irrespective of the specific biotope of each species. This is in agreement with the general trend for climbers to have wide and numerous vessels.

In concluding this chapter on ecological wood anatomy, it can be said that many of the trends observed within the flora of Israel and the adjacent regions agree with those observed in other parts of the world, and that in some cases their tentative functional interpretation is coherent with our limited knowledge of the physiology of xylem sap transport and transpiration rates (cf. Zimmermann, 1983; Baas et al., 1983). At the same time it should be stressed that our interpretation remains unsatisfactory in the absence of data on sapwood proportion and total amount of leaves to be supplied per unit of functioning xylem. Integration of the data discussed here with those on whole-plant physiology, including the functional anatomy of roots and leaves, is a further desideratum. Moreover, due emphasis should be given to the fact that, although there are weak or strong ecological trends based on averages for a given ecological category, each habitat harbours a great wood anatomical diversity among its species, resulting in high degrees of overlap when two individual categories are compared. This applies even when such contrasting groups as desert floras and hydrophyllic species are compared. Such diversity implies that various wood anatomical types are apparently equally successful in a given habitat.

LITERATURE CITED

Airy Shaw H. K. (1966) *Willis' Dictionary of the Flowering Plants and Ferns*, Cambridge University Press, Cambridge.

Baas P. (1973) 'The Wood Anatomical Range of *Ilex* (Aquifoliaceae) and Its Ecological and Phylogenetic Significance', *Blumea*, 21: 193–258.

——(1976) 'Some Functional and Adaptive Aspects of Vessel Member Morphology', in: *Wood Structure in Biological and Technological Research* (eds. P. Baas, A. J. Bolton & D. M. Catling), Leiden Bot. Series No. 3, Leiden University Press, pp.157–181.

——(1982) 'Systematic, Phylogenetic, and Ecological Wood Anatomy. History and Perspectives', in: *New Perspectives in Wood Anatomy* (ed. P. Baas), Nijhoff/Junk, The Hague/Boston, pp.23–58.

——(1985) 'Ecological Patterns in Xylem Anatomy', in: *On the Economy of Plant Form and Function* (ed. T. Givnish), Cambridge University Press, New York (in press).

——, Lee Chenglee, Zhang Xinying, Cui Keming & Deng Yuefen (1984) 'Some Effects of Dwarf Growth on Wood Structure', *IAWA Bull.* n. s., 5: 45–63.

——, E. Werker & A. Fahn (1983) 'Some Ecological Trends in Vessel Characters', *IAWA Bull.* n. s., 4: 141–159.

Bailey I. W. & W. W. Tupper (1918) 'Size Variation in Tracheary Cells. I. A Comparison between the Secondary Xylem of Vascular Cryptogams, Gymnosperms and Angiosperms', *Proc. Amer. Arts Sci.*, 54: 149–204.

Barghoorn E. S. (1940) 'The Ontogenetic and Phylogenetic Specialization of Rays in the Xylem of Dicotyledons. I. The Primitive Ray Structure', *Amer. J. Bot.*, 27: 918–928.

——(1941) 'The Ontogenetic and Phylogenetic Specialization of Rays in the Xylem of Dicotyledons. II. Modification of the Multiseriate and Uniseriate Rays', *Amer. J. Bot.*, 28: 273–282.

Bissing D. R. (1982) 'Variation in Qualitative Anatomical Features of the Xylem of Selected Dicotyledonous Woods in Relation to Water Availability', *Bull. Torrey Bot. Club*, 109: 371–384.

Bruyne A. S. de (1952) 'Wood Structure and Age', *Proc. Koninkl. Nederl. Akad. Wet. Amsterdam*, Series C, 55: 282–286.

Carlquist S. (1966) 'Wood Anatomy of Compositae: A Summary, with Comments on Factors Controlling Wood Evolution', *Aliso,* 6: 25–44.

——(1975) *Ecological Strategies of Xylem Evolution,* University of California Press, Berkeley/Los Angeles/London.

——(1977) 'Ecological Factors in Wood Evolution: A Floristic Approach', *Amer. J. Bot.*, 64: 887–896.

——(1980) 'Further Concepts in Ecological Wood Anatomy, with Comments on Recent Work in Wood Anatomy and Evolution', *Aliso,* 9: 499–553.

Literature Cited

Chalk L. (1983) 'Wood Anatomy, Phylogeny, and Taxonomy', in: C. R. Metcalfe & L. Chalk, *Anatomy of the Dicotyledons,* 2nd ed., Vol. II, Clarendon Press, Oxford.

——& M. M. Chattaway (1937) 'Identification of Woods with Included Phloem', *Trop. Woods*, 50: 1–31.

Chattaway M. M. (1932) 'Proposed Standards for Numerical Values, Used in Describing Woods', *Trop. Woods*, 29: 20–28.

Chudnoff M. (1956) 'Minute Anatomy and Identification of the Woods of Israel', Forest Research Station, *Ilanoth*, 3: 37–52.

Core H. A., W. A. Coté & A. C. Day (1979) *Wood Structure and Identification*, Syracuse Wood Sci. Series 6, Syracuse University Press, 2nd ed.

Cronquist A. (1981) *An Integrated System of Classification of Flowering Plants*, Columbia University Press, New York.

Dahlgren R. M. T. (1980) 'A Revised System of Classification of the Angiosperms', *J. Linn. Soc., Bot.*, 80: 91–124.

Danin A. & I. C. Hedge (1973) 'Contributions to the Flora of Sinai. I. New and Confused Taxa', *Notes Roy. Bot. Gard. Edinburgh*, 32: 259–271.

Desch H. E. (1981) *Timber — Its Structure and Properties* (6th ed., revised by J. M. Dinwoodie), Macmillan, London.

Esau K. (1965) *Plant Anatomy*, 2nd ed., Wiley, New York, London & Sydney.

Esser P. & M. van der Westen (1983) 'Wood Anatomy and Classification of the Oleaceae', *IAWA Bull.* n. s., 4: 71–72.

Fahn A. (1953) 'Annual Wood Ring Development in Maquis Trees of Israel', *Palestine J. Bot. Jerusalem Ser.*, 6: 1–26.

——(1955) 'The Development of the Growth Ring in Wood of *Quercus infectoria* and *Pistacia lentiscus* in the Hill Region of Israel', *Trop. Woods*, 101: 52–59.

——(1958) 'Xylem Structure and Annual Rhythm of Development in Trees and Shrubs of the Desert. I. *Tamarix aphylla, T. jordanis* var. *negevensis, T. gallica* var. *maris-mortui*', *Trop. Woods*, 109: 81–94.

——(1959) 'Xylem Structure and Annual Rhythm of Development in Trees and Shrubs of the Desert. II. *Acacia tortilis* and *A. raddiana*', *Bull. Res. Counc. Israel, Bot.*, 7D: 23–28.

——(1964) 'Some Anatomical Adaptations of Desert Plants', *Phytomorphology*, 14: 93–102.

——(1979) *Secretory Tissues in Plants*, Academic Press, London.

——(1982) *Plant Anatomy*, 3rd ed., Pergamon Press, Oxford.

——& N. Arnon (1963) 'The Living Wood Fibres of *Tamarix aphylla* and the Changes Occurring in Them in the Transition from Sapwood to Heartwood', *New Phytol.*, 62: 99–104.

——& B. Leshem (1963) 'Wood Fibres with Living Protoplasts', *New Phytol.*, 62: 91–98.

——& C. Sarnat (1963) 'Xylem Structure and Annual Rhythm of Development in Trees and Shrubs of the Desert. IV. Shrubs', *Bull. Res. Counc. Israel, Bot.*, 11D: 198–209.

——& Y. Shchori (1968) 'The Organization of the Secondary Conducting Tissues in Some Species of the Chenopodiaceae', *Phytomorphology*, 17: 147–154.

——, E. Werker & P. Ben-Tzur (1979) 'Seasonal Effects of Wounding and Growth Substances on Development of Traumatic Resin Ducts in *Cedrus libani*', *New Phytol.*, 82: 537–544.

Feinbrun-Dothan N. (1978) *Flora Palaestina*, Vol. III, The Israel Academy of Sciences and Humanities, Jerusalem.

Wood Anatomy

Graaff N. A. van der & P. Baas (1974) 'Wood Anatomical Variation in Relation to Latitude and Altitude', *Blumea*, 22: 101–121.

Greguss P. (1955) *Xylotomische Bestimmung der heute lebenden Gymnospermen*, Akadémiai Kiadó, Budapest.

——(1959) *Holzanatomie der europäischen Laubhölzer und Sträucher*, Akadémiai Kiadó, Budapest.

——(1972) *Xylotomy of the Living Conifers*, Akadémiai Kiadó, Budapest.

Grosser D. (1977) *Die Hölzer Mitteleuropas*, Springer, Berlin/Heidelberg/New York.

Grundwag M. & E. Werker (1976) 'Comparative Wood Anatomy as an Aid to Identification of *Pistacia* L. Species', *Israel J. Bot.*, 25: 152–167.

International Association of Wood Anatomists (1964) *Multilingual Glossary of Terms Used in Wood Anatomy*, Committee on Nomenclature, IAWA.

——(1981) *Standard List of Characters Suitable for Computerized Hardwood Identification*, Committee of the IAWA.

Jacquiot C. (1955) *Altas d'Anatomie des Bois des Conifères*, 2 vols., Centre Technique du Bois, Paris.

——, Y. Trenard & D. Dirol (1973) *Atlas d'Anatomie des Bois des Angiospermes (Essences Feuillues)*, 2 vols., Centre Technique du Bois, Paris.

Jane F. W. (1970) *The Structure of Wood* (revised by K. Wilson & D. J. B. White), 2nd ed., A. & C. Black, London.

Kribs D. A. (1935) 'Salient Lines of Structural Specialization in the Wood Rays of Dicotyledons', *Bot. Gaz.*, 96: 547–557.

——(1950) *Commercial Foreign Woods on the American Market*, Edward Bros. Inc. Ann. Arbor, Michigan.

——(1968) *Commercial Foreign Woods on the American Market* (revised "Dover edition"), Dover Publ. Inc., New York.

Melchior H. (1964) *A. Engler's Syllabus der Pflanzenfamilien*, Vol. II, Borntraeger, Berlin/Nikolassee.

Metcalfe C. R. & L. Chalk (1950) *Anatomy of the Dicotyledons*, Clarendon Press, Oxford.

——&——(1979) *Anatomy of the Dicotyledons*, Vol. I, 2nd ed., Clarendon Press, Oxford.

——&——(1983) *Anatomy of the Dicotyledons*, Vol. II, 2nd ed., Clarendon Press, Oxford.

Miller H. J. (1975) 'Anatomical Characteristics of Some Woody Plants of the Angmagssalik District in Southeast Greenland', *Meddr Grønland*, 198(6): 1–30.

Miller R. B. (1978) 'Potassium Calcium Sulphate Crystals in the Secondary Xylem of *Capparis*', *IAWA Bull.*, 1978 (2 & 3): 50.

Moll J. W. & H. H. Janssonius (1906–1936) *Mikrographie des Holzes der auf Java vorkommenden Baumarten*, Vols. I–VI, Brill, Leiden.

Oever L. van den, P. Baas & M. Zandee (1981) 'Comparative Wood Anatomy of *Symplocos* and Latitude and Altitude of Provenance', *IAWA Bull.* n. s., 2: 3–24.

Page V. M. (1981) 'Dicotyledonous Wood from the Upper Cretaceous of California. III. Conclusions', *J. Arn. Arbor.*, 62: 437–455.

Panshin A. J. & C. de Zeeuw (1980) *Textbook of Wood Technology*, 4th ed., McGraw-Hill, New York.

Parameswaran N. & H. Conrad (1982) 'Wood and Bark Anatomy of *Balanites aegyptiaca* in Relation to Ecology and Taxonomy', *IAWA Bull.* n. s., 3: 75–88.

Pfeiffer H. (1926) 'Das abnorme Dickenwachstum', in: K. Linsbauer, *Handbuch der Pflanzenanatomie*, Bd. 9, Lief. 15, Gebr. Borntraeger, Berlin.

Phillips E. W. J. (1948) 'Identification of Softwoods by Their Microscopic Structure', *For. Prod. Res. Bull.*, 22: 1–56.

Robbertse P. J., G. Venter & H. Janse van Rensburg (1980) 'The Wood Anatomy of the South African Acacias', *IAWA Bull.* n. s., 1: 93–103.

Schweingruber F. H. (1978) *Mikroskopische Holzanatomie*, Kommissionsverlag Zürcher AG, Zug.

Solereder H. (1908) *Systematic Anatomy of the Dicotyledons*, Clarendon Press, Oxford.

Tomlinson P. B. (1961) *Anatomy of the Monocotyledons*, Vol. II. Palmae, Clarendon Press, Oxford.

Webber I. E. (1936) 'The Wood of Sclerophyllous and Desert Shrubs and Desert Plants of California', *Amer. J. Bot.*, 23: 181–188.

Werker E. & A. Fahn (1969) 'Resin Ducts of *Pinus halepensis* Mill.: Their Structure, Development and Pattern of Arrangement', *J. Linn. Soc., Bot.*, 62: 379–411.

Zimmermann M. H. (1983) *Xylem Structure and the Ascent of Sap,* Springer, Berlin/Heidelberg/New York/Tokyo.

Zohary M. (1962) *Plant Life of Palestine — Israel and Jordan, Chronica Botanica*, n. s. 33, Ronald Press, New York.

——(1966, 1972) *Flora Palaestina*, Vols. I, II, The Israel Academy of Sciences and Humanities, Jerusalem.

——(1973) *Geobotanical Foundations of the Middle East*, Vols. I, II, Fischer, Stuttgart, and Swets & Zeitlinger, Amsterdam.

SELECTED REFERENCES ON WOODS FROM THE REGION

Baas P. & E. Werker (1981) 'A New Record of Vestured Pits in Cistaceae', *IAWA Bull.* n. s., 2: 41–42.

Bernstein Z. & A. Fahn (1960) 'The Effect of Annual Pruning on the Seasonal Changes in Xylem Formation in the Grapevine', *Ann. Bot.* n. s., 24: 159–171.

Bridgwater S. D. & P. Baas (1978) 'Wood Anatomy of the Punicaceae', *IAWA. Bull.*, 1978 (1): 3–6.

Cambini A. (1967a) 'Micrografia comparata dei legni del genere *Quercus*', *Consiglio Maxionale delle Ricerche*, 1: 9–49.

——(1967b) 'Riconoscimento microscopico del legno delle querce italiane', *ibid.*: 53–70.

Carlquist S. (1961) 'Wood Anatomy of the Inuleae (Compositae)', *Aliso*, 5: 21–37.

——(1966a) 'Wood Anatomy of Anthemideae, Ambrosieae, Calenduleae, and Arctotideae (Compositae)', *Aliso*, 6: 1–23.

——(1966b) 'Wood Anatomy of Compositae: A Summary, with Comments on Factors Controlling Wood Evolution', *Aliso*, 6: 25–44.

Chudnoff M. (1956) 'Minute Anatomy and Identification of the Woods of Israel', *Ilanoth*, 3: 37–52.

Fahn A. (1953) 'Annual Wood Ring Development in Maquis Trees of Israel', *Palestine J. Bot. Jerusalem Ser.*, 6: 1–26.

——(1955) 'The Development of the Growth Ring in Wood of *Quercus infectoria* and *Pistacia lentiscus* in the Hill Region of Israel', *Trop. Woods*, 101: 52–59.

——(1958) 'Xylem Structure and Annual Rhythm of Development in Trees and Shrubs of the Desert. I. *Tamarix aphylla, T. jordanis* var. *negevensis, T. gallica* var. *maris-mortui*', Trop. Woods, 109: 81–94.

——(1959) 'Xylem Structure and Annual Rhythm of Development in Trees and Shrubs of the Desert. II. *Acacia tortilis* and *A. raddiana*', *Bull. Res. Counc. Israel, Bot.*, 7D: 23–28.

——, J. Burley, K. A. Longman, A. Mariaux & P. B. Tomlinson (1981) 'Possible Contributions of Wood Anatomy to the Determination of the Age of Tropical Trees', in: *Age and Growth Rate of Tropical Trees* (eds. F. H. Bormann & G. Berlyn), School of Forestry & Environmental Studies, Bull. 94, Yale University, New Haven, pp. 31–54.

——& C. Sarnat (1963) 'Xylem Structure and Annual Rhythm of Development in Trees and Shrubs of the Desert. IV. Shrubs', *Bull. Res. Counc. Israel, Bot.*, 11D: 198–209.

Giannuoli S. (1949) 'Ciclo di accresimento e differenziazione delle gemme in piante perenni nel territorio di Bari. VI. Istologia e ritmo di accrescimento del legno di *Ficus carica* L.', *Nuovo Giorn. Bot.* n. s., 56: 188–197.

Gindel I. (1952) 'Some Anatomical Features of the Indigenous Woody Vegetation of Israel', *Bull. Res. Counc. Israel, Bot.*, Suppl. 2: 1–16.

Selected References

Gottwald H. (1983) 'Wood Anatomical Studies of Boraginaceae (s. l.). I. Cordioideae', *IAWA Bull.* n. s., 4: 161–178.

Greguss P. (1955) *Xylotomische Bestimmung der heute lebenden Gymnospermen*, Akadémiai Kiadó, Budapest.

——(1959) *Holzanatomie der europäischen Laubhölzer und Sträucher*, Akadémiai Kiadó, Budapest.

——(1972) *Xylotomy of the Living Conifers*, Akadémiai Kiadó, Budapest.

Grosser D. (1977) *Die Hölzer Mitteleuropas. Ein mikrophotographischer Lehratlas*, Springer-Verlag, Berlin.

Grundwag M. & E. Werker (1976) 'Comparative Wood Anatomy as an Aid to Identification of *Pistacia* L. Species', *Israel J. Bot.*, 25: 152–167.

Huber B. & C. Rouschal (1954) *Mikrophotographischer Atlas mediterraner Hölzer*, Fritz Haller Verlag, Berlin-Grunewald.

Jacquiot C. (1955) *Atlas d'Anatomie des Bois des Conifères*, Centre Technique de Bois, Paris.

——, Y. Trenard & D. Dirol (1973) *Atlas d'Anatomie des Bois des Angiospermes (Essences Feuillues)*, 2 vols., Centre Technique de Bois, Paris.

Liphschitz N. & Y. Waisel (1970) 'The Effect of Water Stresses on Radial Growth of *Populus euphratica* Oliv.', *La-Yaaran*, 1970 (3): 53–84.

Messeri A. (1938) 'Studio anatomico-ecologico del legno secondario di alcune piante del Fezzan', *Nuovo Giorn. Bot. It.* n. s., 45: 1–89.

——(1948) 'L'evoluzione della cerchia legnosa in *Pinus halepensis* Mill. in Bari', *Nuovo Giorn. Bot. It.* n. s., 55: 111–132.

——(1951) 'Anomalie nella differenziazione del legno in rami di olivo defoliati a varie epoche', *Nuovo Giorn. Bot. It.* n. s., 58: 337–354.

Minervini I. (1949) 'Ciclo di accrescimento e differenziazione delle gemme in piante perenni nel territorio di Bari. IV. L'evoluzione della cerchia legnosa in *Viburnum tinus* L. dal Dicembre 1946 al Novembre 1947 a Bari', *Nuovo Giorn. Bot. It.* n. s., 55: 433–445.

Mullan D. P. (1932–1933) 'Observations on the Biology and Physiological Anatomy of Some Indian Halophytes', *J. Ind. bot. Soc.*, 11: 103–118, 285–302; 12: 165–182, 235–253.

Nassonov V. (1933) 'The Anatomy of *Pistacia vera* L.', *Trudy Prikl. Bot. Gen. Selek.*, Ser. 3, 4: 113–134 (Russian, English summary).

Niloufari P. (1961) *Textbook of Wood Technology. I. Structure, Identification and Defects of the Iranian Timbers with Notes on Commercial Timbers of the World*, University of Teheran, Publ. 738.

Nobile M. T. (1947) 'Ricerche sulla istologia dei legno secondario delle Gymnospermae. VIII. Studio istologico del legno secondario di *Pinus halepensis* Mill.', *Nuovo Giorn. Bot. It.* n. s., 54: 1–10.

Oppenheimer H. R. (1945) 'Cambial Wood Production in Stems of *Pinus halepensis*', *Palestine J. Bot, Rehovot*, Ser. 5: 22–51.

Paolis D. de (1949) 'Ciclo di accrescimento e differenziazione delle gemme in piante perenni nel territorio di Bari. V. L'evoluzione della cerchia legnosa in *Rhamnus alaternus* L. dal Dicembre 1946 al Marzo 1949', Nuovo *Giorn. Bot. It.* n. s., 56: 328–338.

Puth M. & N. K. Abbas (1968) 'A Lens Key to Important Hardwoods Growing in and Being Imported into Iraq', *Mesopot. J. Agric.*, 3: 54–64.

Robbertse P. J., G. Venter & H. Janse van Rensburg (1980) 'The Wood Anatomy of the South African Acacias', *IAWA Bull.* n. s., 1: 93–103.
Rohweder O. & K. Urmi-König (1975) 'Centrospermen Studien 8', *Bot. Jahrb.*, 96: 375–409.
Saya I. (1957, 1959) 'Contributo alla conoscenza del legno dei principali arbusti mediterranei', *Ann. Accad. ital. Sci. for.*, 6: 299–312; 8: 309–326.
Schweingruber F. H. (1978) *Mikroskopische Holzanatomie. Anatomie Microscopique du Bois. Microscopical Wood Anatomy. Structural Variability of Stem and Twigs in Recent and Subfossil Woods from Central Europe*, Swiss Fed. Inst. For. Res., Birmensdorf, Edition Zürcher, Zug, Switzerland.
Sieber M. & L. J. Kučera (1980) 'On the Stem Anatomy of *Clematis vitalba* L.', *IAWA Bull.* n. s., 1: 49–54.
Werker E. & P. Baas (1981) 'Trabeculae of Sanio in Secondary Tissues of *Inula viscosa* (L.) Desf. and *Salvia fruticosa* Mill.', *IAWA Bull.* n. s., 2: 69–76.
Zamski E. (1976) 'The Mode of Secondary Growth and the Three-Dimensional Structure of the Phloem in *Avicennia*', *Bot. Gaz.*, 140: 67–76.

INDEX OF PLANT NAMES

Page numbers in italics indicate detailed description of the wood anatomy; bold type indicates plate number.

PLATES

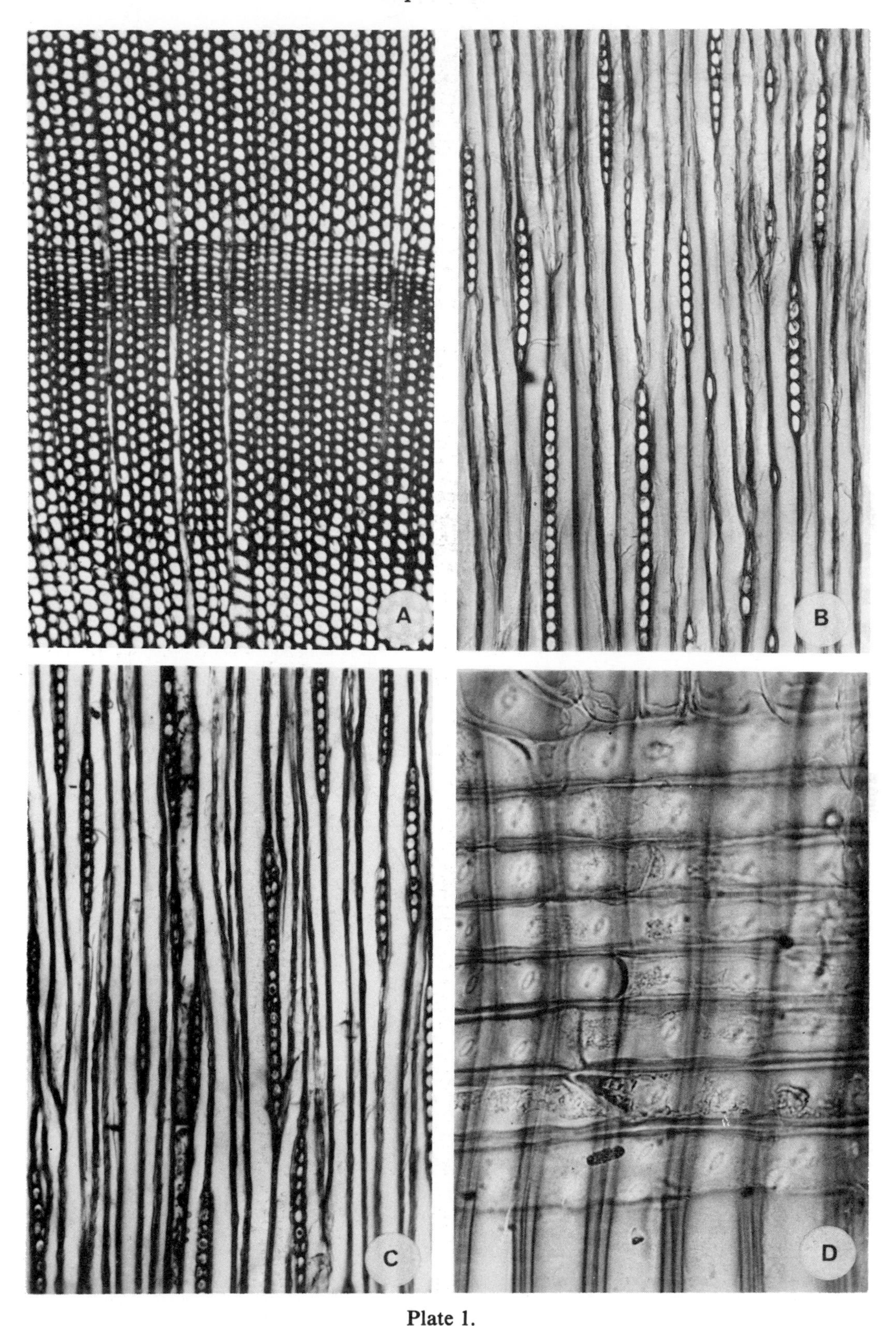

Plate 1.

Cupressus sempervirens. A and B — var. *horizontalis*, ×125. C and D — var. *pyramidalis*; C, ×125; D, showing more or less smooth end walls of ray cells, ×530

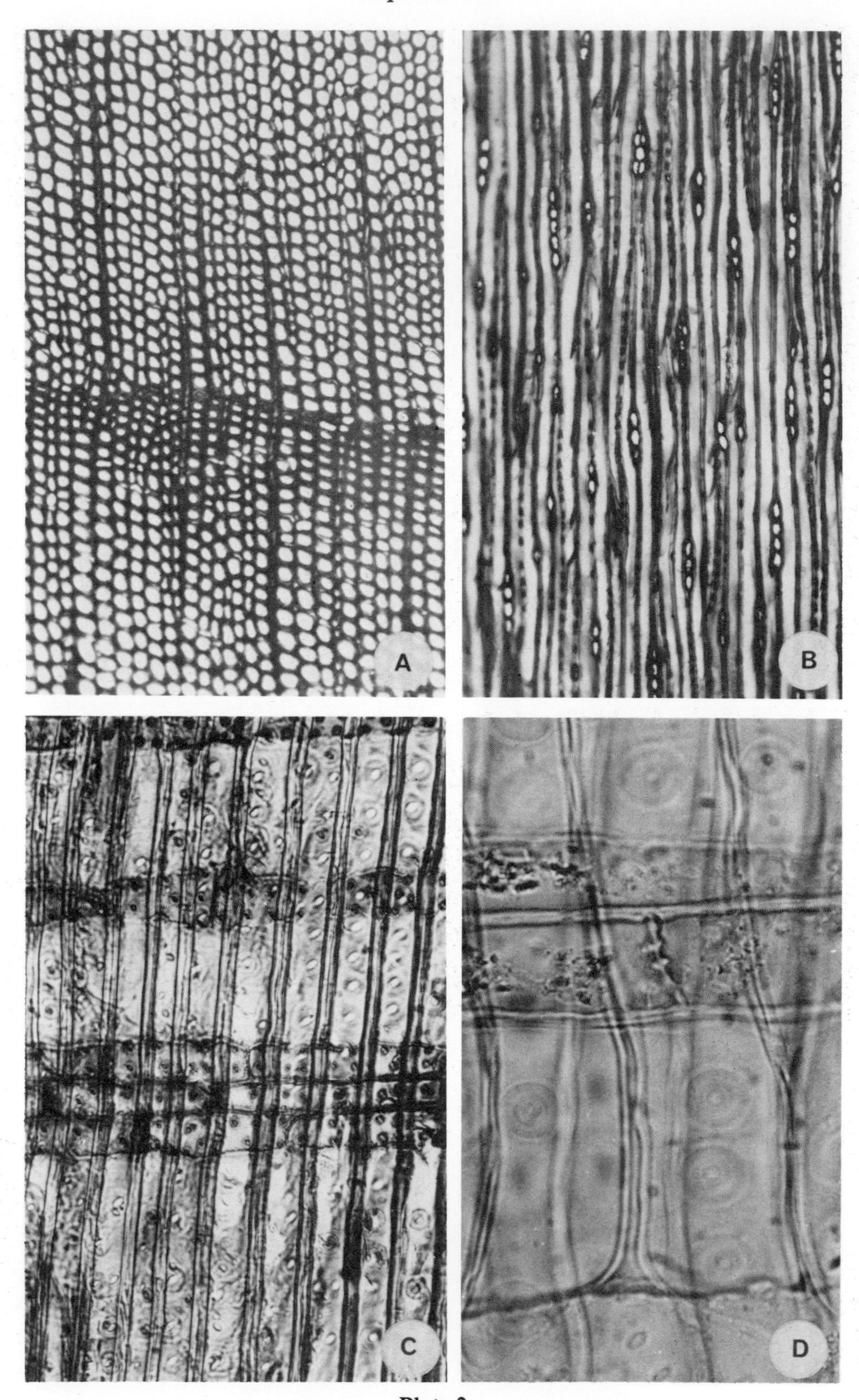

Plate 2.
Juniperus. A–C — *J. phoenicea*; A and B, ×125; C, ×300. D — *J. oxycedrus*, showing a nodular end wall of a ray cell, ×850

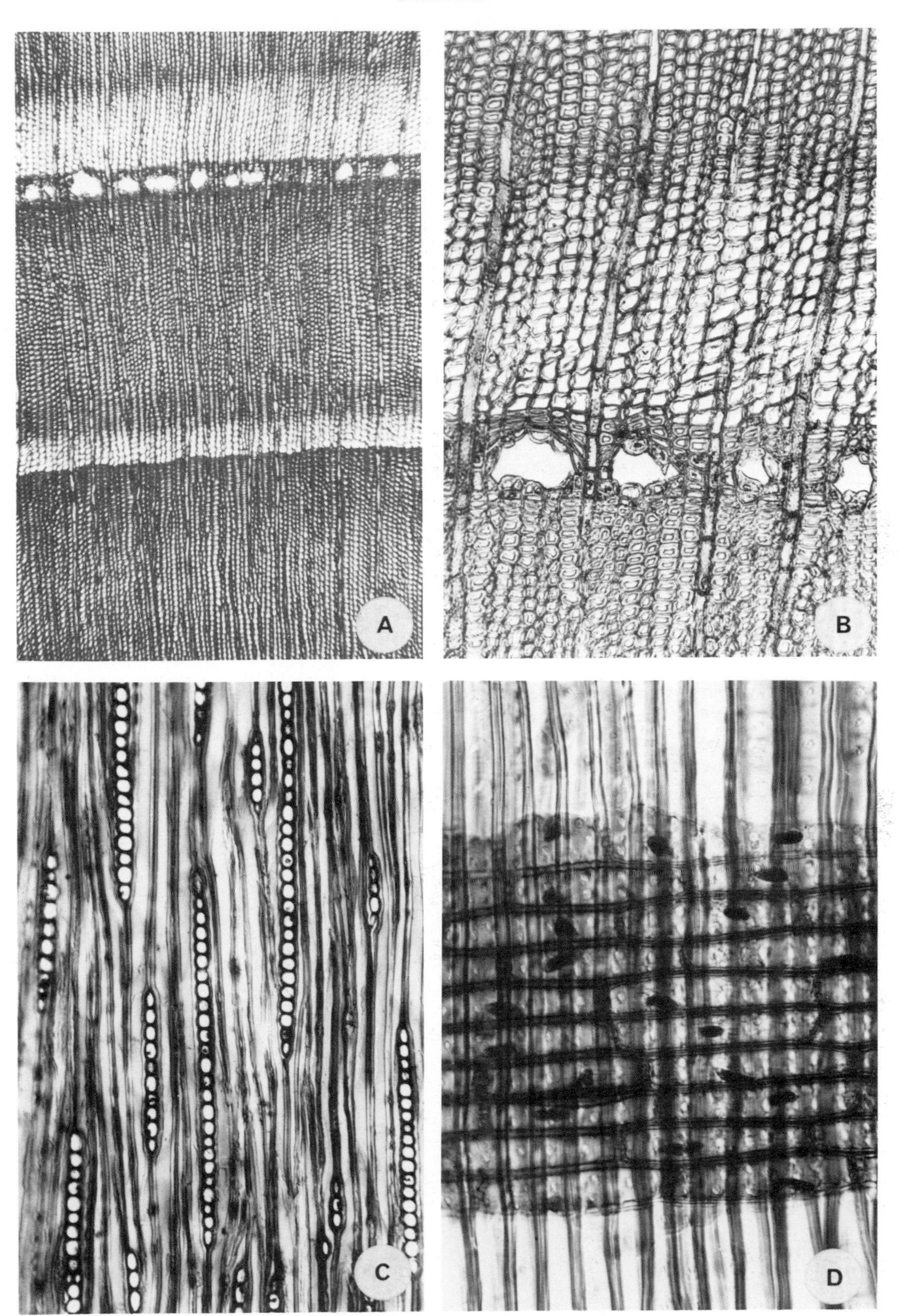

Plate 3.
Cedrus libani. A, ×42; B and C, ×125; D, ×300

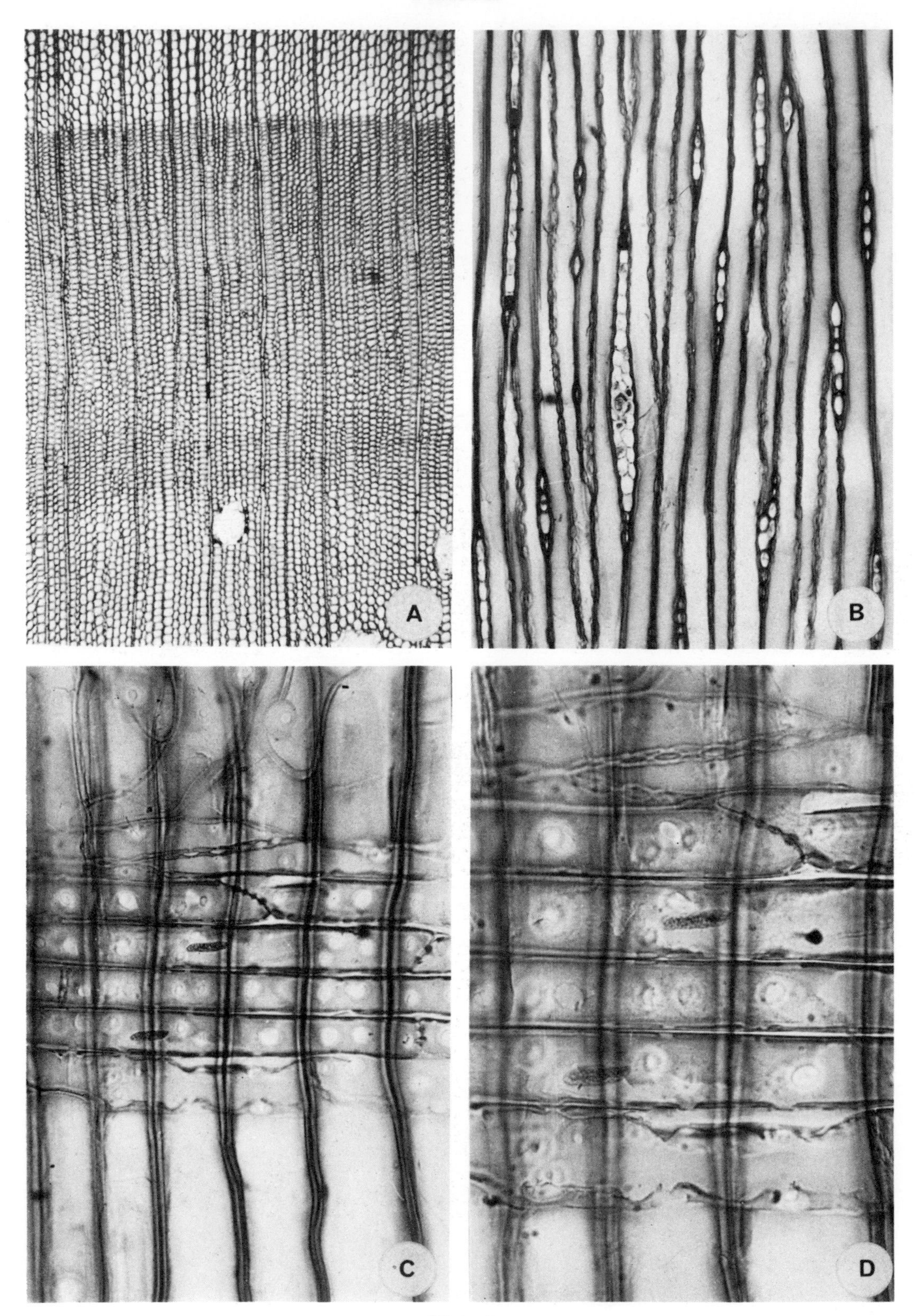

Plate 4
Pinus halepensis. A, ×42; B, ×125; C×300; D, ×365

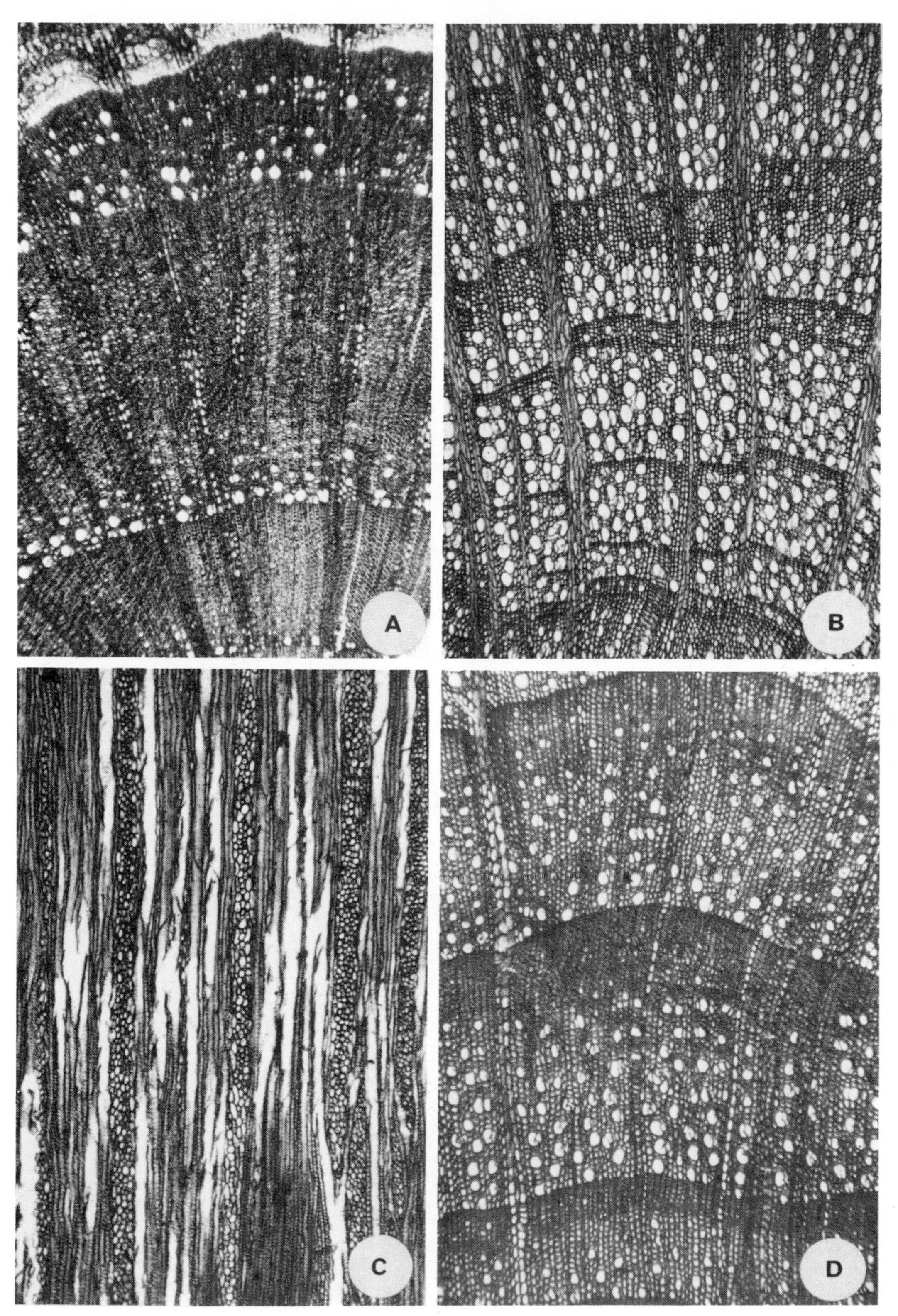

Plate 5.
Ephedra, ×42. A — *E. alata*. B and C — *E. campylopoda*. D — *E. foliata*

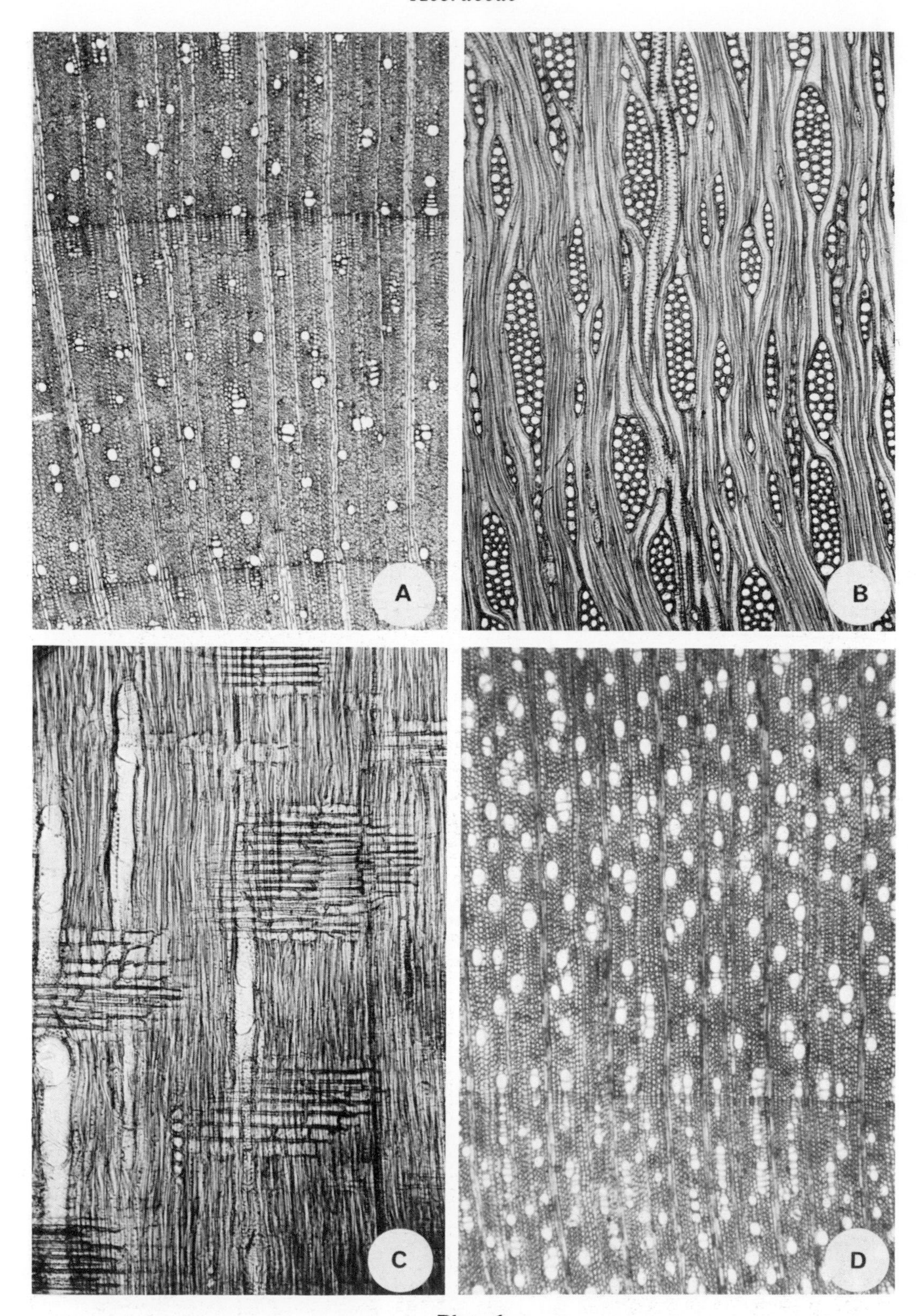

Plate 6.
Acer. A–C — *A. hermoneum*; A, ×48; B and C, ×120. D — *A. obtusifolium*, ×48

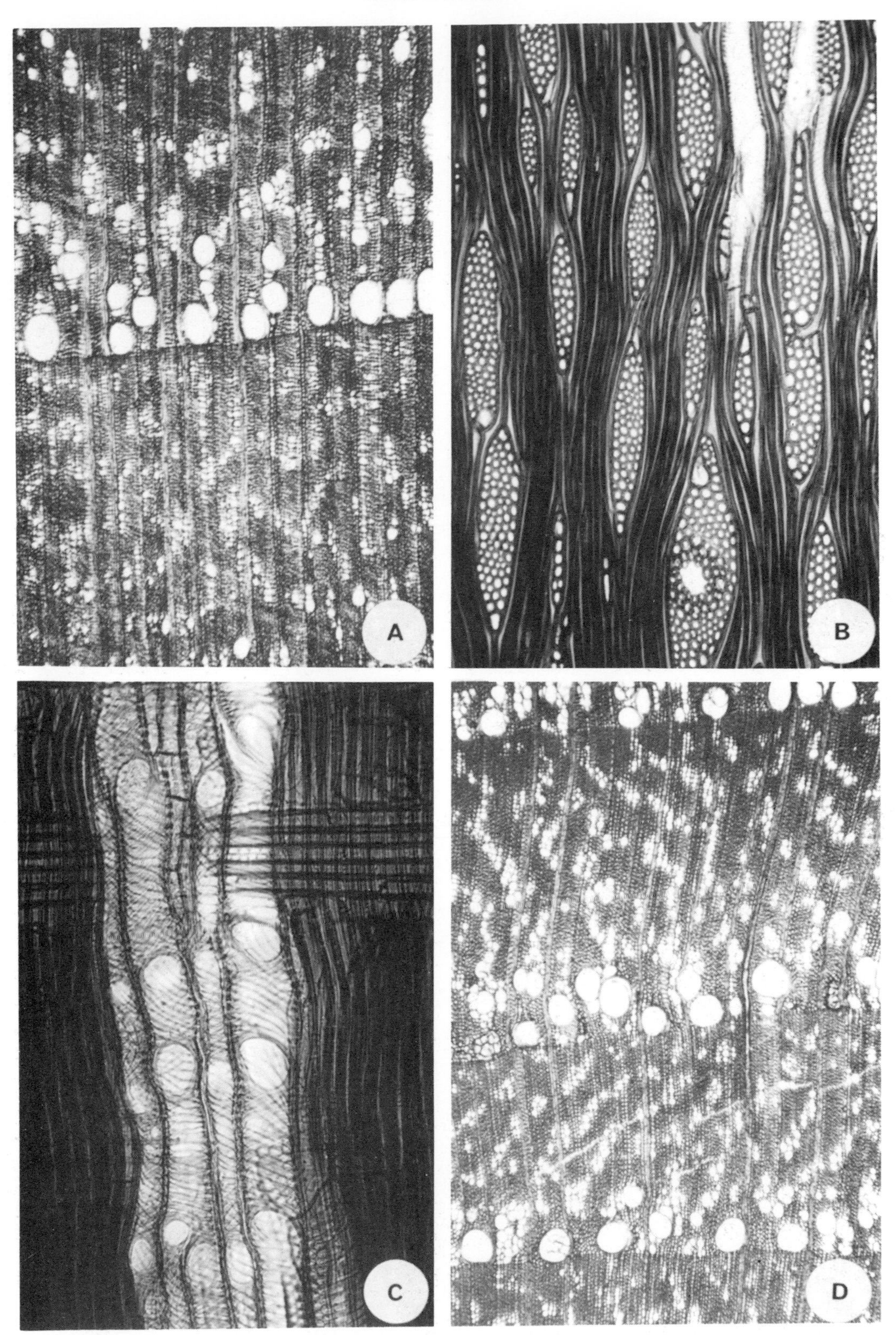

Plate 7.
Pistacia. A–C — *P. atlantica*; A, ×42; B, ×120; C, ×300. D — *P. khinjuk*, ×42

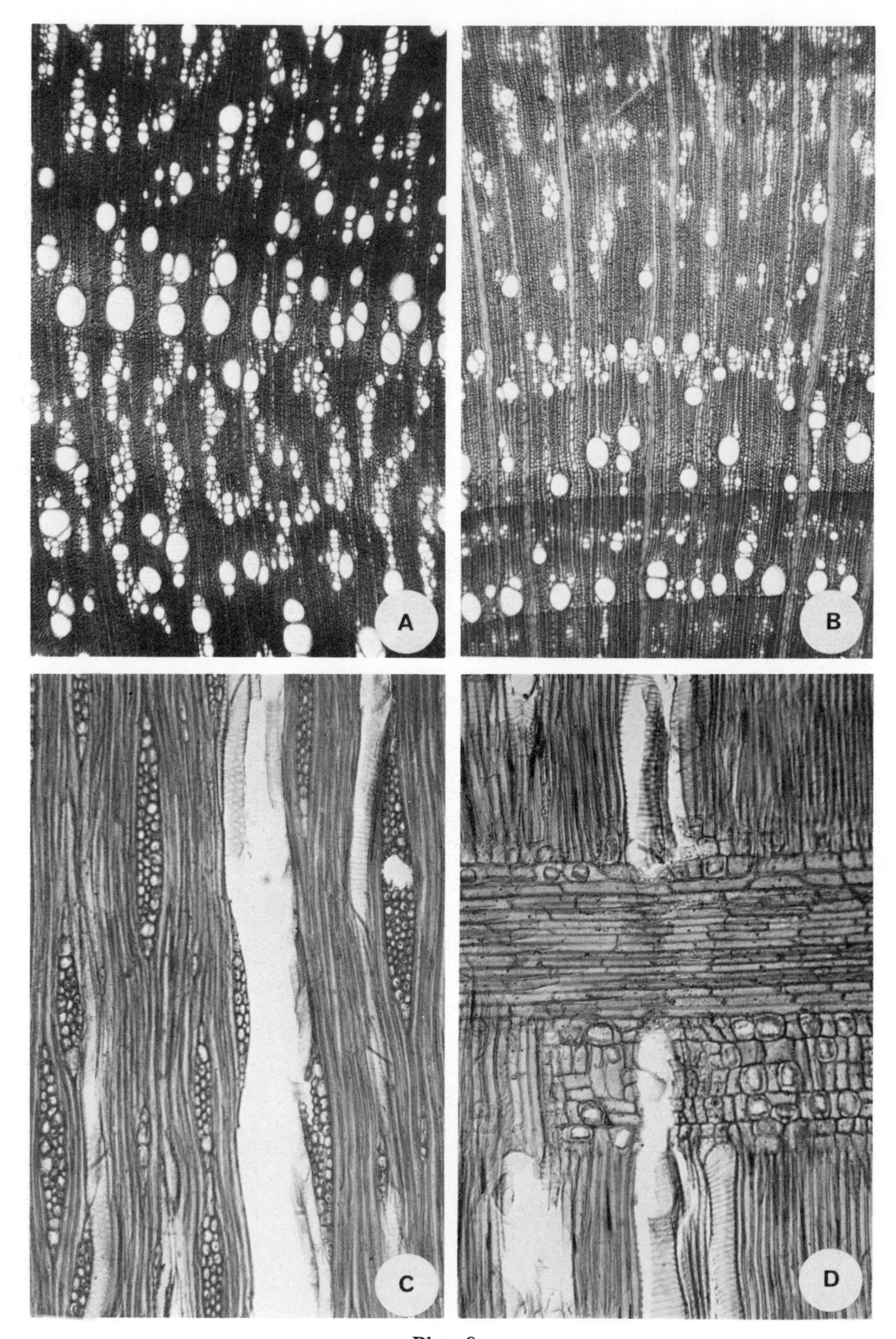

Plate 8

Pistacia. A — *P. lentiscus*, ×42. B–D — *P. palaestina*; B, ×42; C and D, ×126

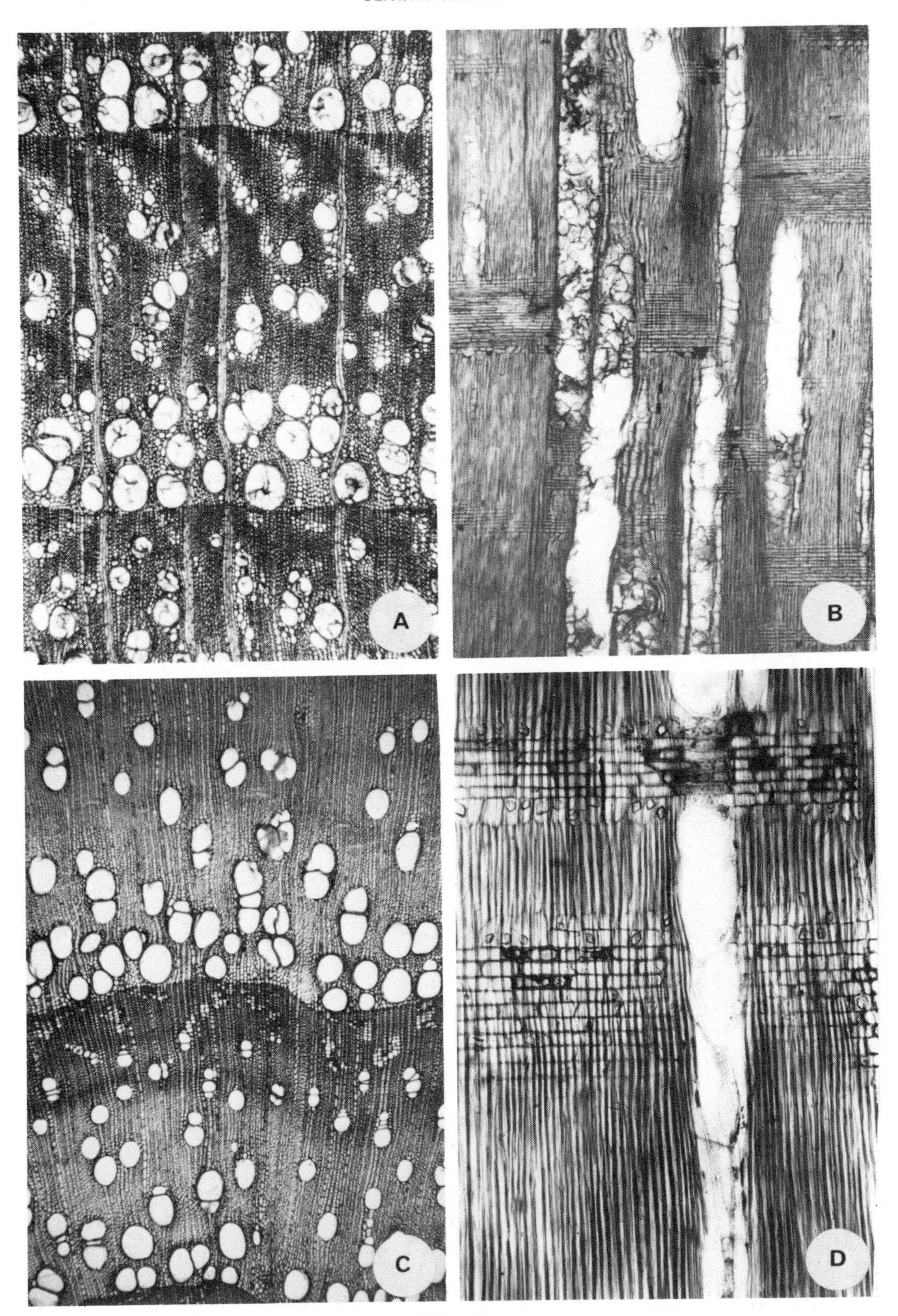

Plate 9.
A and B — *Pistacia vera*, ×42. C and D — *Rhus coriaria*; C, ×42; D, ×126

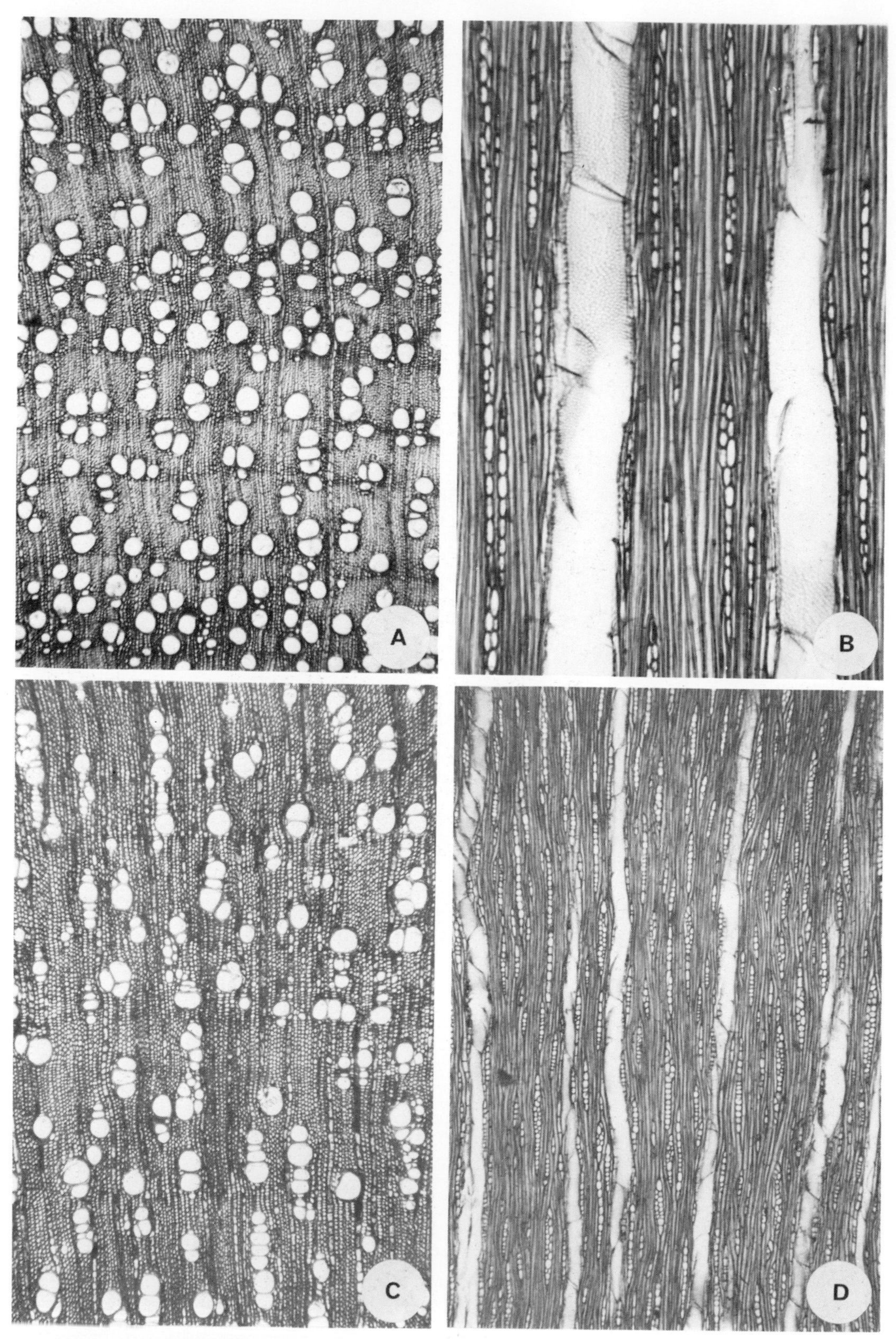

Plate 10.
Rhus. A and B — *R. pentaphylla*; A, ×42; B, ×126. C and D — *R. tripartita*, ×42

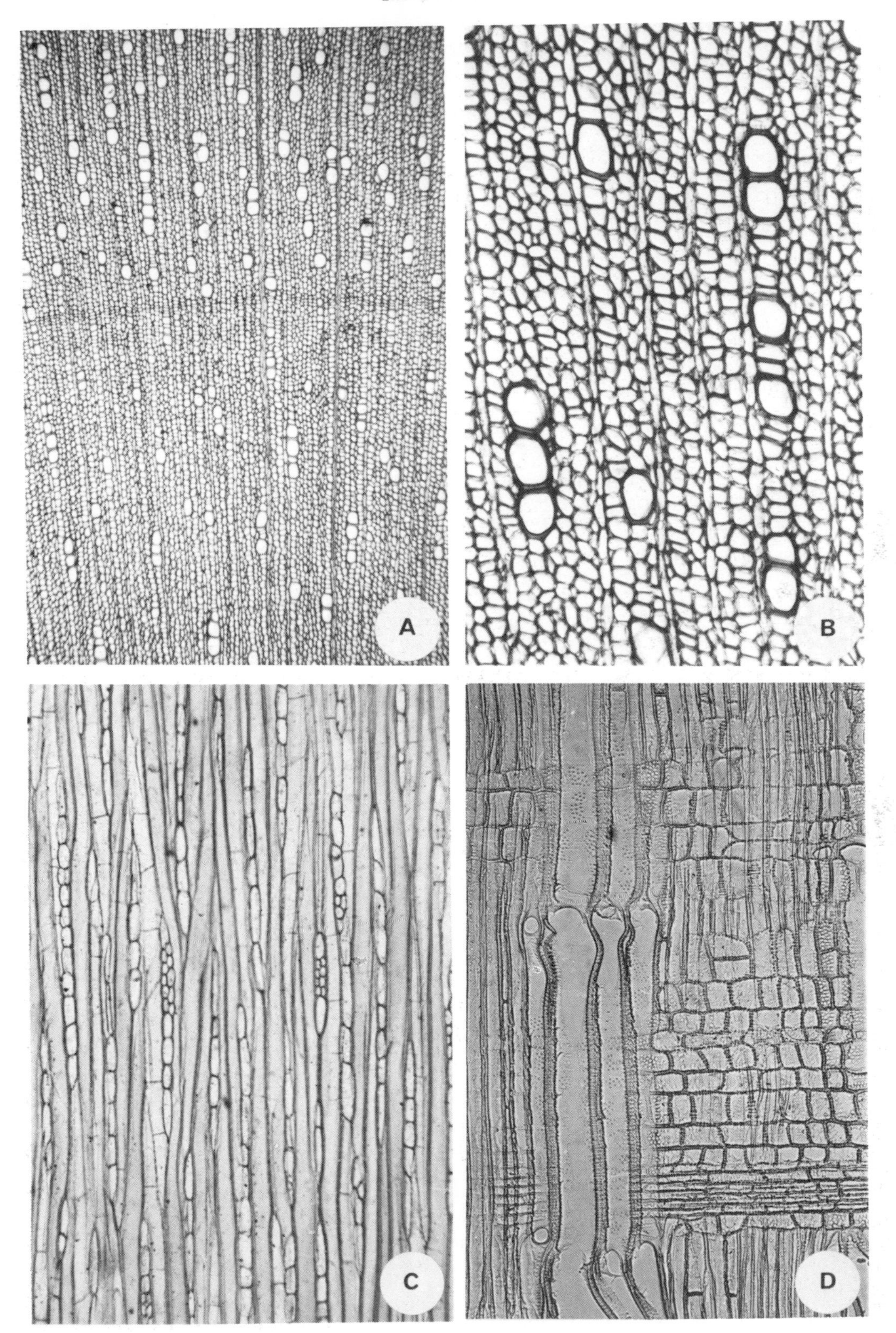

Plate 11.
Nerium oleander. A, ×42; B–D, ×125

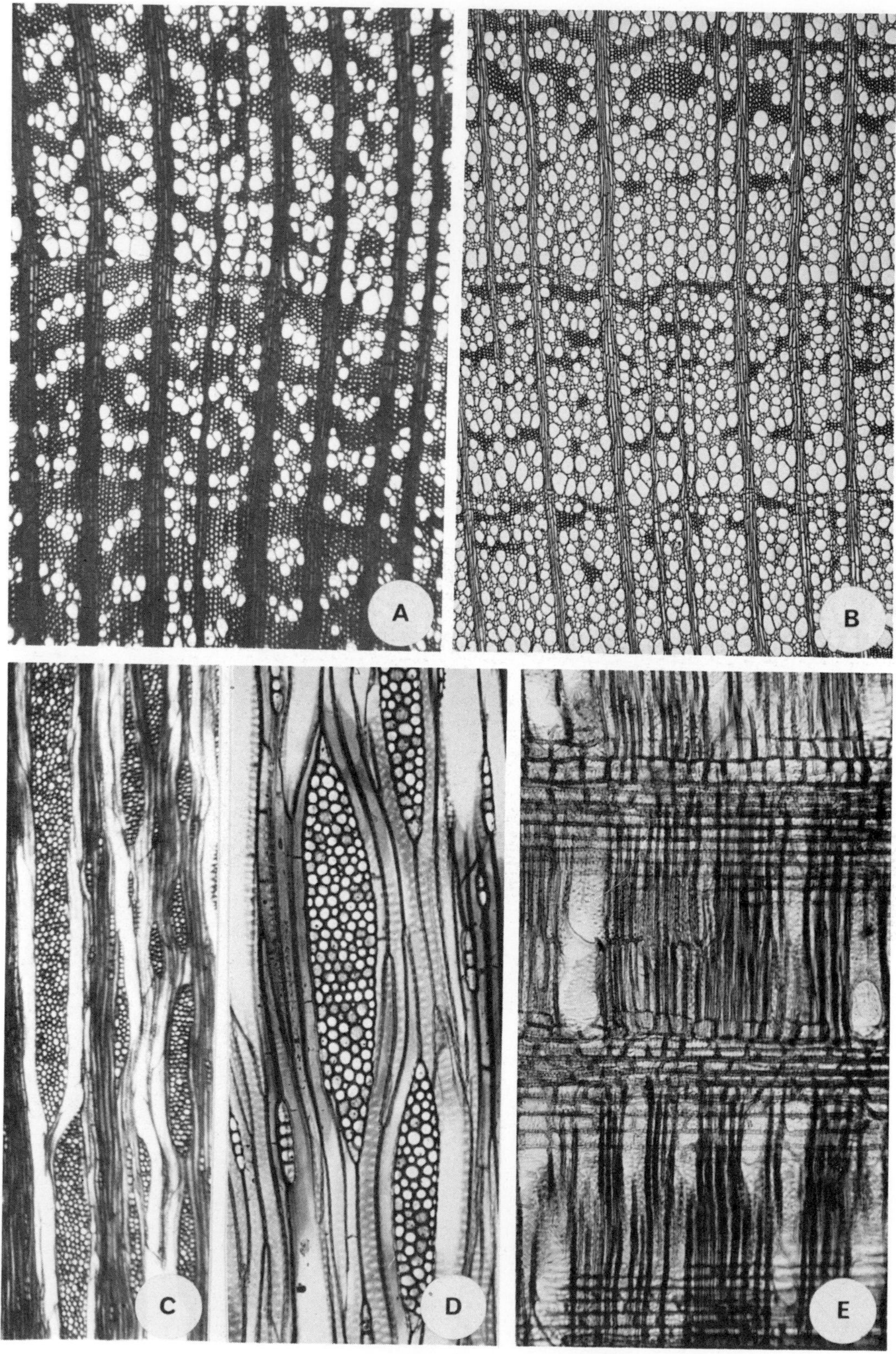

Plate 12.
Hedera helix. A, ×42; B, ×30; C, ×42; D and E, ×120

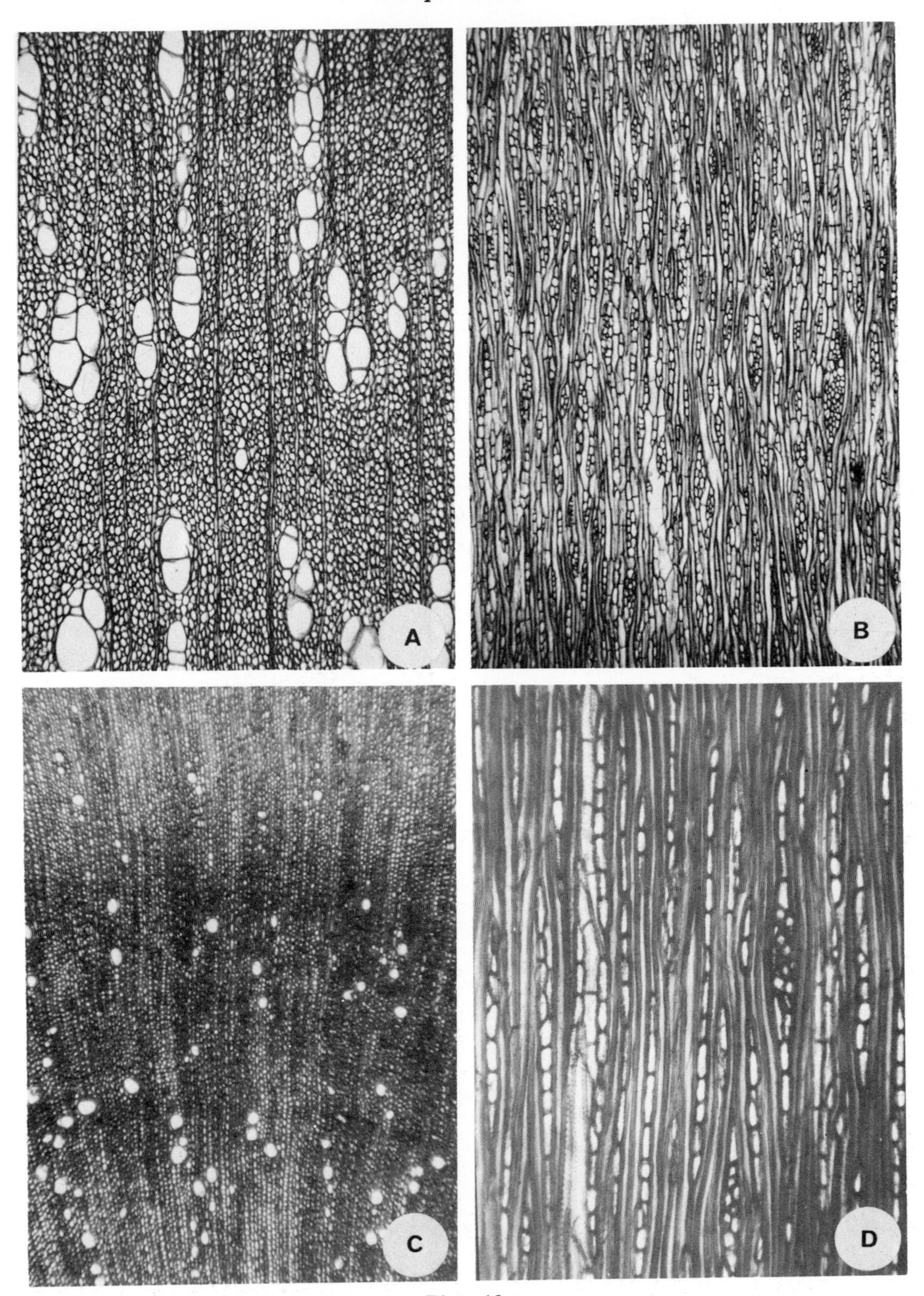

Plate 13.
A and B — *Calotropis procera*, ×42. C and D — *Periploca aphylla*; C, ×42; D, ×126

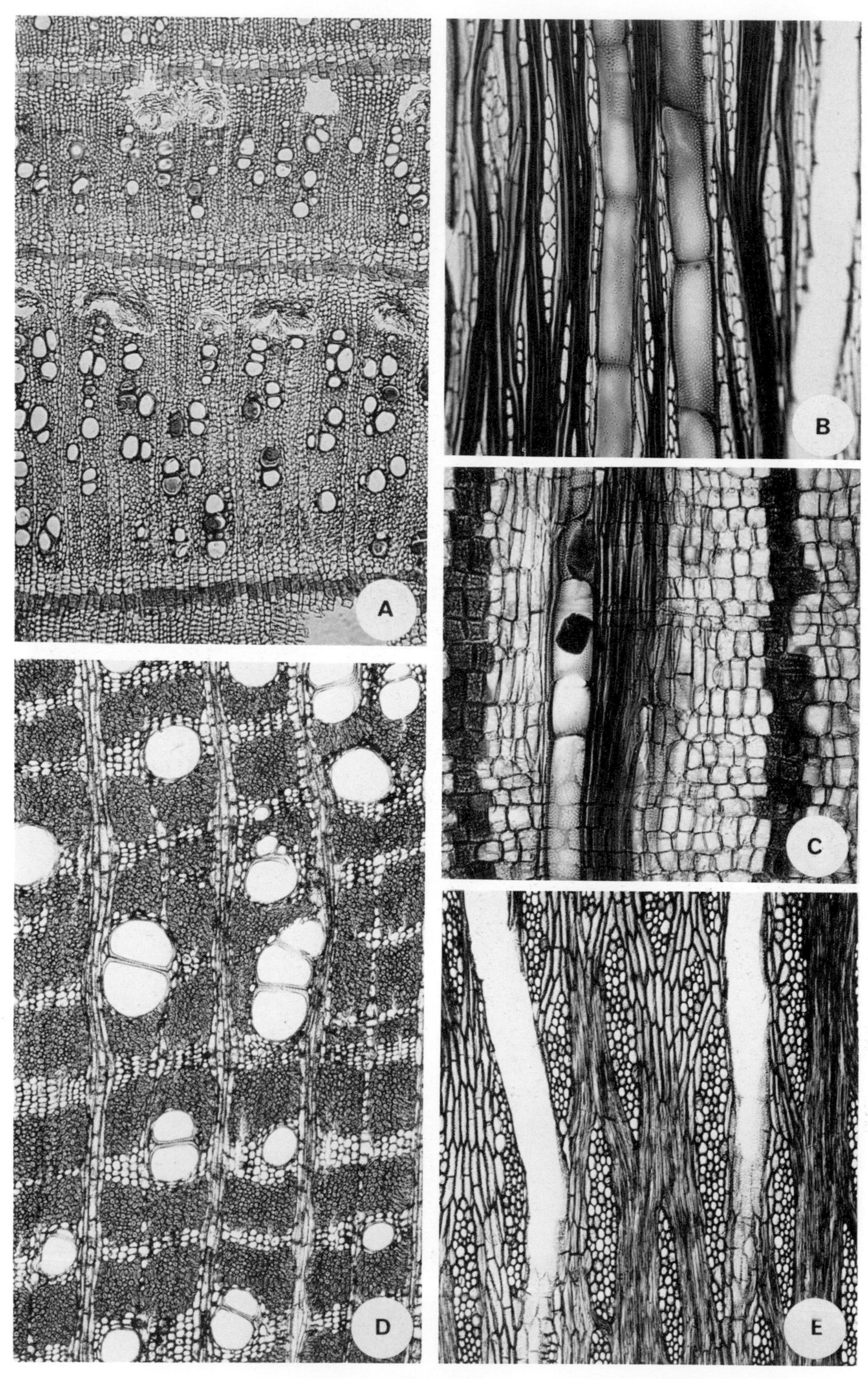

Plate 14.
A–C — *Avicennia marina*; A, ×48; B and C, ×120. D and E — *Cordia sinensis*, ×51

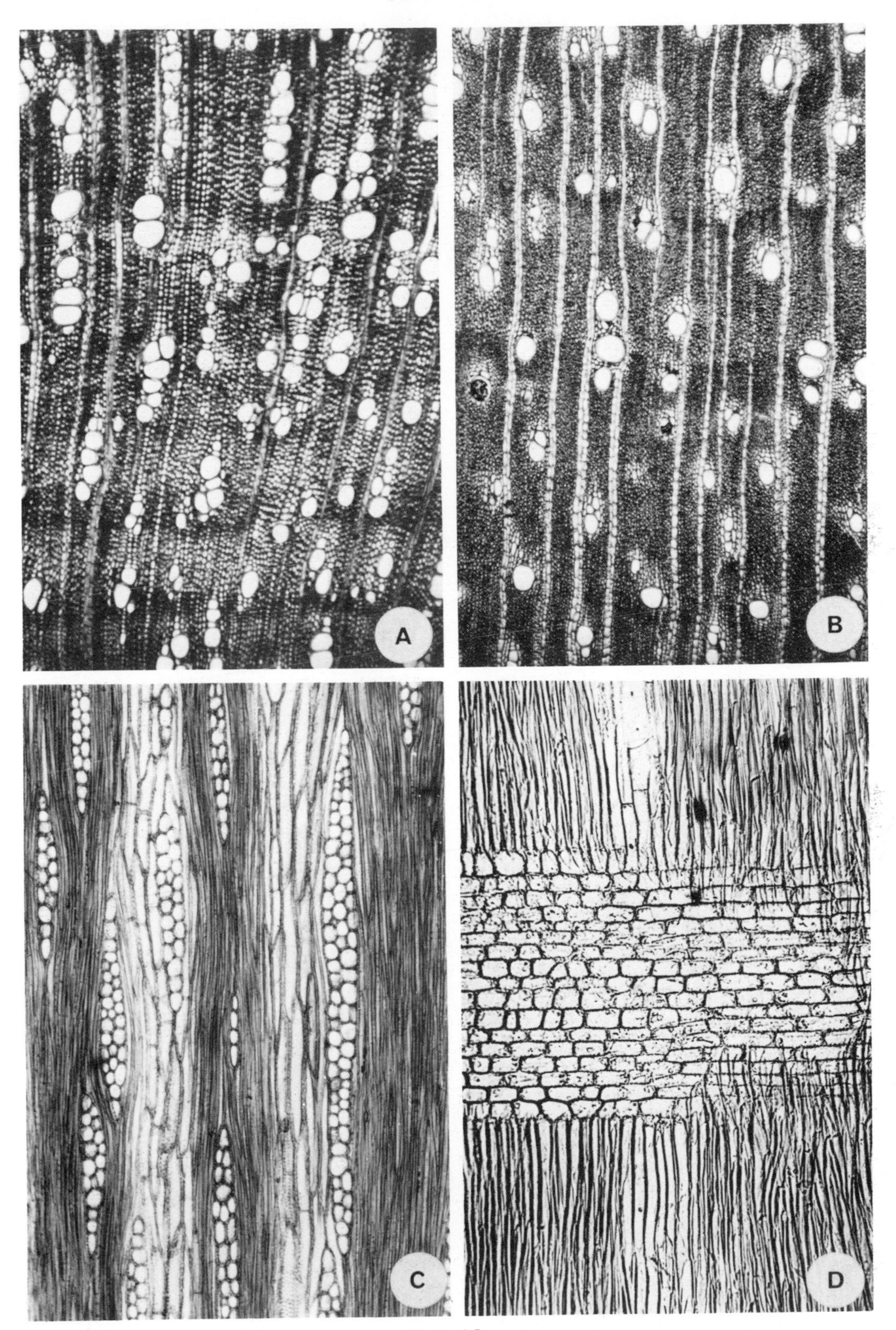

Plate 15.
Capparis. A — *C. cartilaginea*, ×42. B–D — *C. decidua;* B, ×42; C and D, ×126

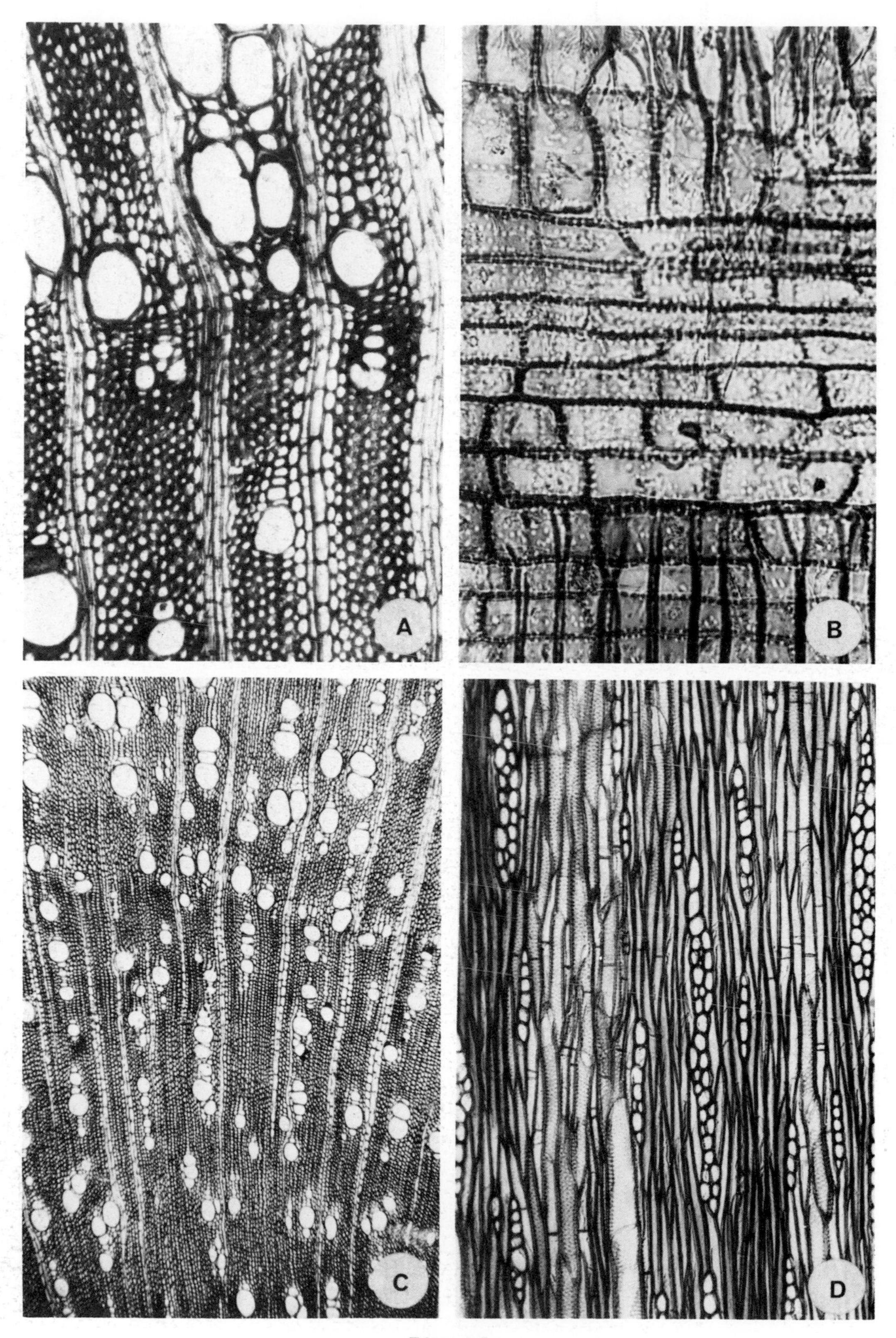

Plate 16.
Capparis. A and B — *C. ovata*; A, ×125; B, ×300. C and D — *C. spinosa*; C, ×42; D, ×125

Plate 17.
A — *Cleome droserifolia*, ×42. B–D — *Maerua crassifolia*; B, ×42; C, ×126; D, ×630

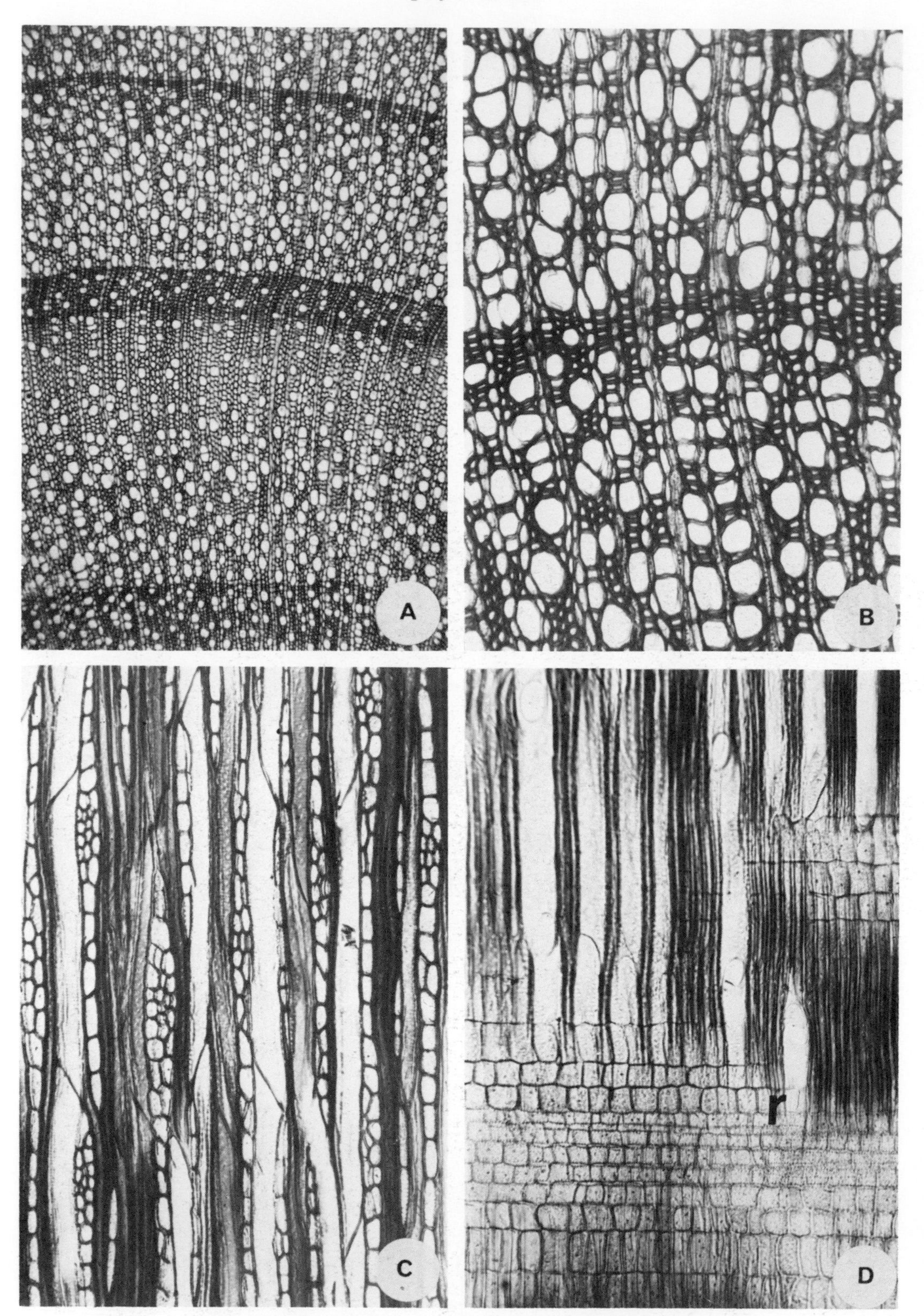

Plate 18.
Lonicera etrusca; A, ×42; B–D, ×126

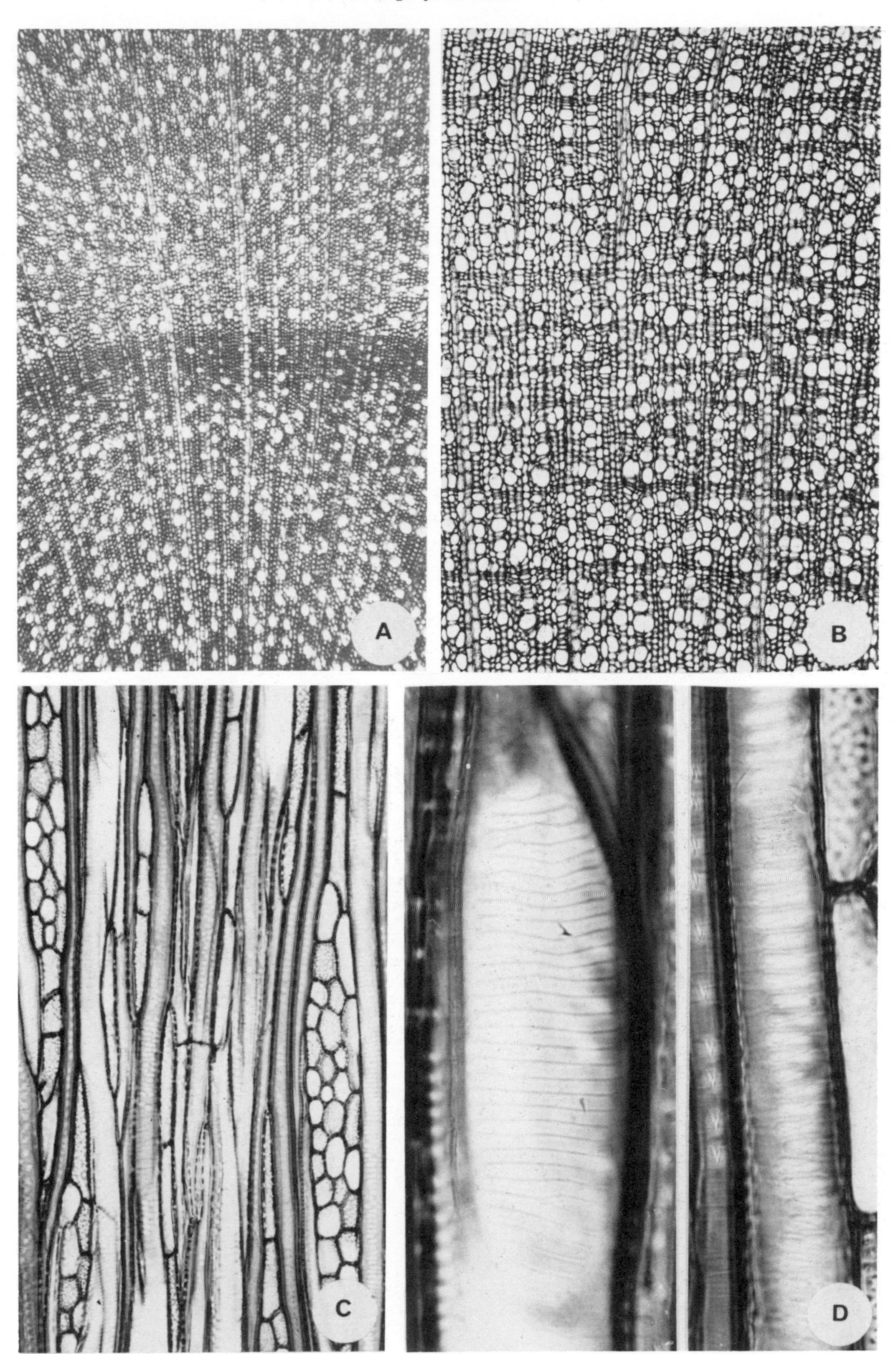

Plate 19.
Viburnum tinus. A, ×42; B, ×48; C, ×120; D, ×430

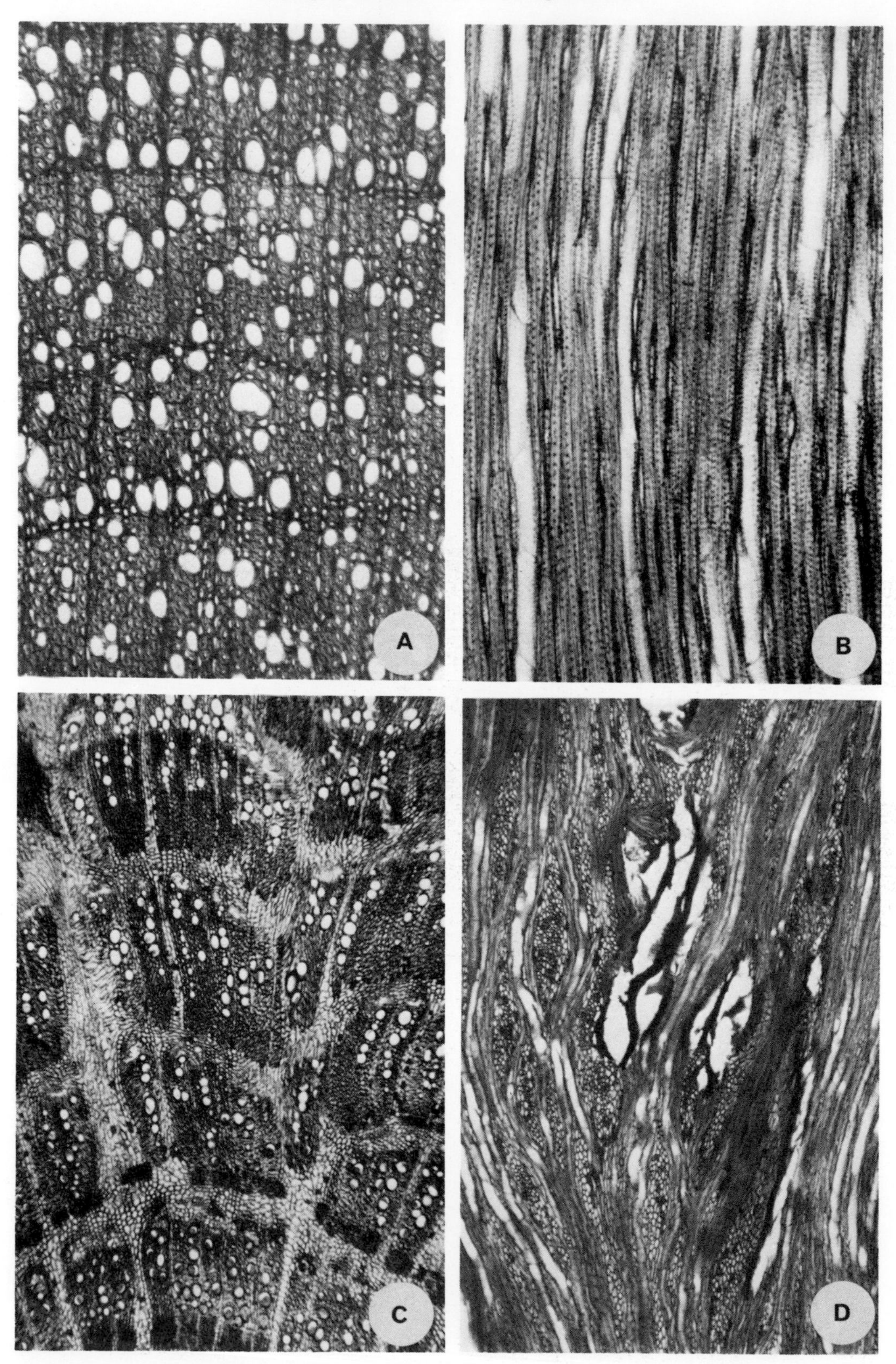

Plate 20.
A and B — *Gymnocarpos decandrum*, ×126. C and D — *Aellenia lancifolia*, ×42

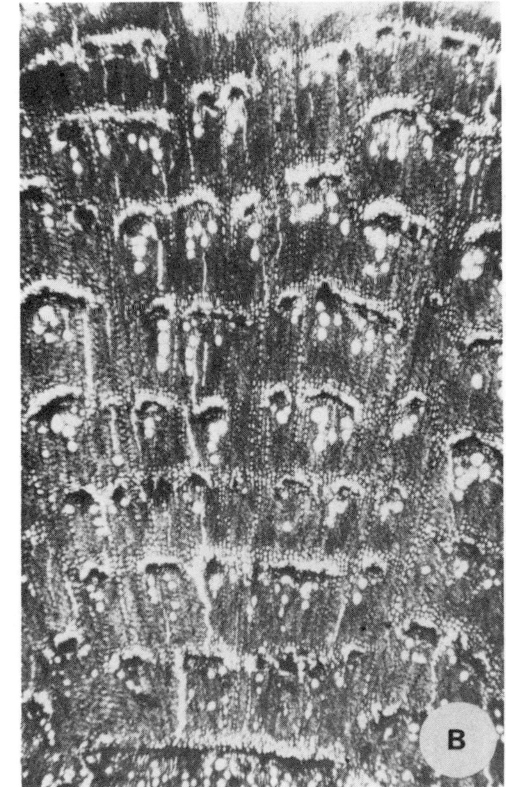

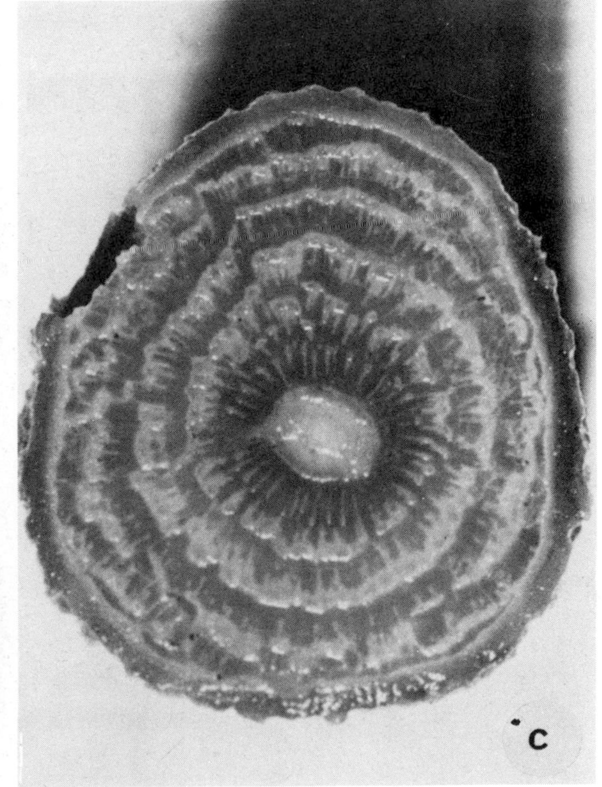

Plate 21.
Anabasis. A — *A. articulata*, ×8. B — *A. setifera*, ×42. C — *A. syriaca*, ×10

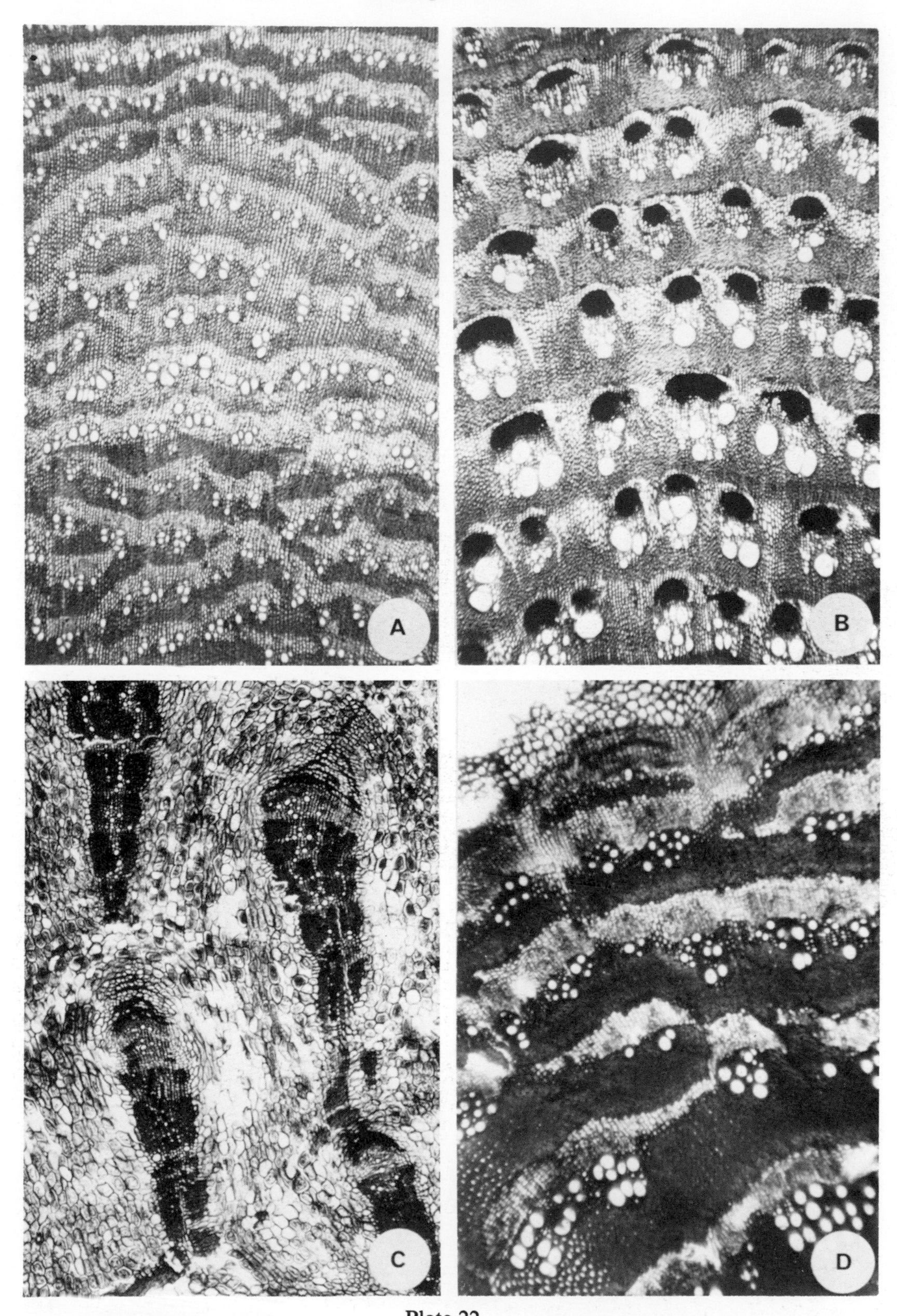

Plate 22.
A — *Arthrocnemum macrostachyum*, ×42. B — *Atriplex halimus*, ×42. C — *Chenolea arabica*, ×48. D — *Halogeton alopecuroides*, ×42

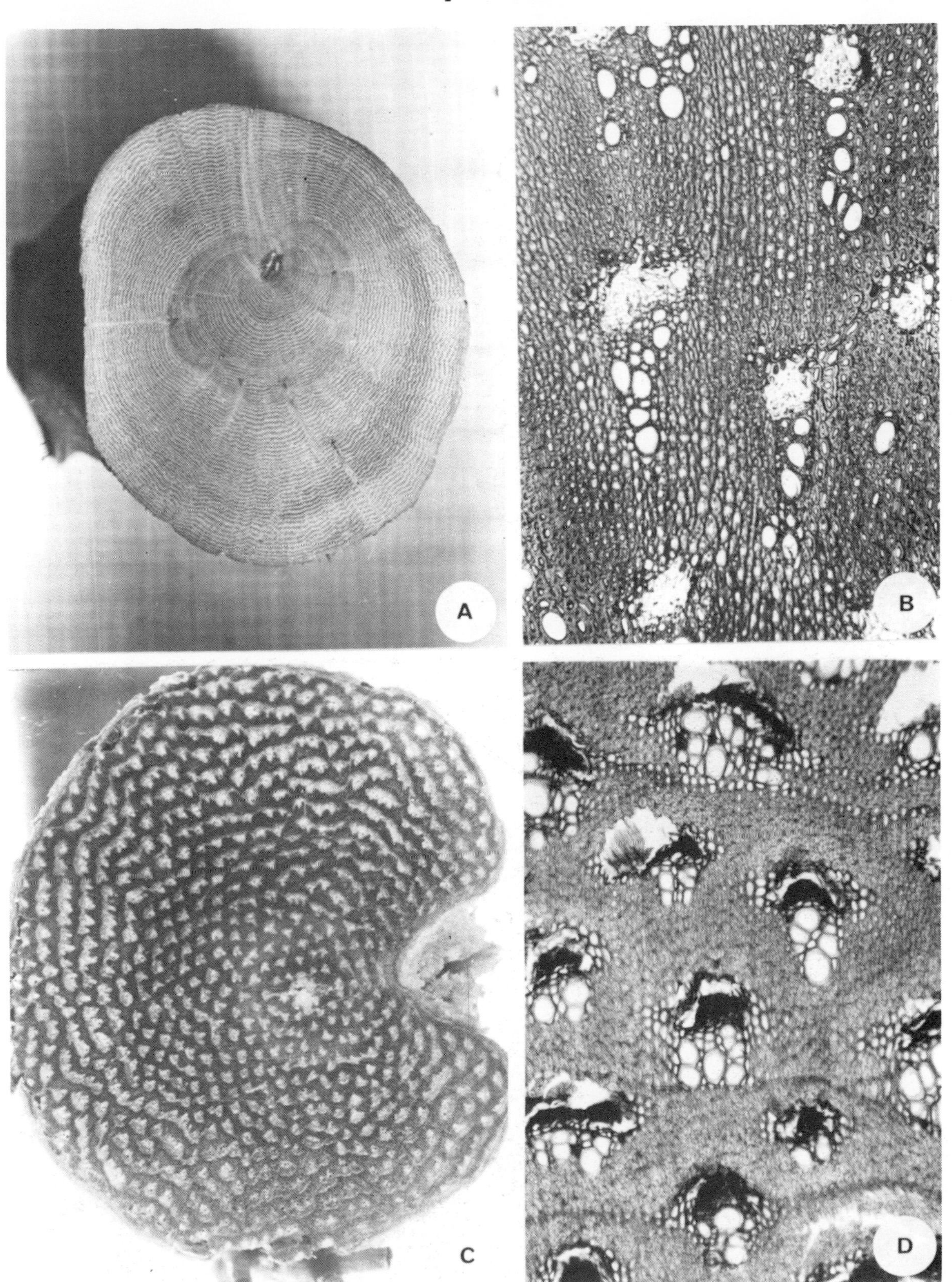

Plate 23.
A — *Haloxylon persicum*, ×1.1. B–D — *Hammada*. B — *H. negevensis*, ×126.
C and D — *H. scoparia*; C, ×8; D, ×126

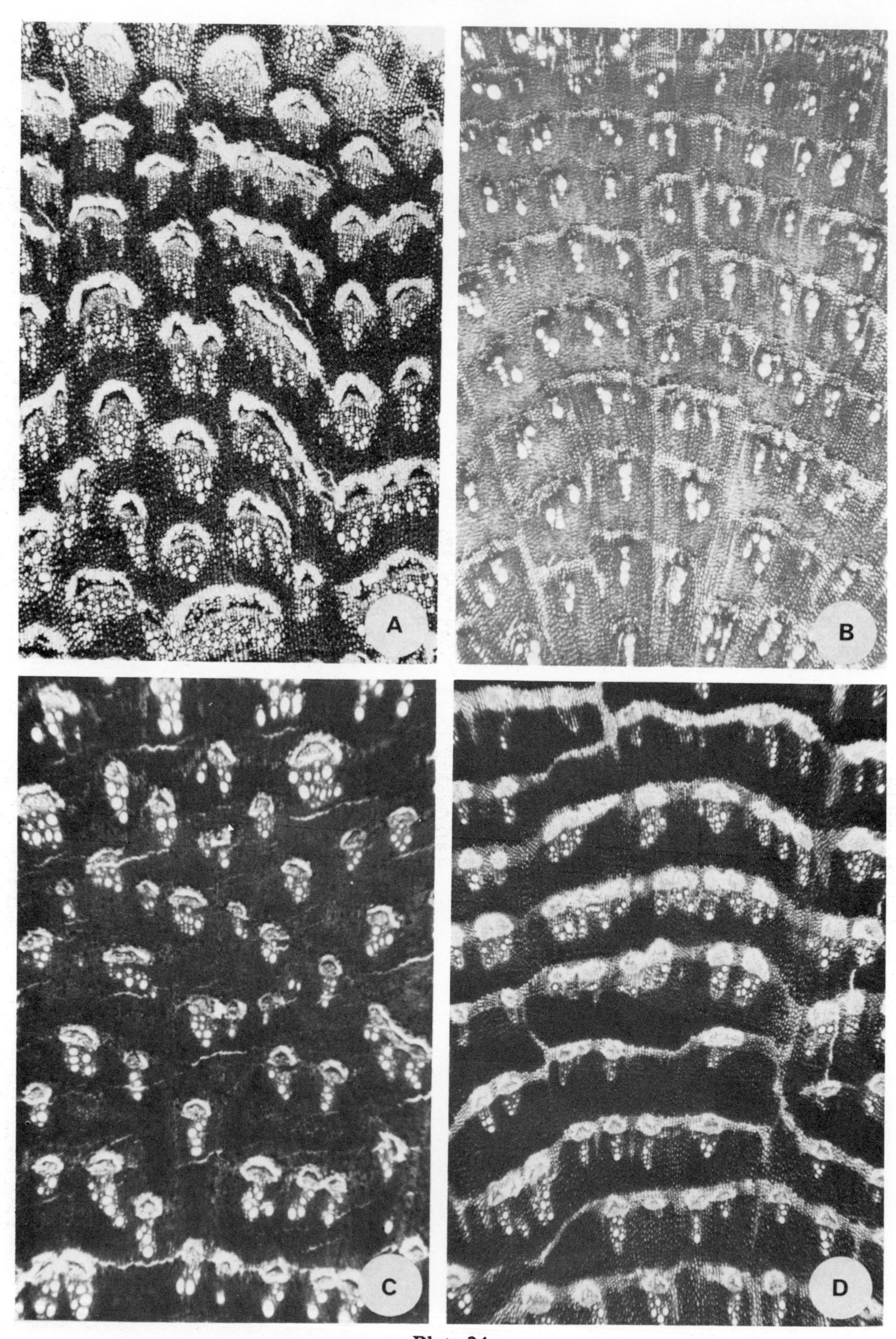

Plate 24.
A — *Noaea mucronata*, ×48. B–D — *Salsola*, ×42. B — *S. baryosma*. C — *S. tetrandra*. D — *S. vermiculata*

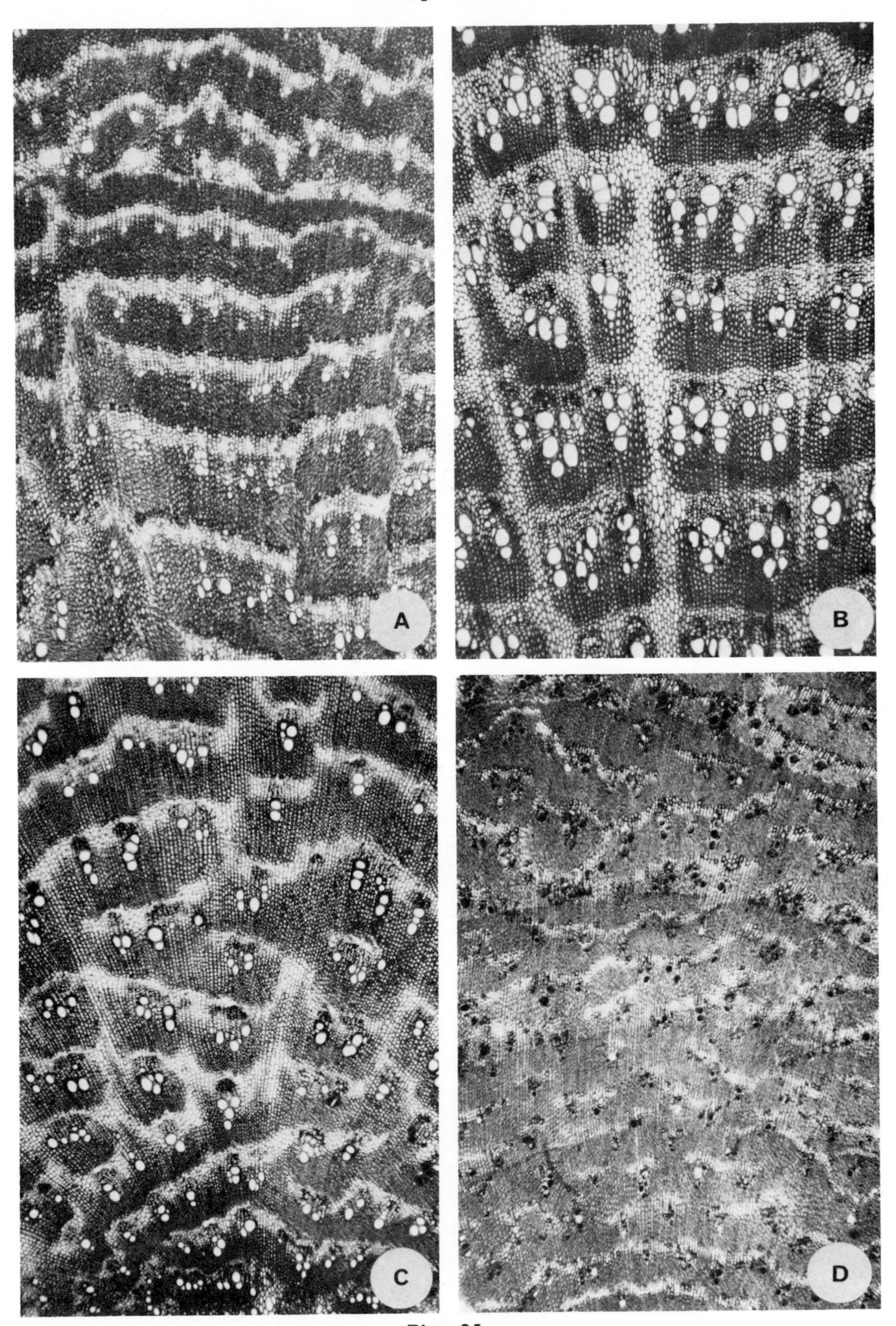

Plate 25.
Suaeda, ×42. A —*S. asphaltica*. B — *S. fruticosa*. C — *S. monoica*. D — *S. palaestina*

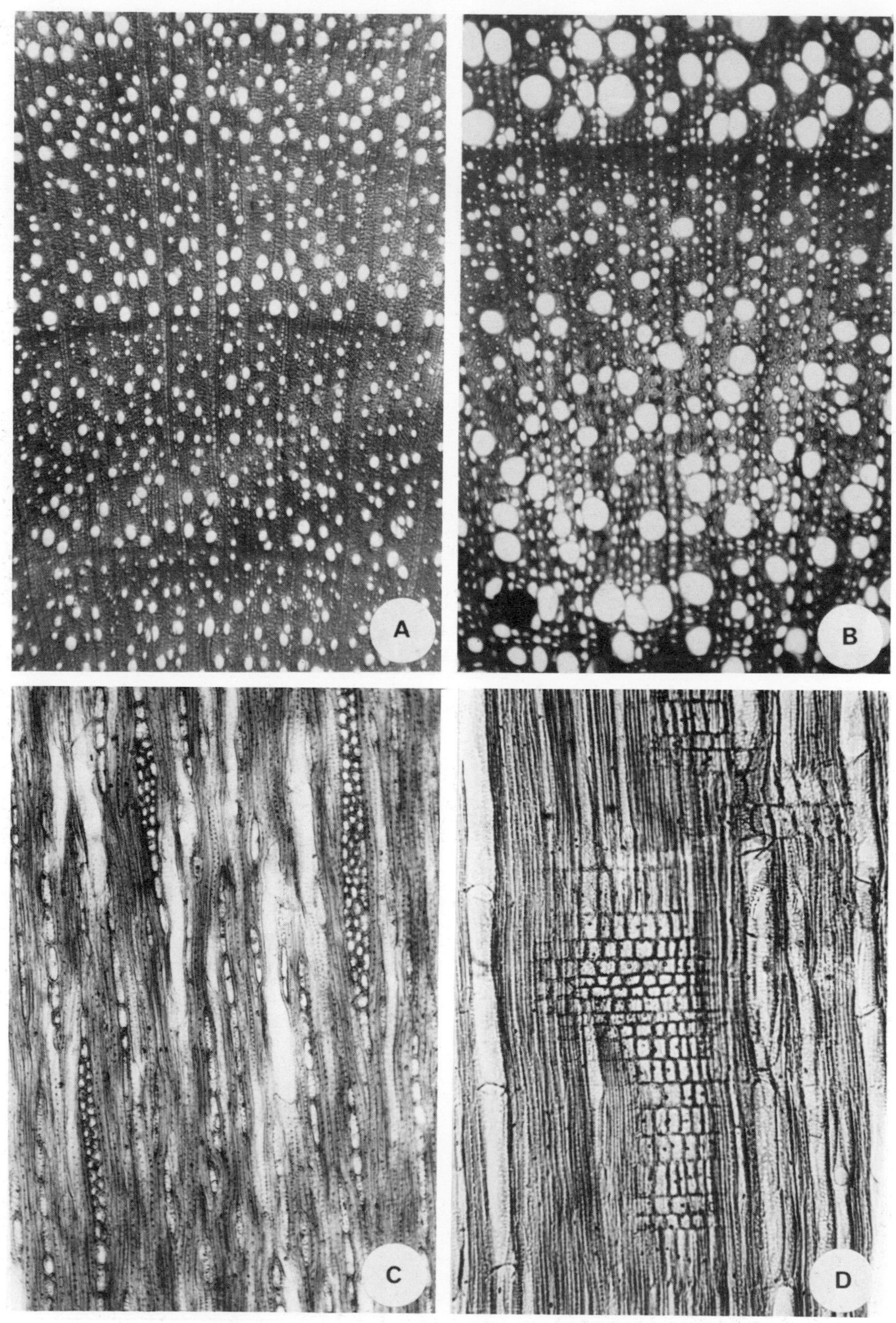

Plate 26.
Cistus. A — *C. creticus*, ×42. B–D — *C. salvifolius*, × 126

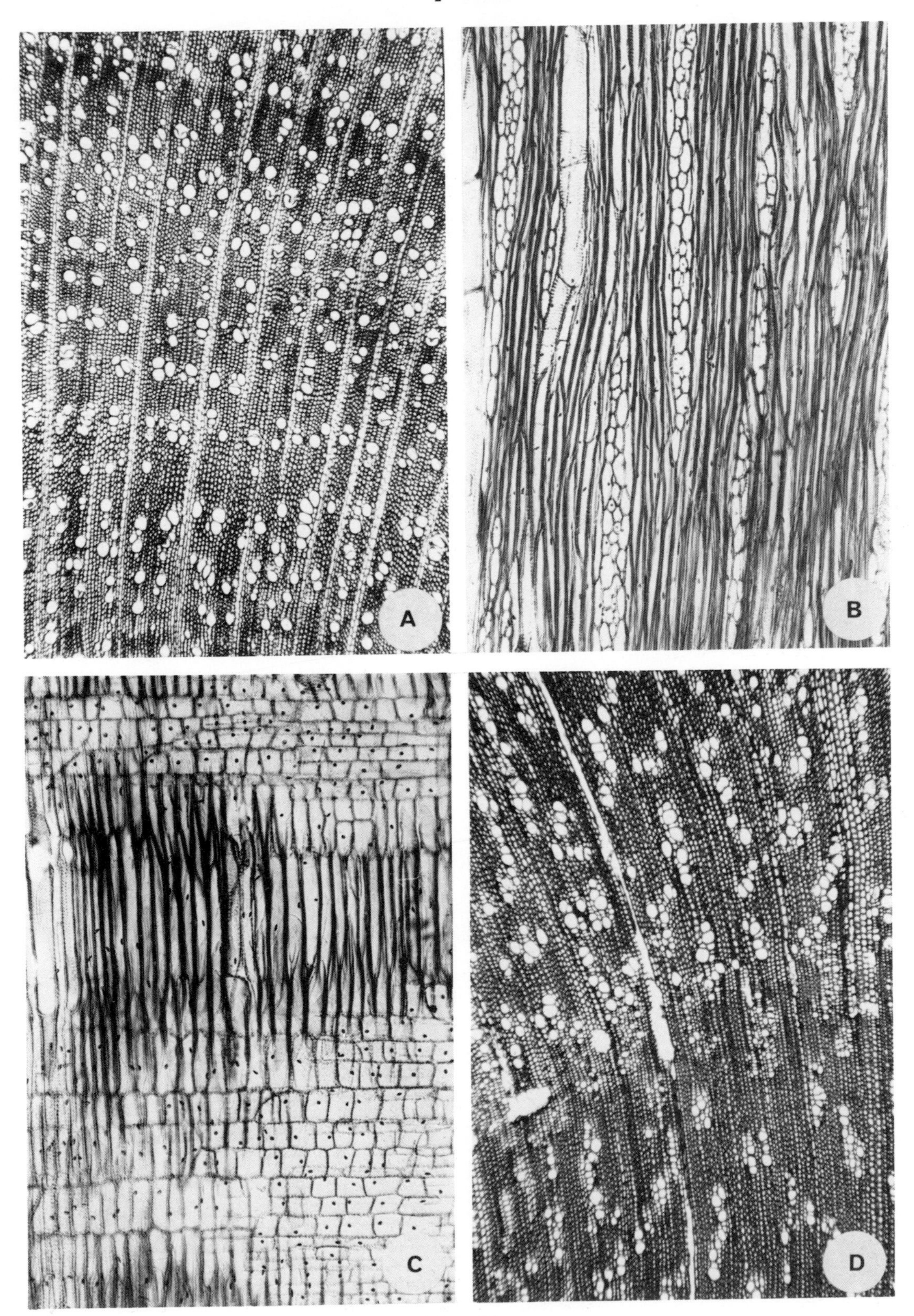

Plate 27.
Artemisia. A–C — *A. arborescens*; A, ×42; B and C, ×126. D — *A. monosperma*, ×42

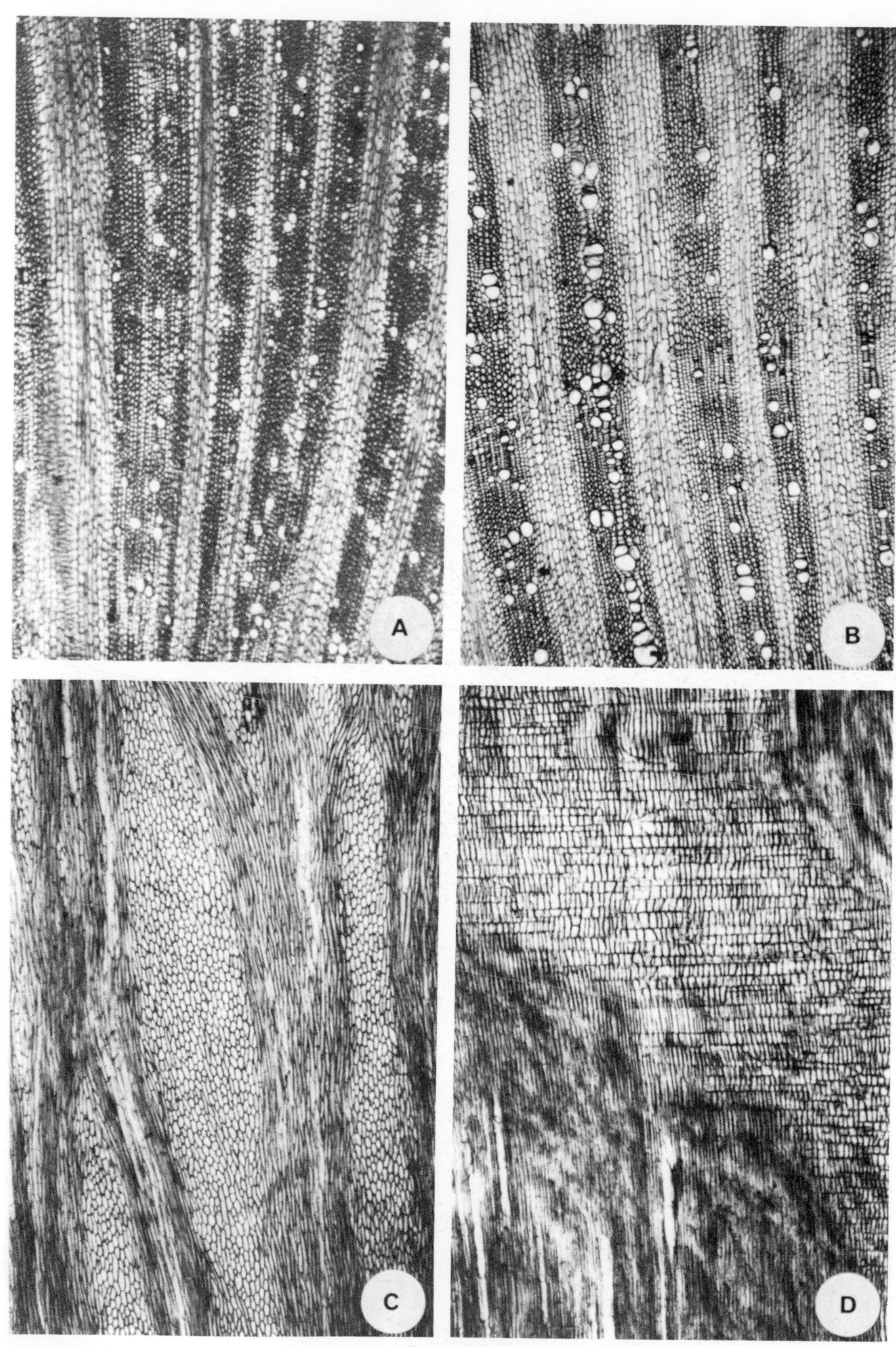

Plate 28.
Inula, ×42. A — *I. crithmoides*. B–D — *I. viscosa*

Plate 29.
A and B — *Pluchea dioscoridis*, ×42. C and D — *Zilla spinosa*, ×42

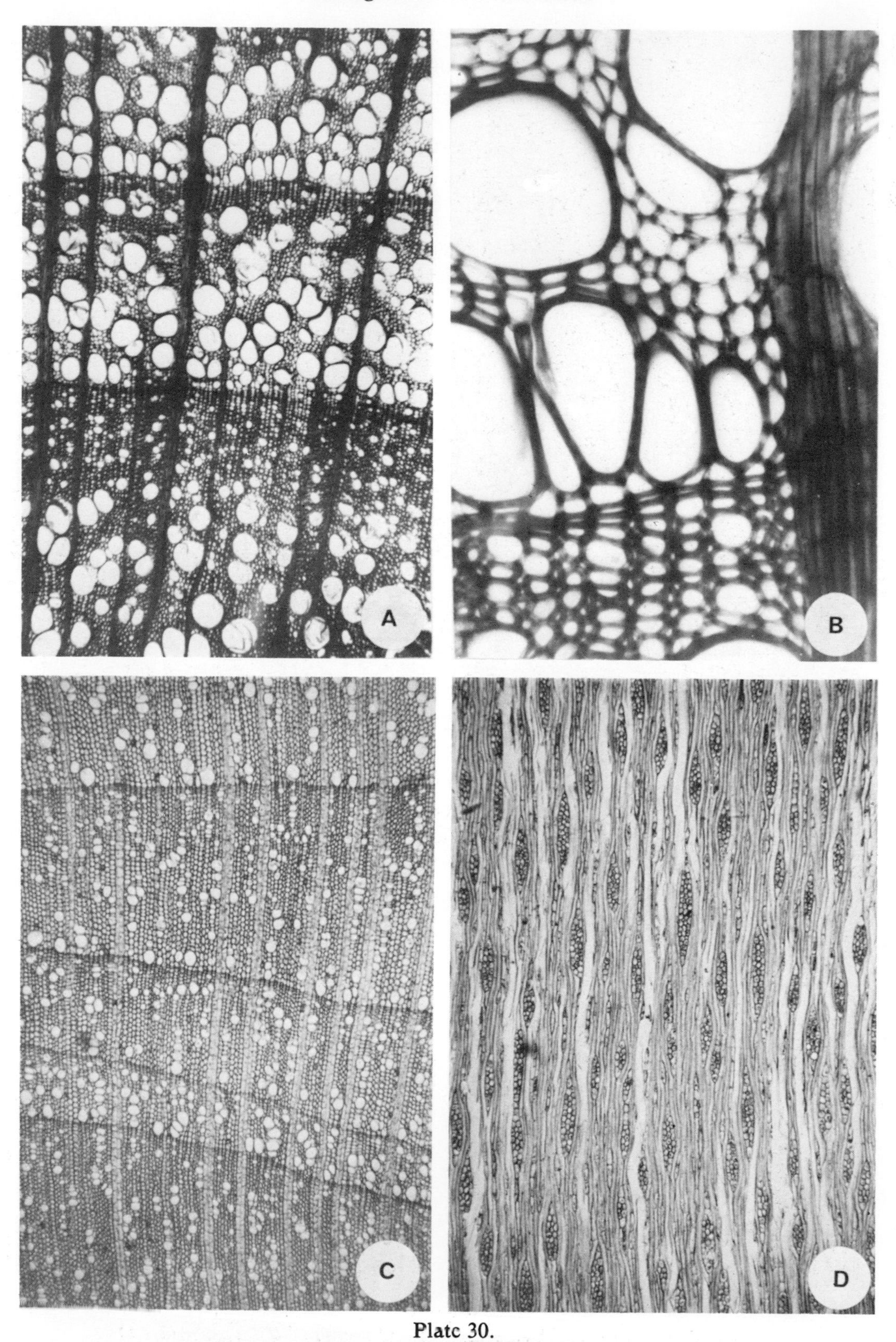

Plate 30.
A and B — *Elaeagnus angustifolia*; A, ×42; B, ×300. C and D — *Arbutus andrachne*, ×42

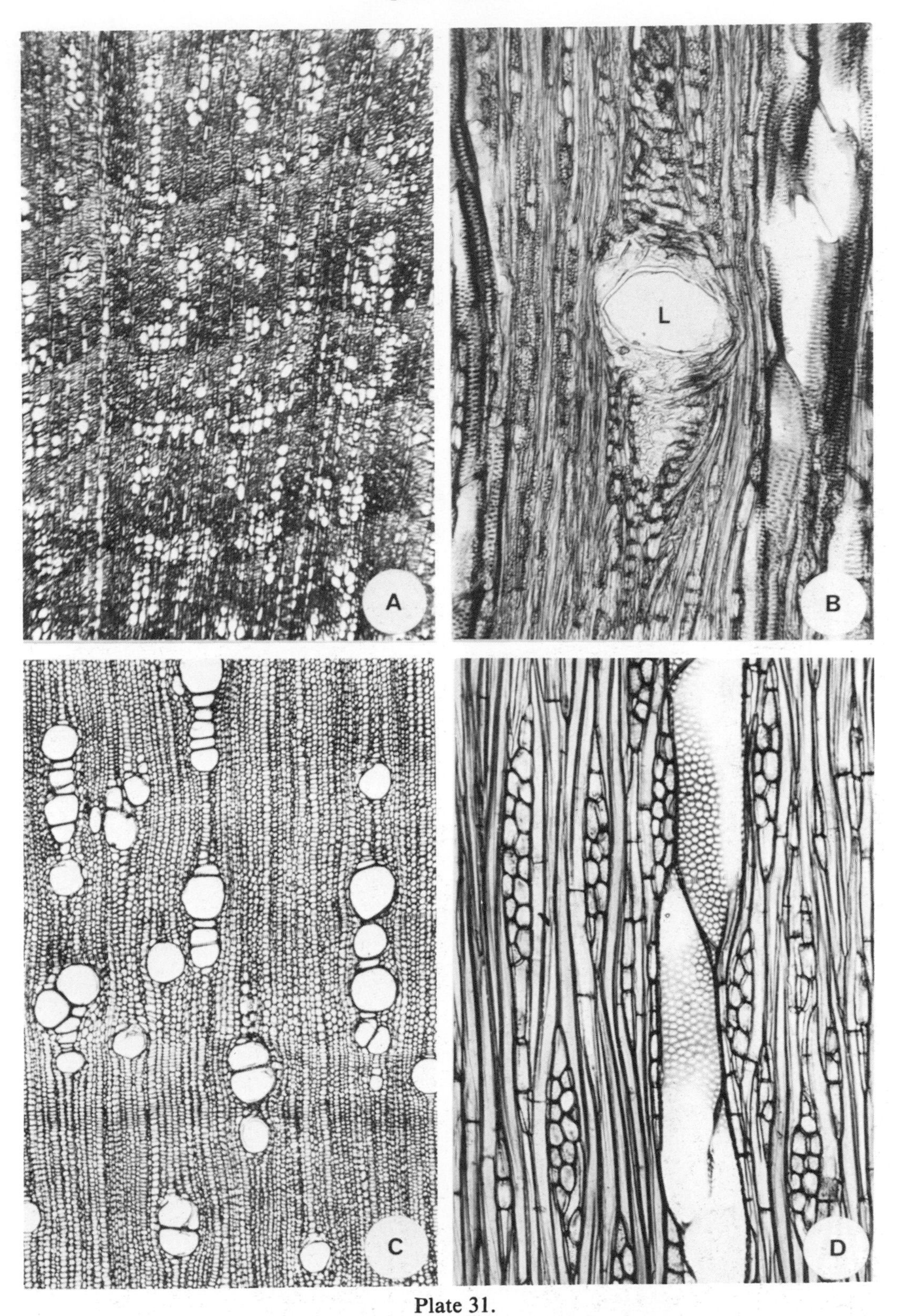

Plate 31.
A and B — *Euphorbia hierosolymitana*; A, ×42; B, showing a laticifer (L) in a ray, ×126.
C and D — *Ricinus communis*; C, ×48; D, ×120

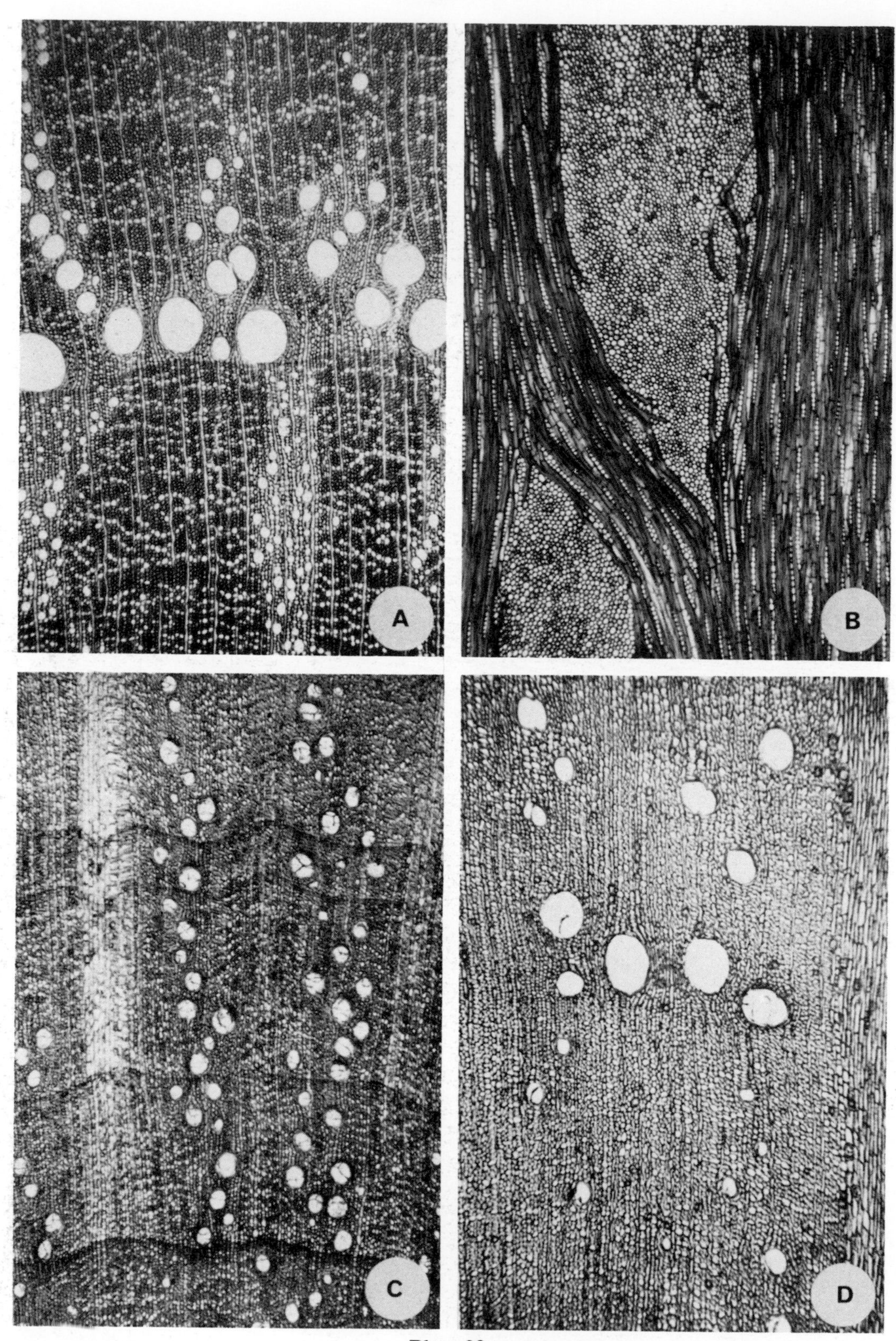

Plate 32.
Quercus, ×42. A and B — *Q. boissieri*. C — *Q. calliprinos*. D — *Q. ithaburensis*

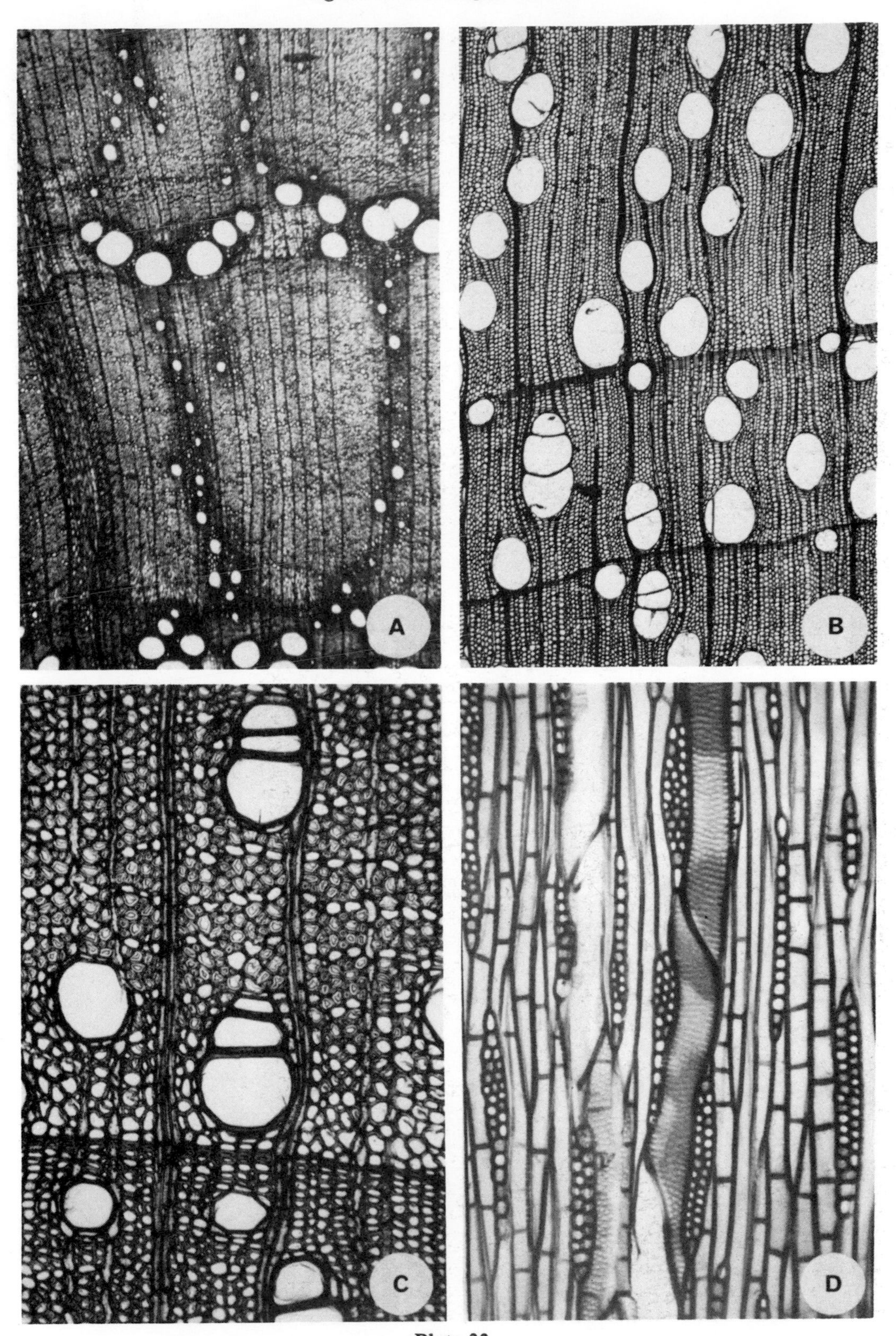

Plate 33.

A — Quercus libani, ×42. B–D — *Juglans regia*; B, ×30; C and D, ×126

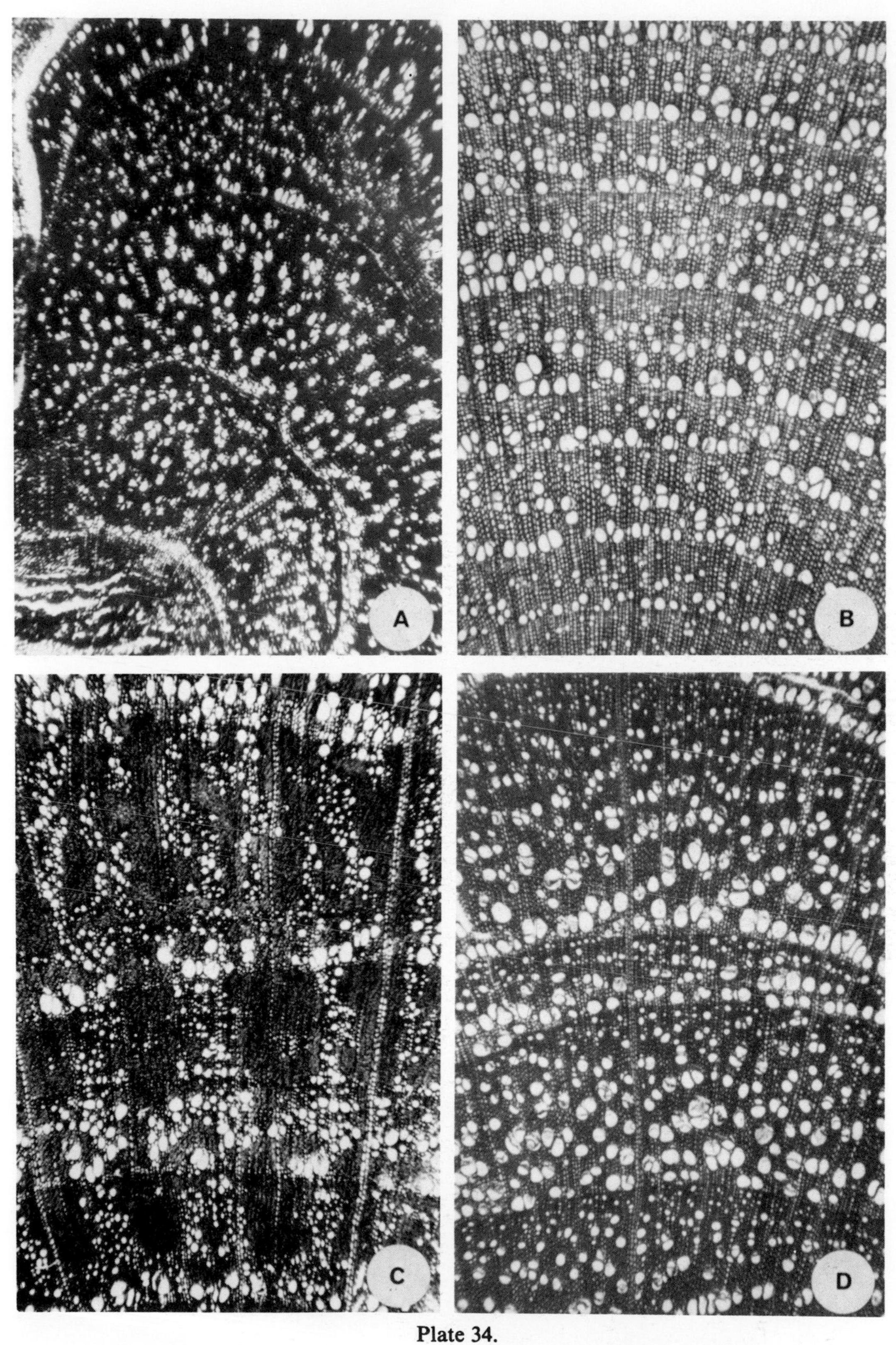

Plate 34.
A — *Coridothymus capitatus*, ×42. B — *Prasium majus*, ×42.
C — *Rosmarinus officinalis*, ×42. D — *Salvia fruticosa*, ×42

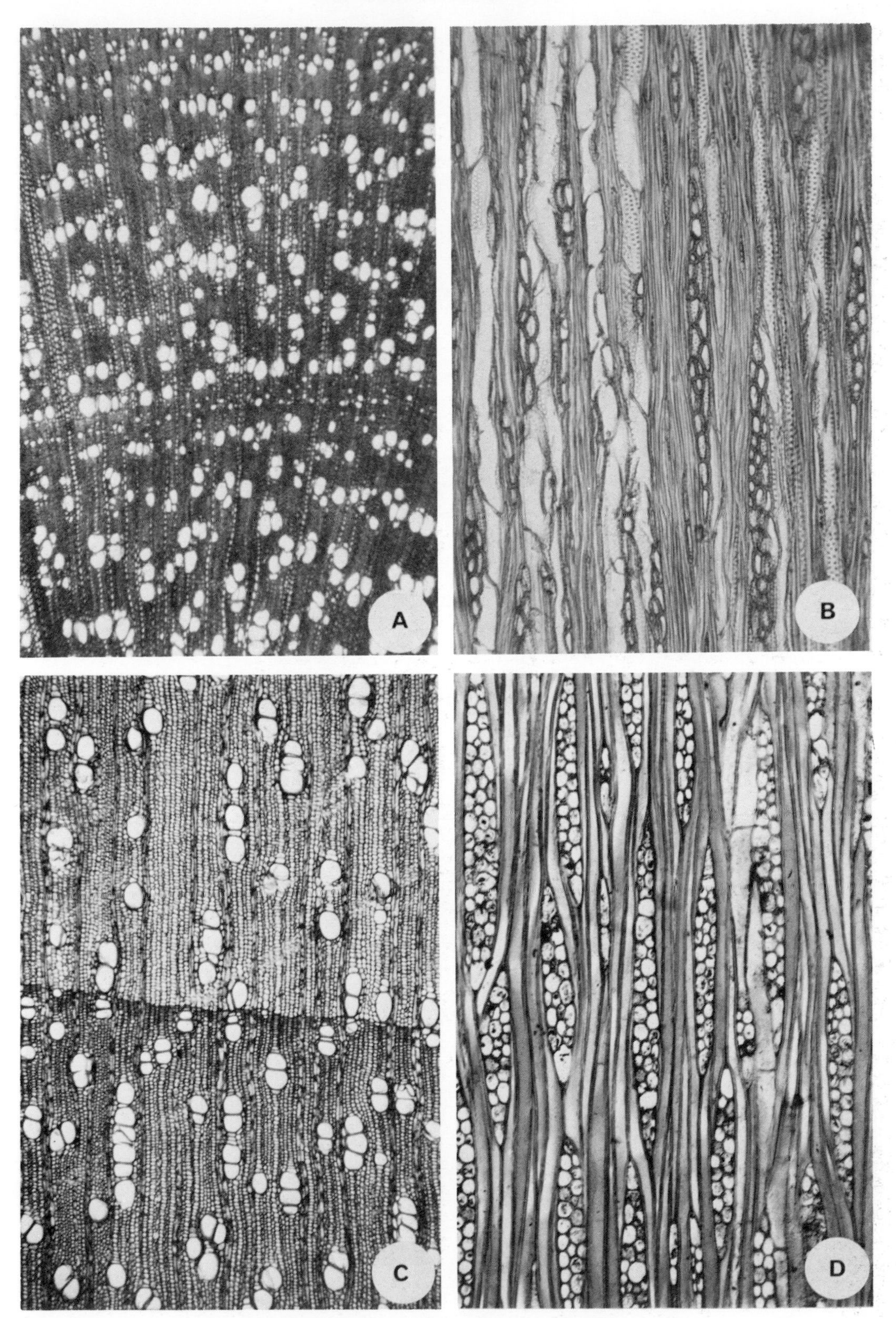

Plate 35.
A and B — *Teucrium creticum*; A, ×42; B, ×126.
C and D — *Laurus nobilis*; C, ×42; D, ×126

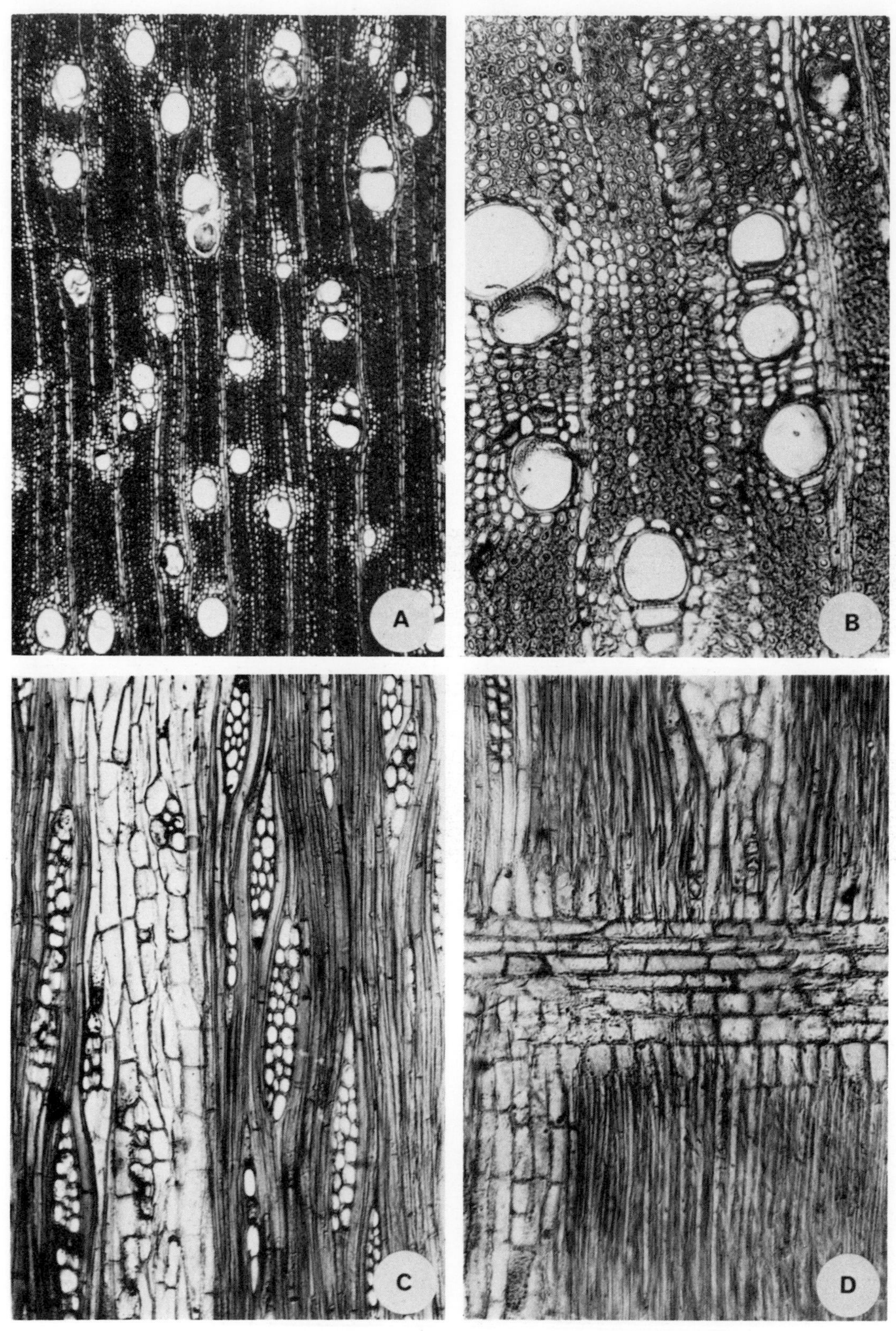

Plate 36.
Ceratonia siliqua. A, ×42; B–D, ×126

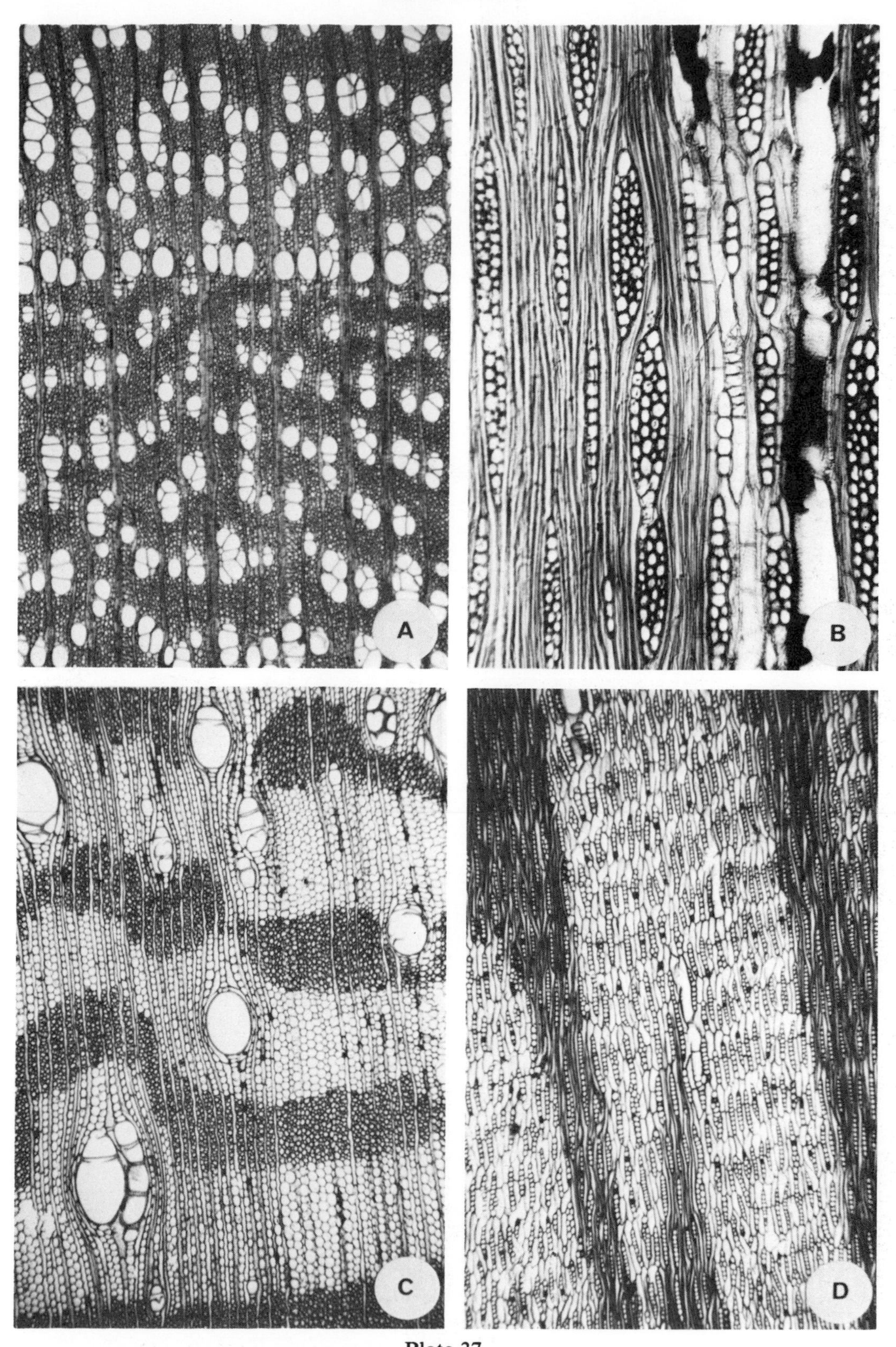

Plate 37.
A and B — *Cercis siliquastrum*; A, ×42; B, ×126. C and D — *Acacia albida*, ×42

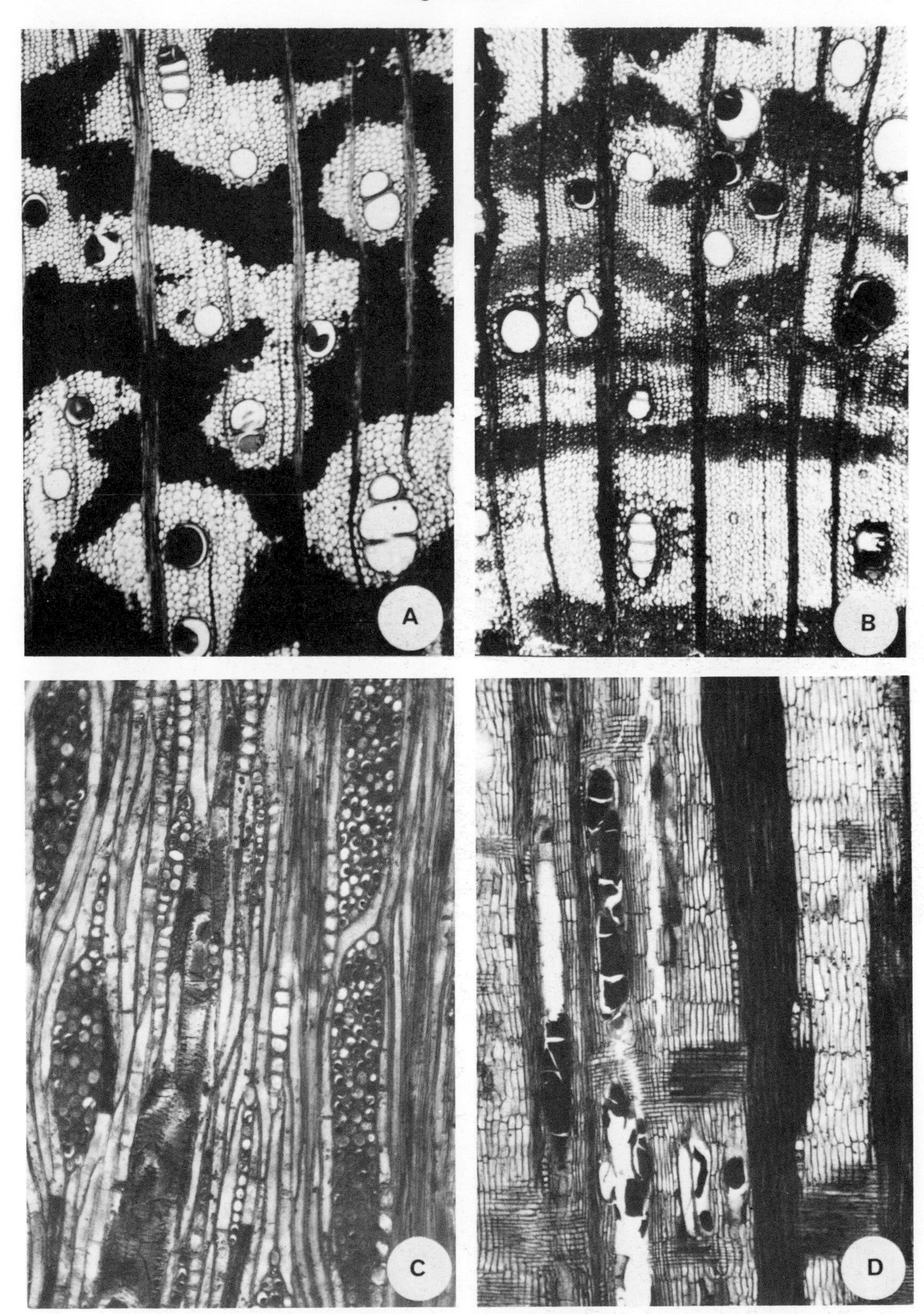

Plate 38.
Acacia gerrardii subsp. *negevensis*; A and B, ×42; C, ×126; D, ×42

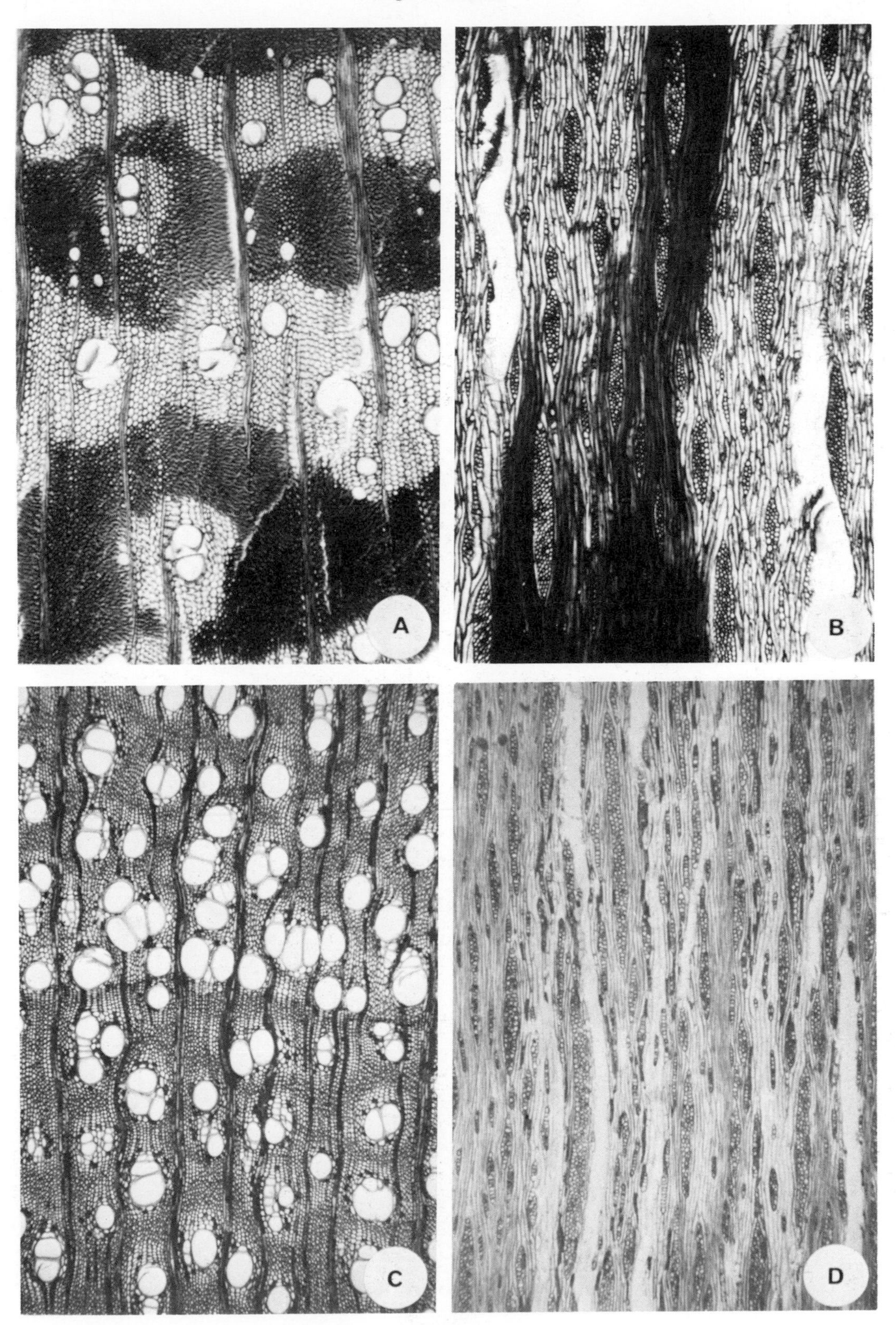

Plate 39.
A and B — *Acacia raddiana*, ×42. C and D — *Prosopis farcta*, ×42

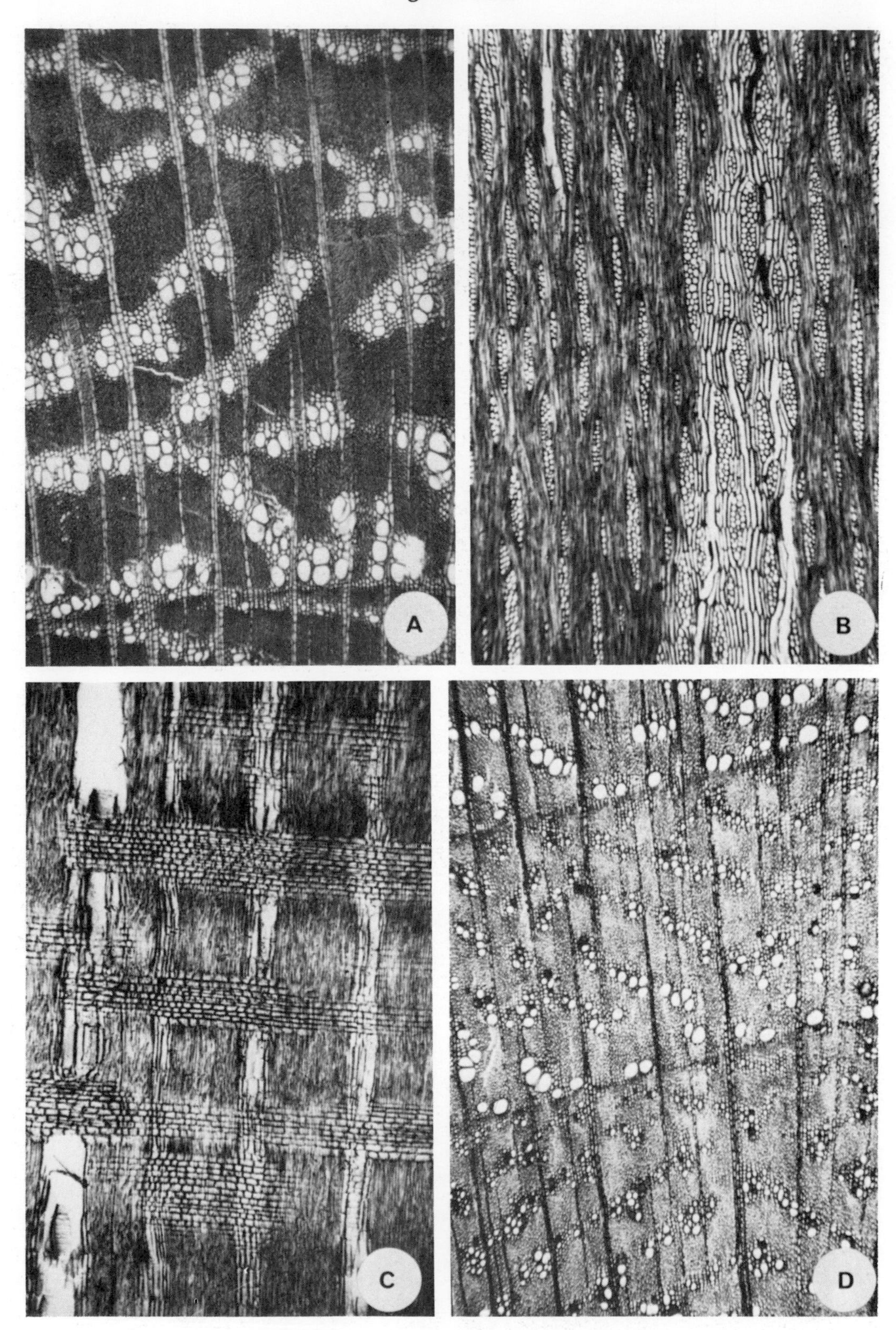

Plate 40.
A–C — *Anagyris foetida*, ×42. D — *Calycotome villosa*, ×42

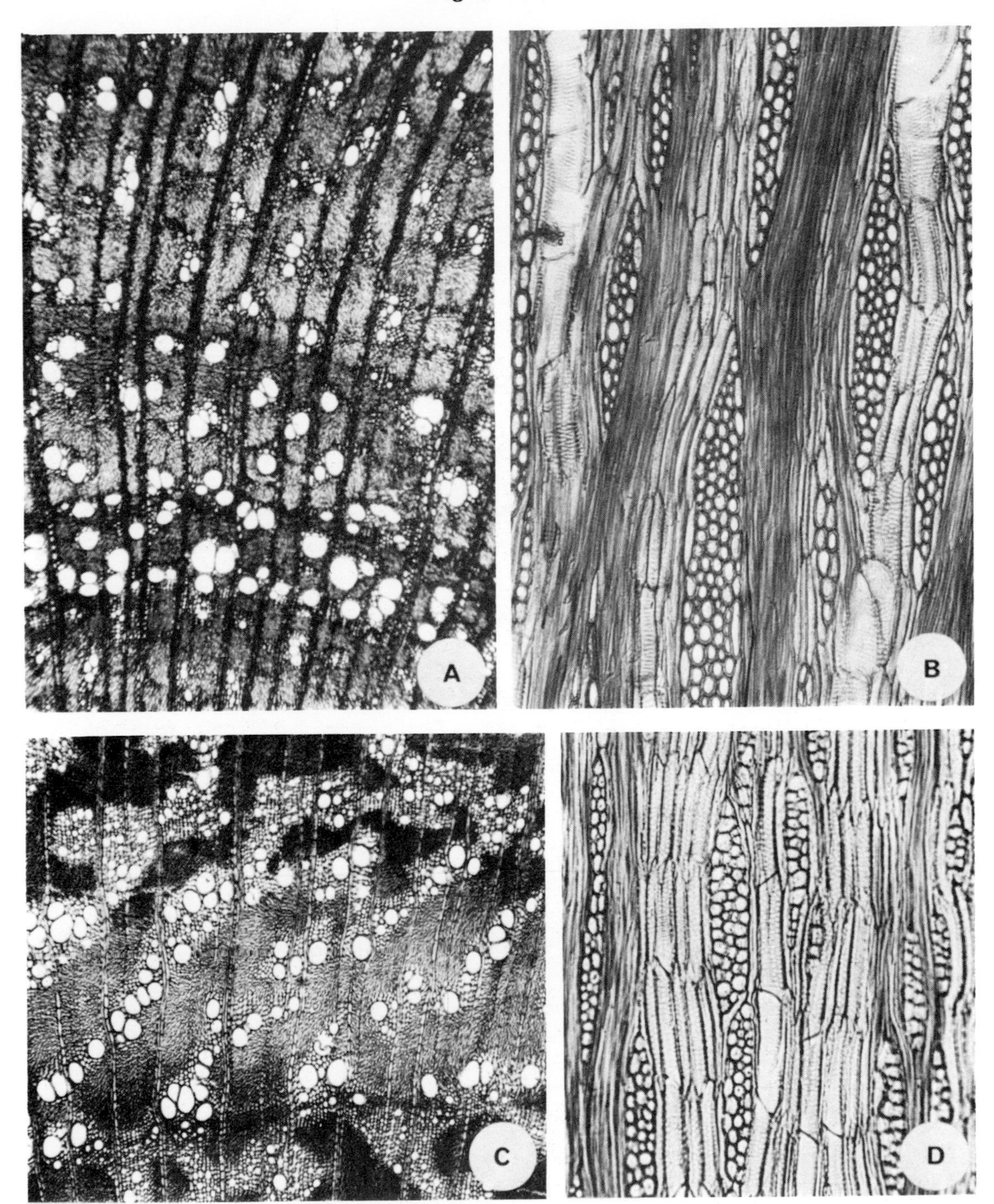

Plate 41.
A and B — *Colutea cilicica*; A, ×42; B, ×126. C and D — *Genista fasselata*; C, ×42; D, ×126

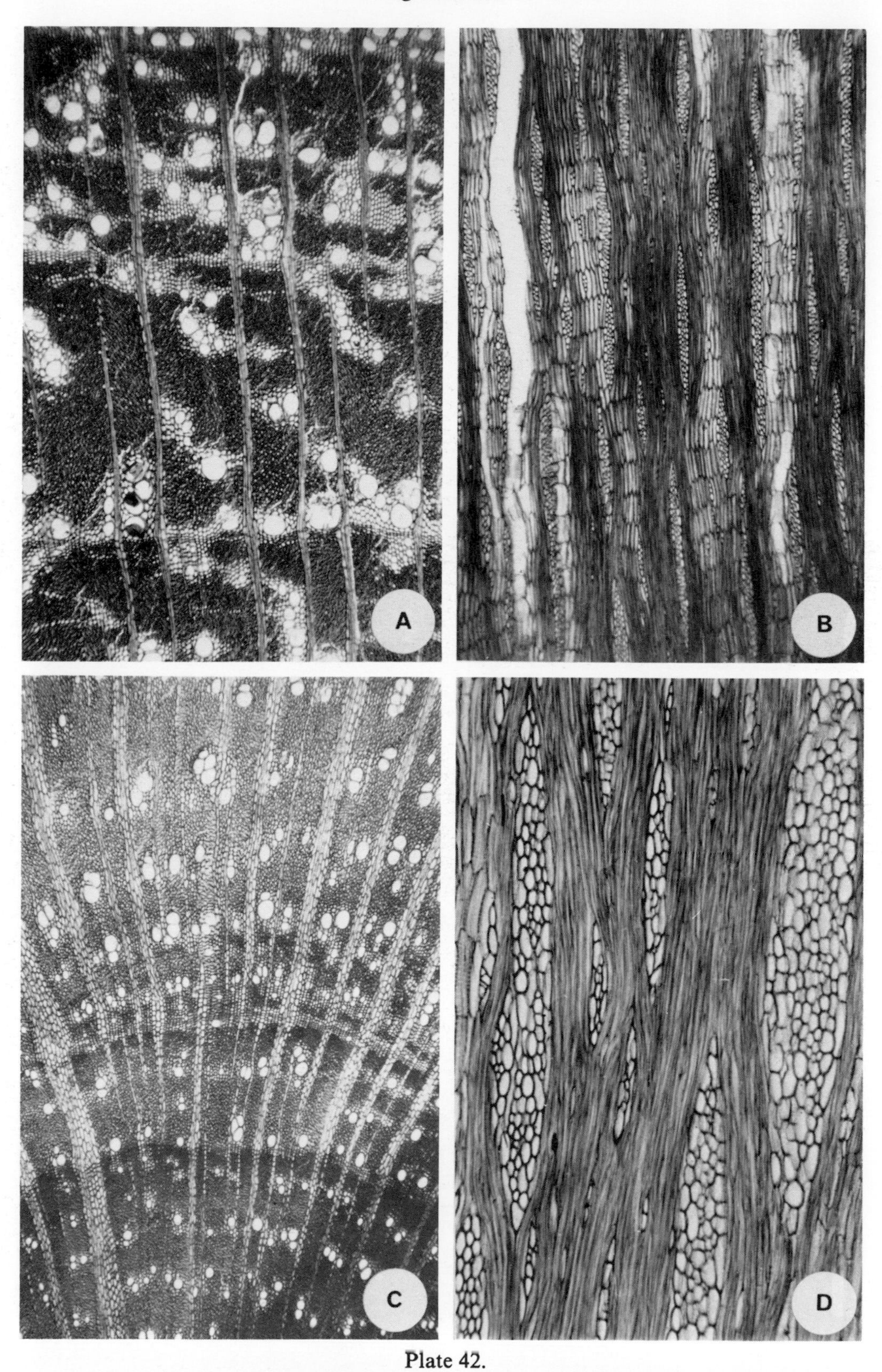

Plate 42.
A and B — *Gonocytisus pterocladus*, ×42. C and D — *Ononis natrix*; C, ×42; D, ×126

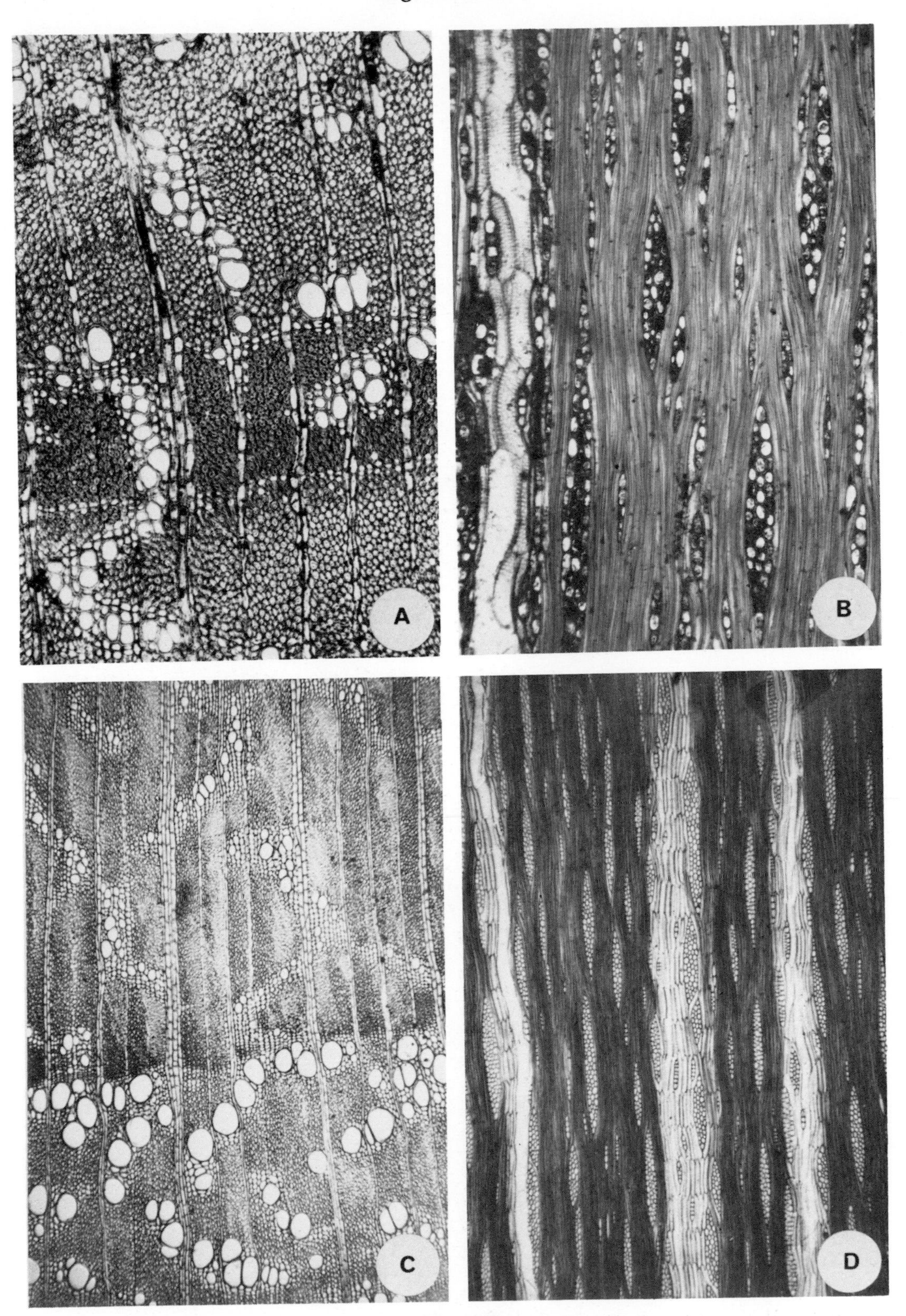

Plate 43.
A and B — *Retama raetam*, ×126. C and D — *Spartium junceum*, ×42

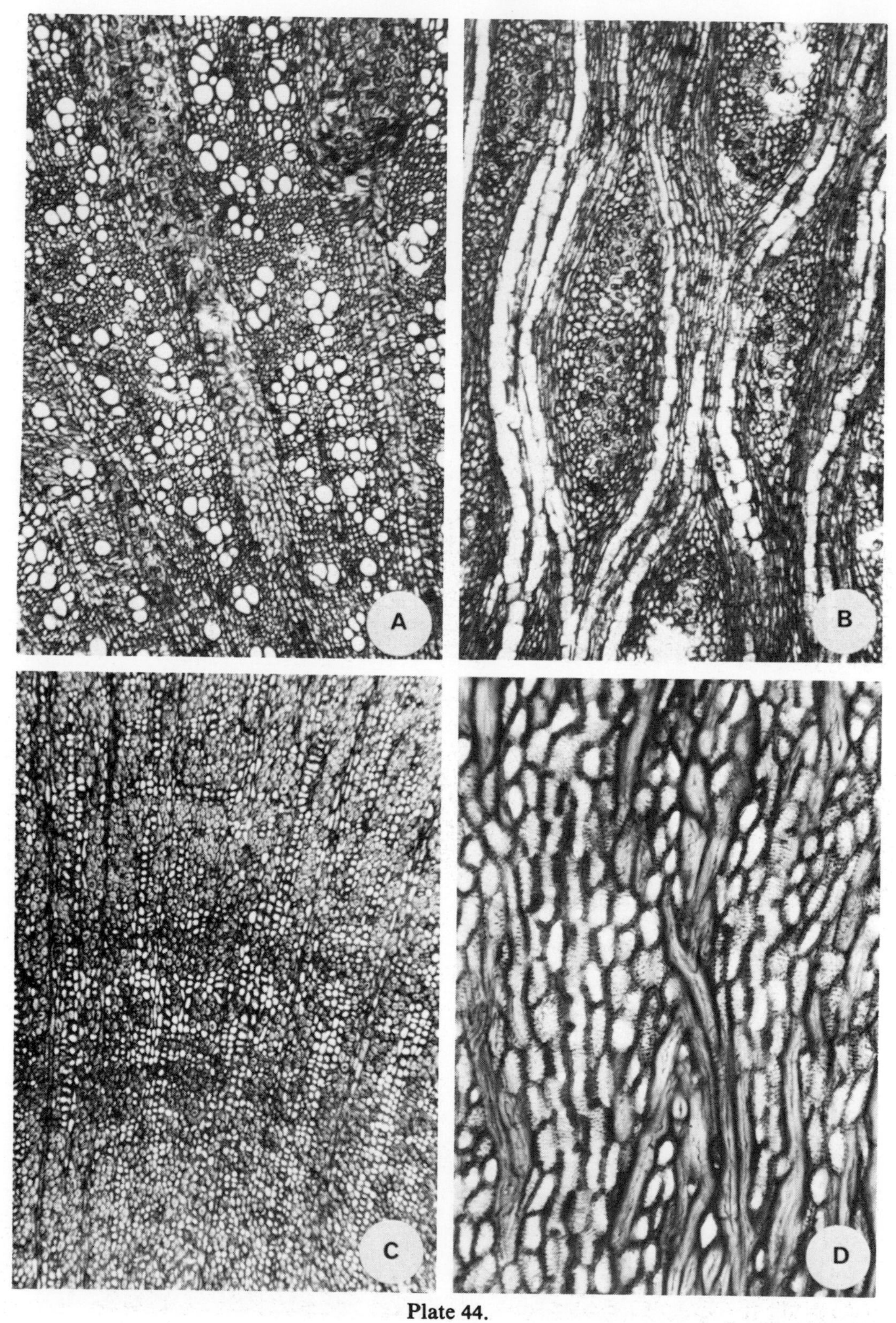

Plate 44.
A and B — *Loranthus acaciae*, ×42. C and D — *Viscum cruciatum*; C, ×42; D, ×126

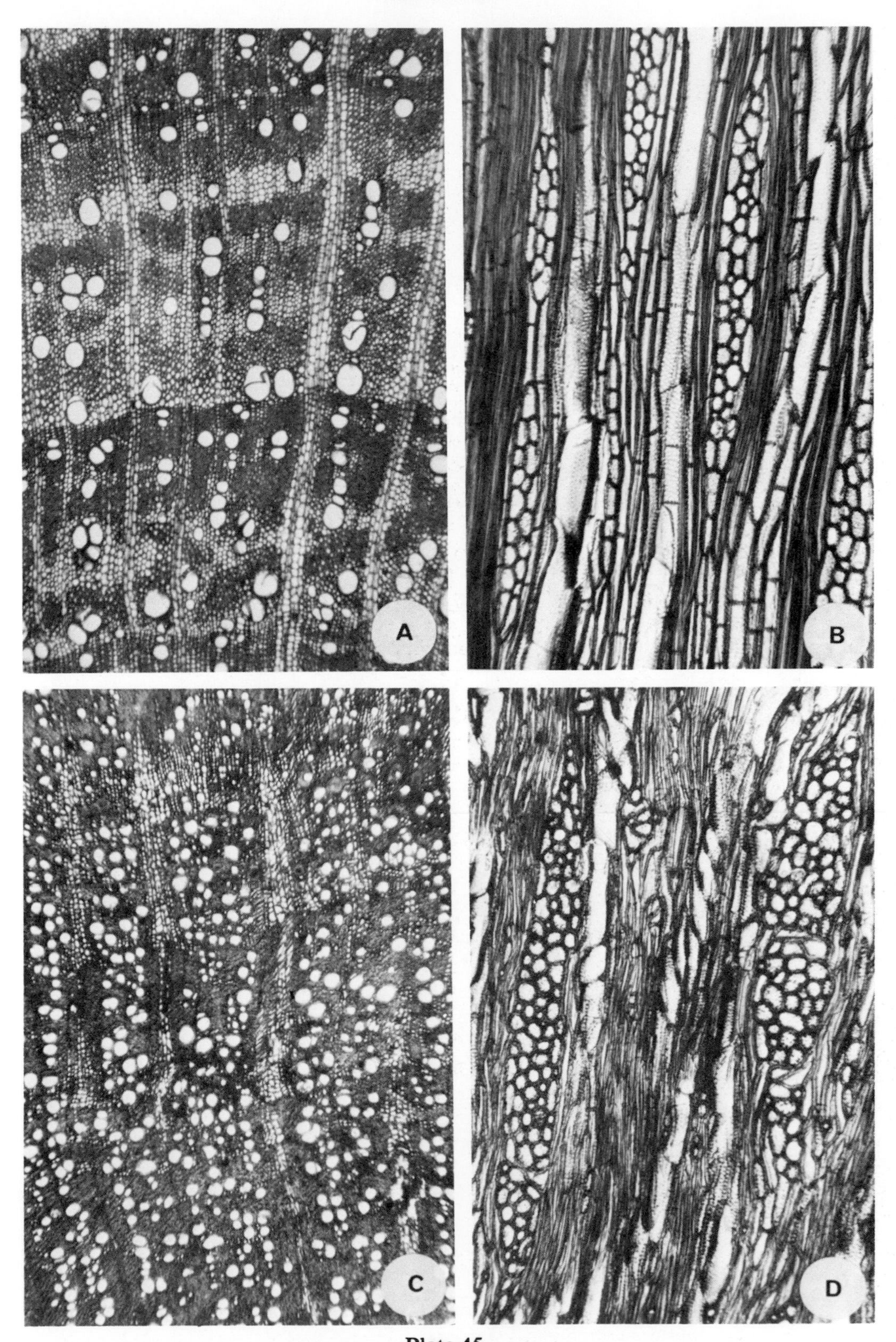

Plate 45.
A and B — *Abutilon pannosum*; A, ×42; B, ×126. C and D — *Hibiscus micranthus*; C, ×42; D, ×126

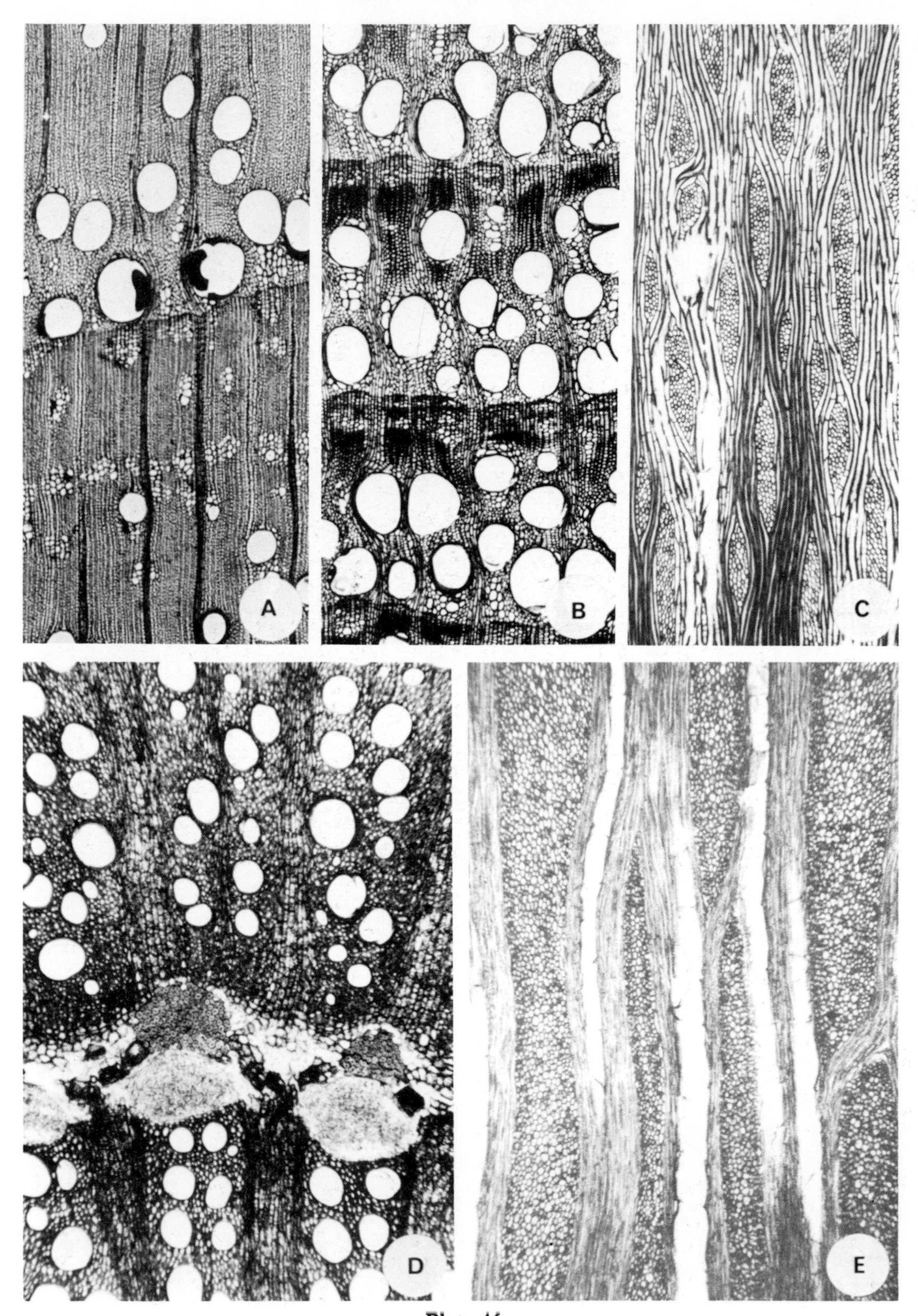

Plate 46.
A–C — *Melia azedarach*; A and B, ×30; C, ×42. D and E — *Cocculus pendulus*, ×42

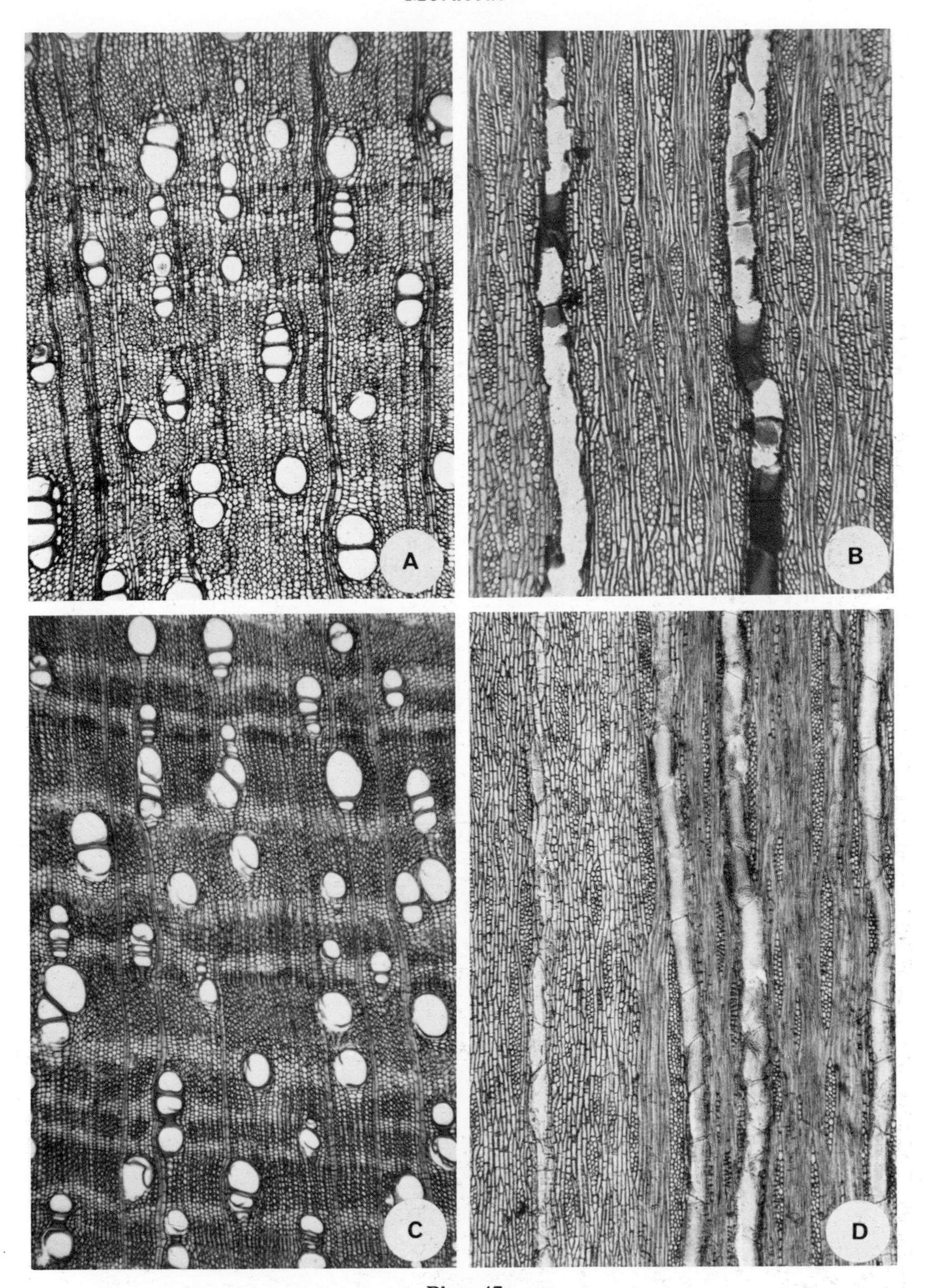

Plate 47.
Ficus, ×42. A and B — *F. carica.* C and D — *F. pseudo-sycomorus*

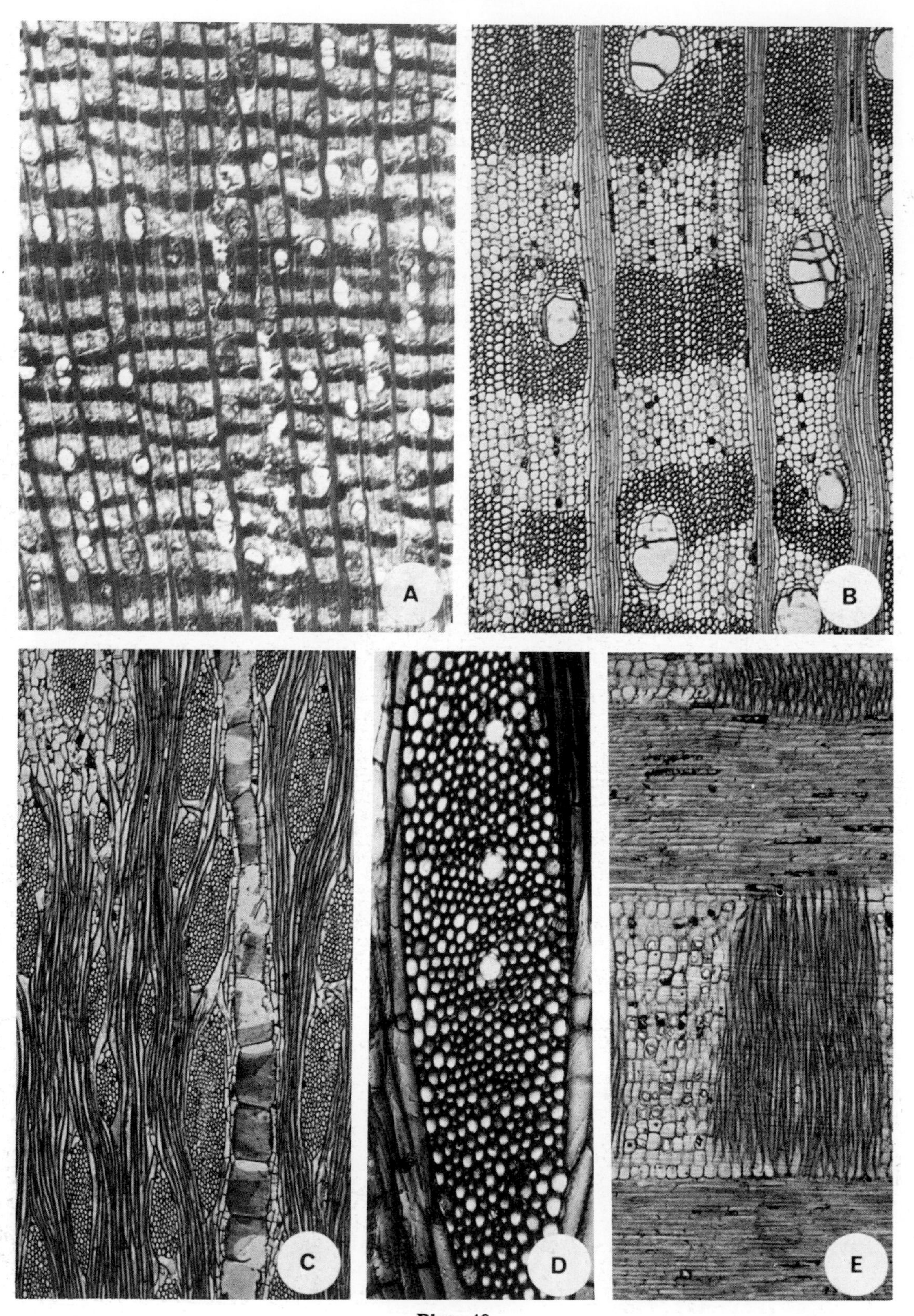

Plate 48.
Ficus sycomorus. A, ×24; B and C, ×48; D, ×300; E, ×48

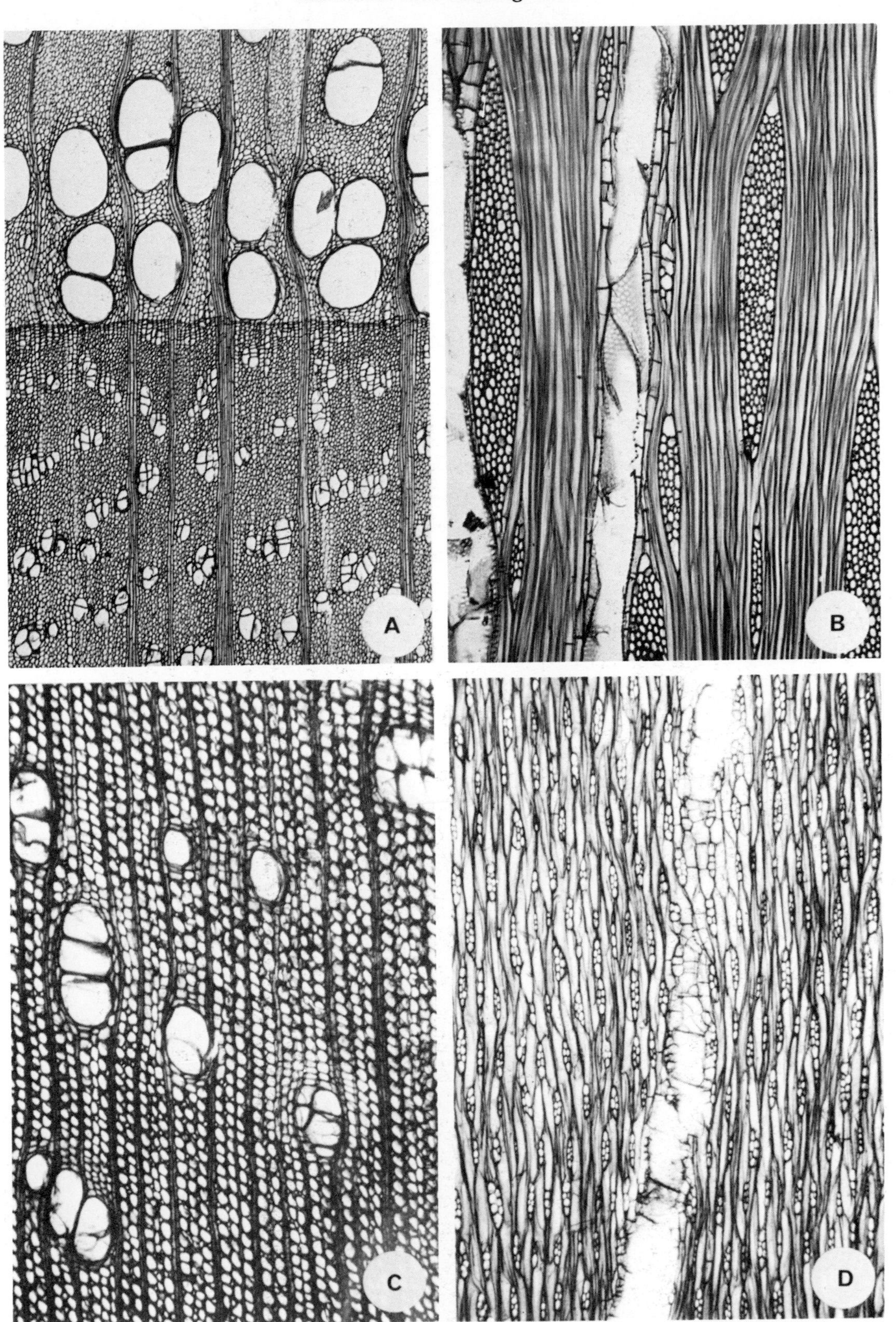

Plate 49.
A and B — *Morus alba*; A, ×48; B, ×120. C and D — *Moringa peregrina*, ×42

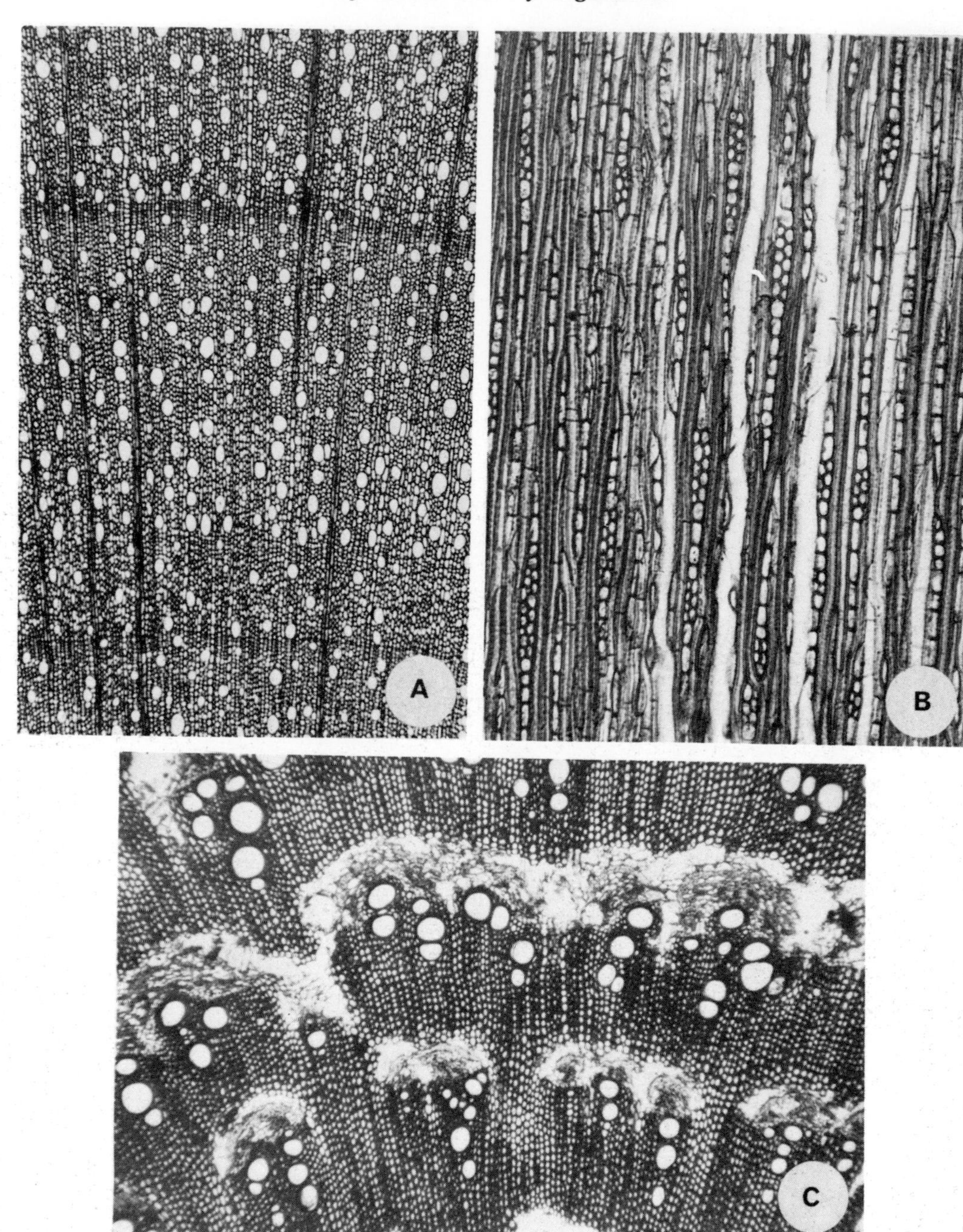

Plate 50.
A and B — *Myrtus communis*; A, ×48; B, ×126. C — *Commicarpus africanus*, ×42

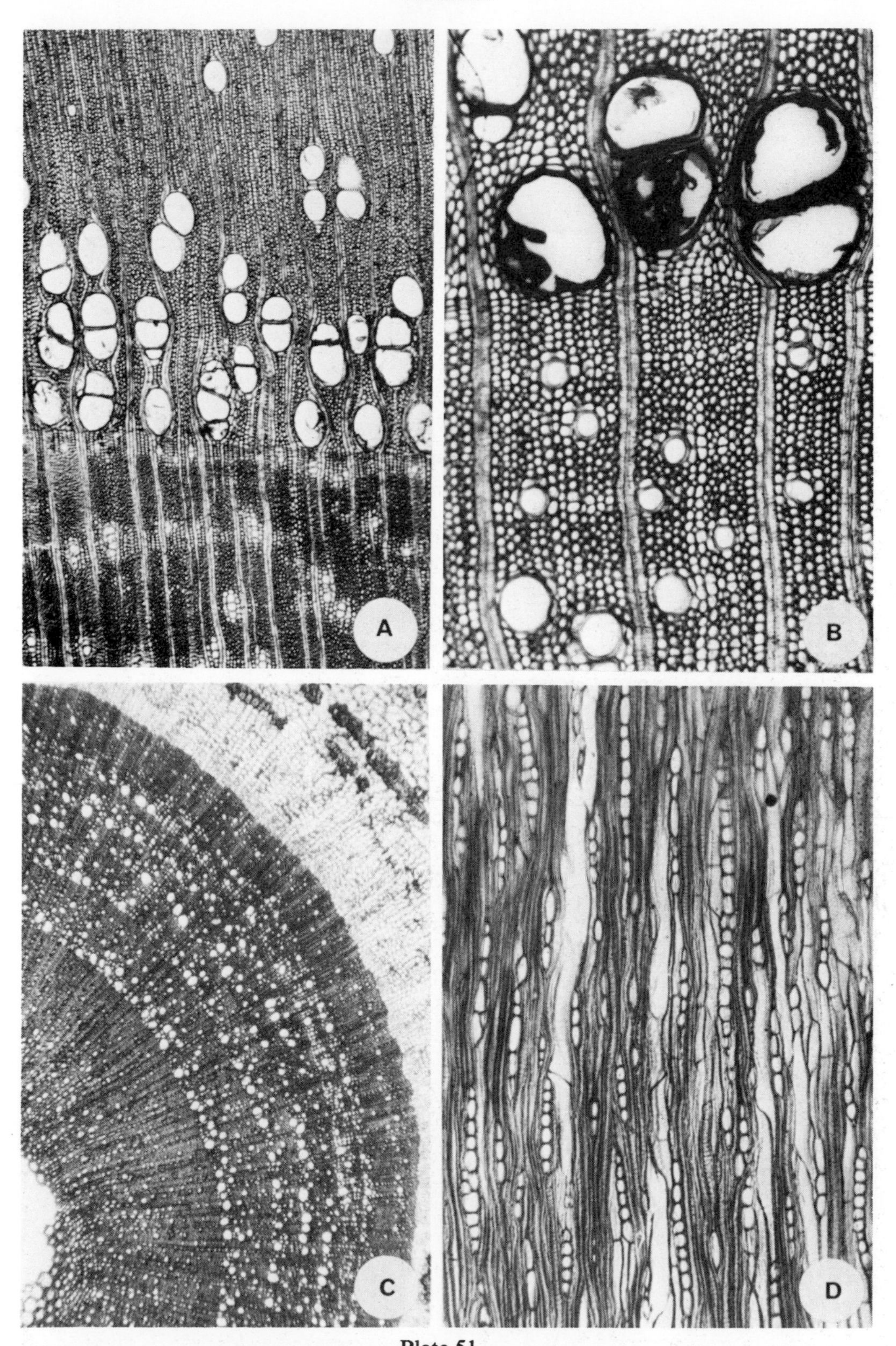

Plate 51.
A and B — *Fraxinus syriaca*; A, ×30; B, ×126. C and D — *Jasminum fruticans*; C, ×42; D, ×126

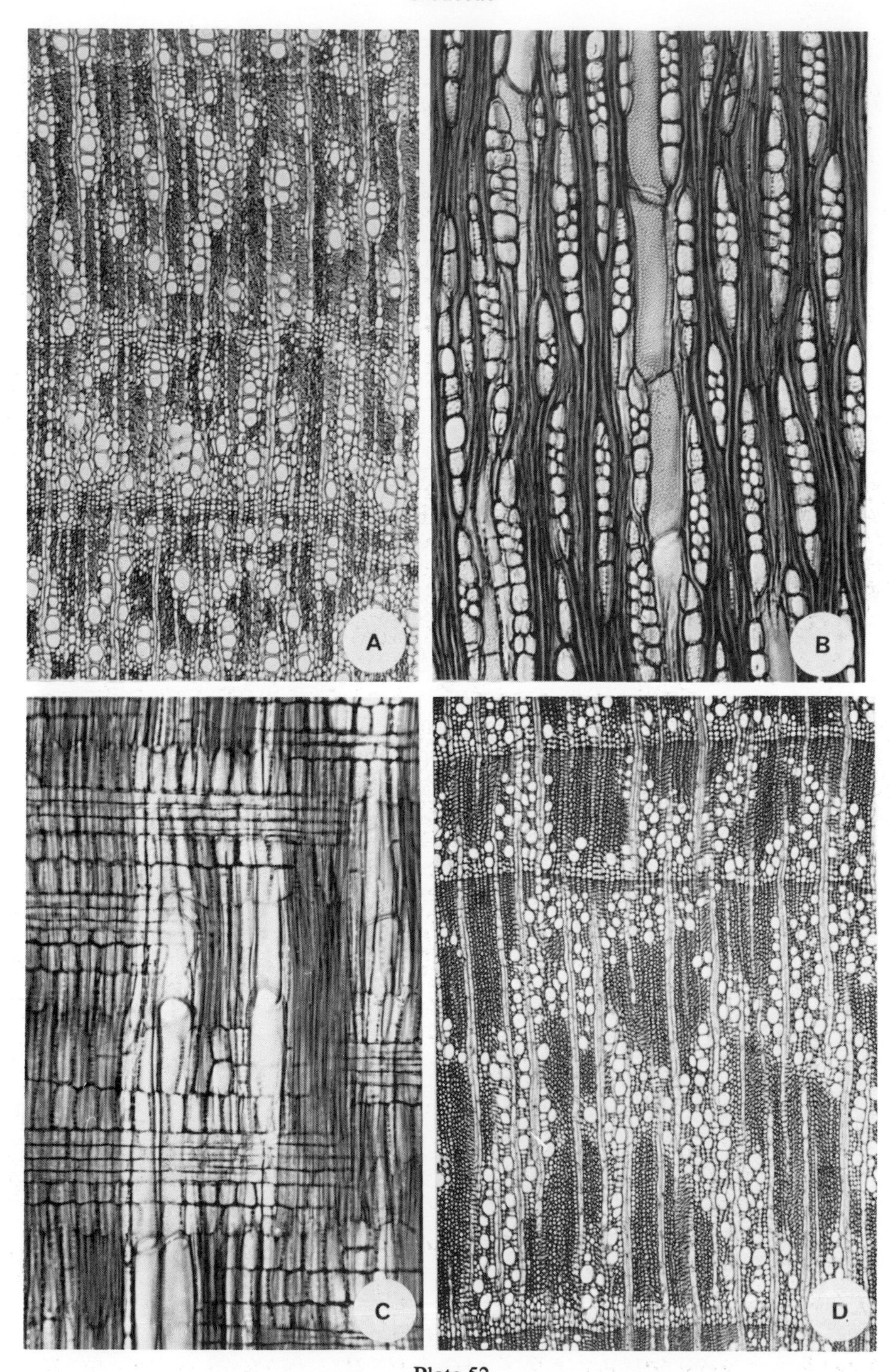

Plate 52.
A–C — *Olea europaea*; A, ×48; B and C, ×120. D — *Phillyrea latifolia*, ×48

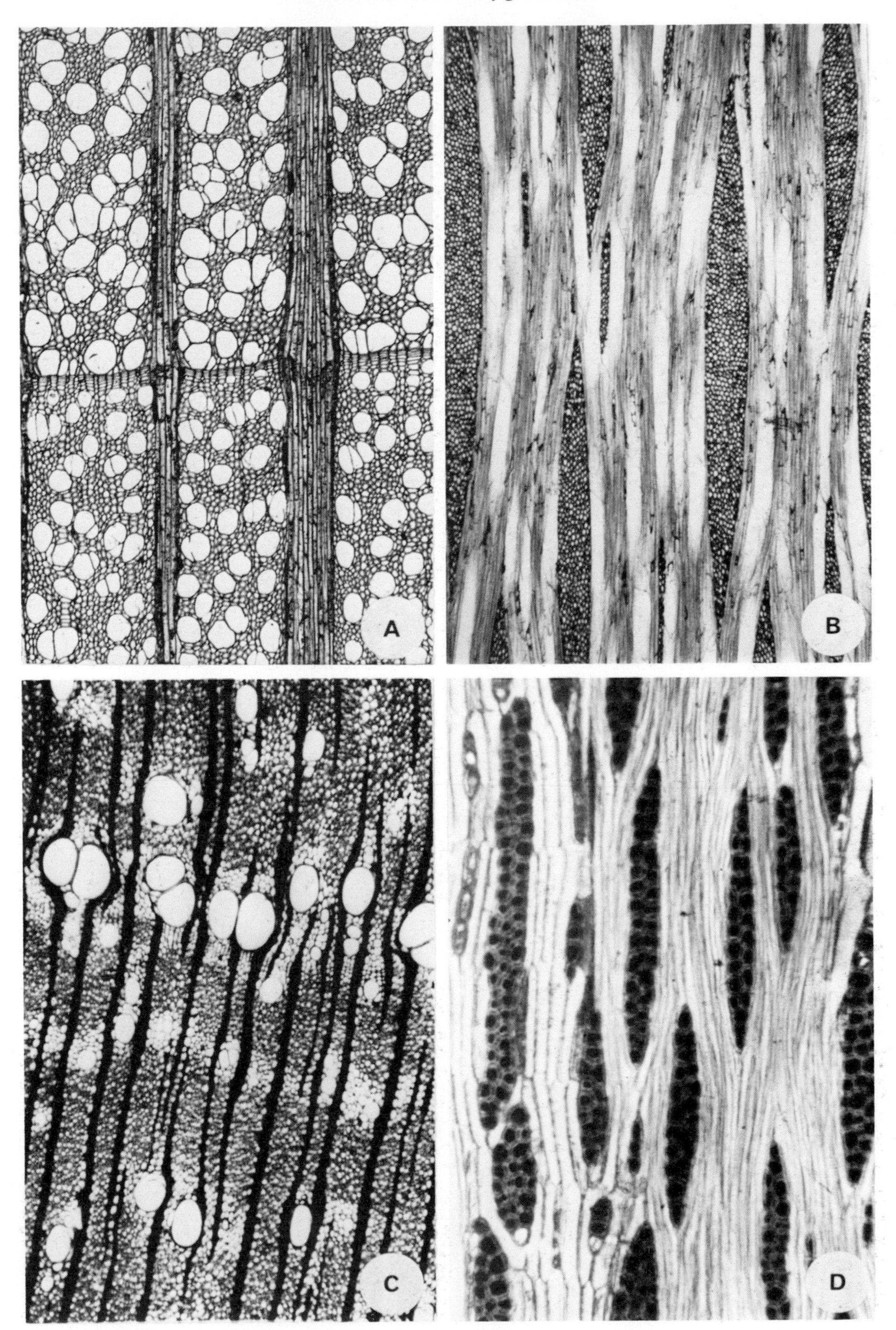

Plate 53.
A and B — *Platanus orientalis*; A, ×48; B, ×42. C and D — *Calligonum comosum*; C, ×42. D, ×126

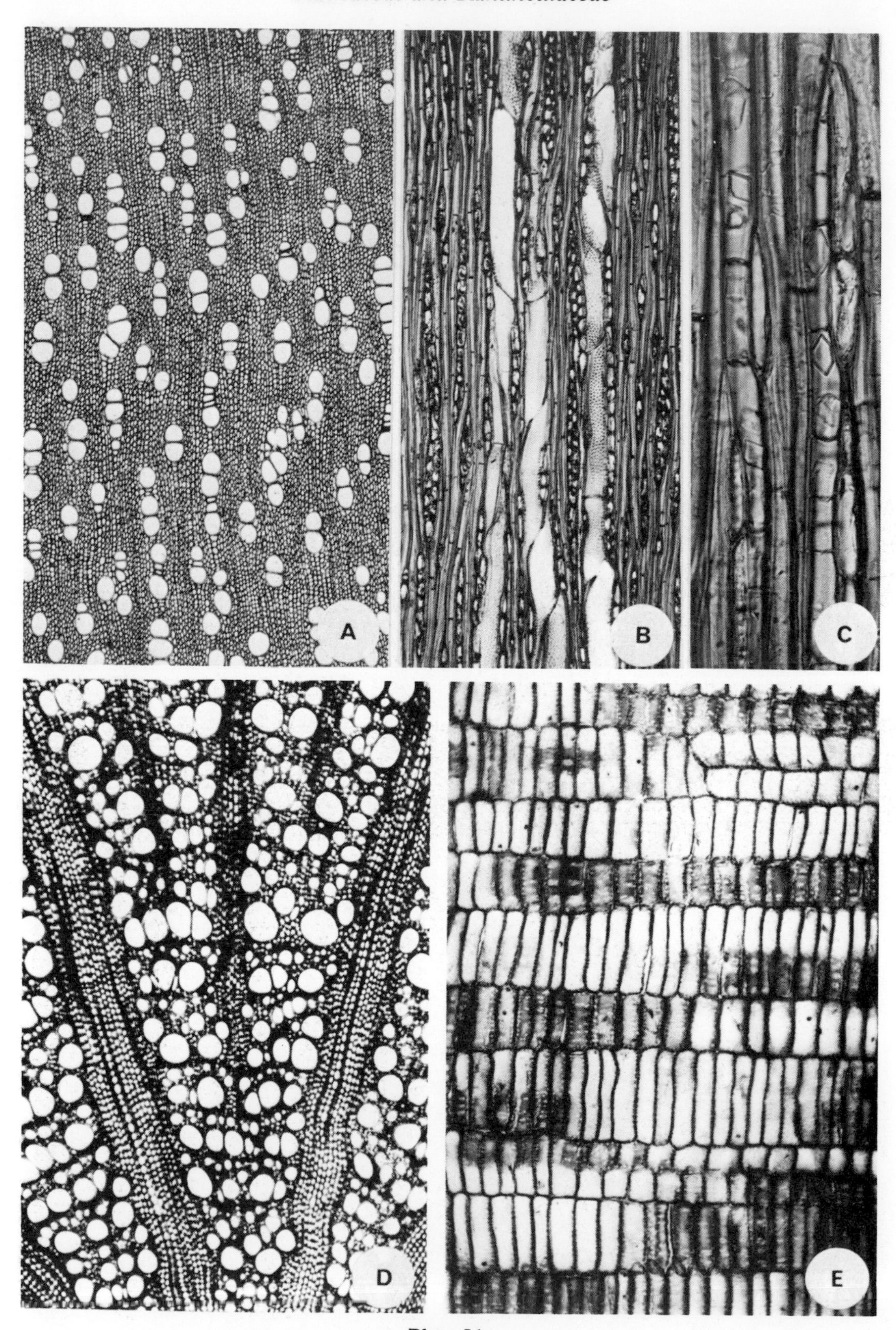

Plate 54.
A–C — *Punica granatum*; A, ×36; B, ×120; C, ×430. D and E — *Clematis cirrhosa*; D, ×42; E, ×126

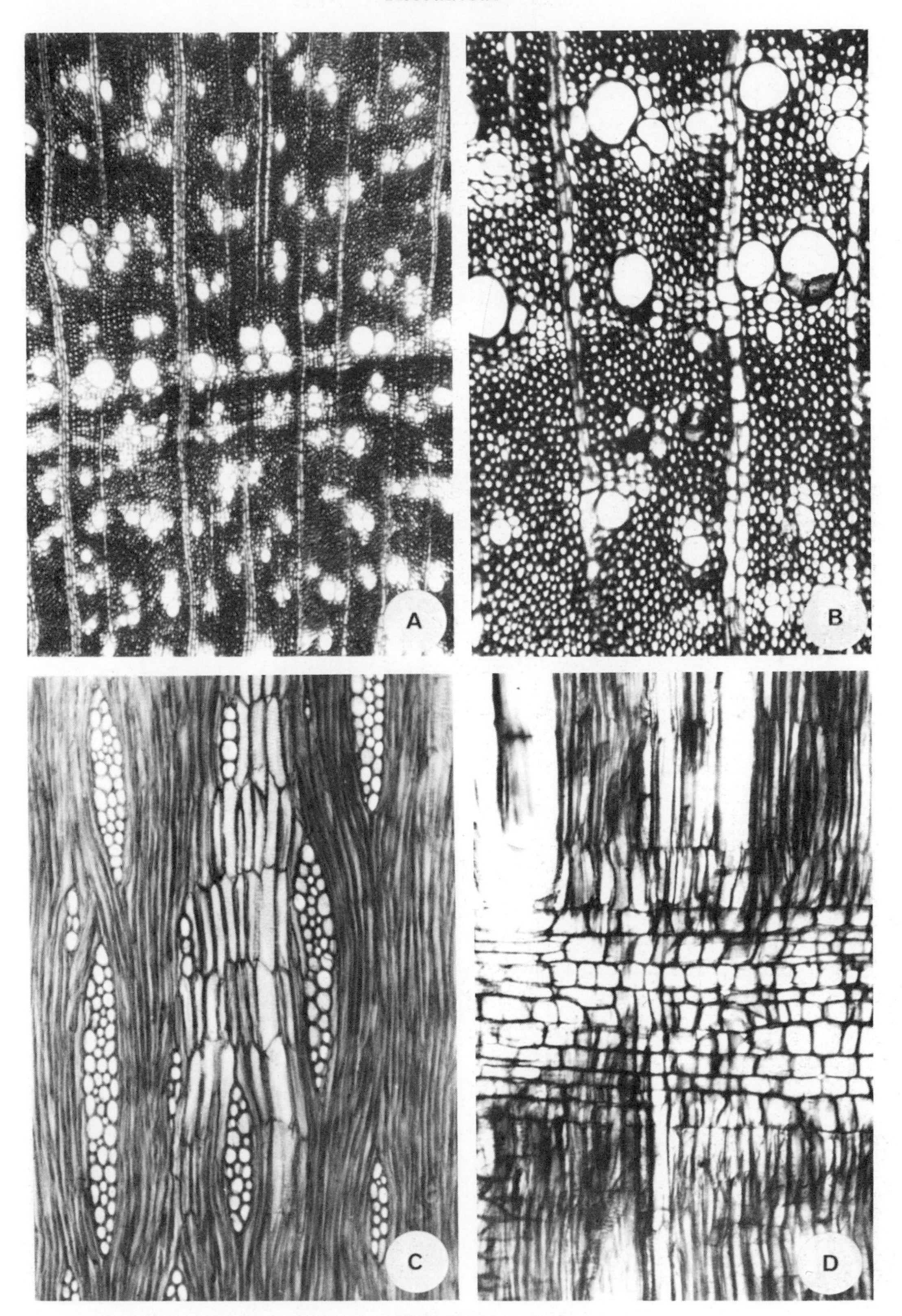

Plate 55.
Ochradenus baccatus; A, ×42; B, ×126; C and D, ×120

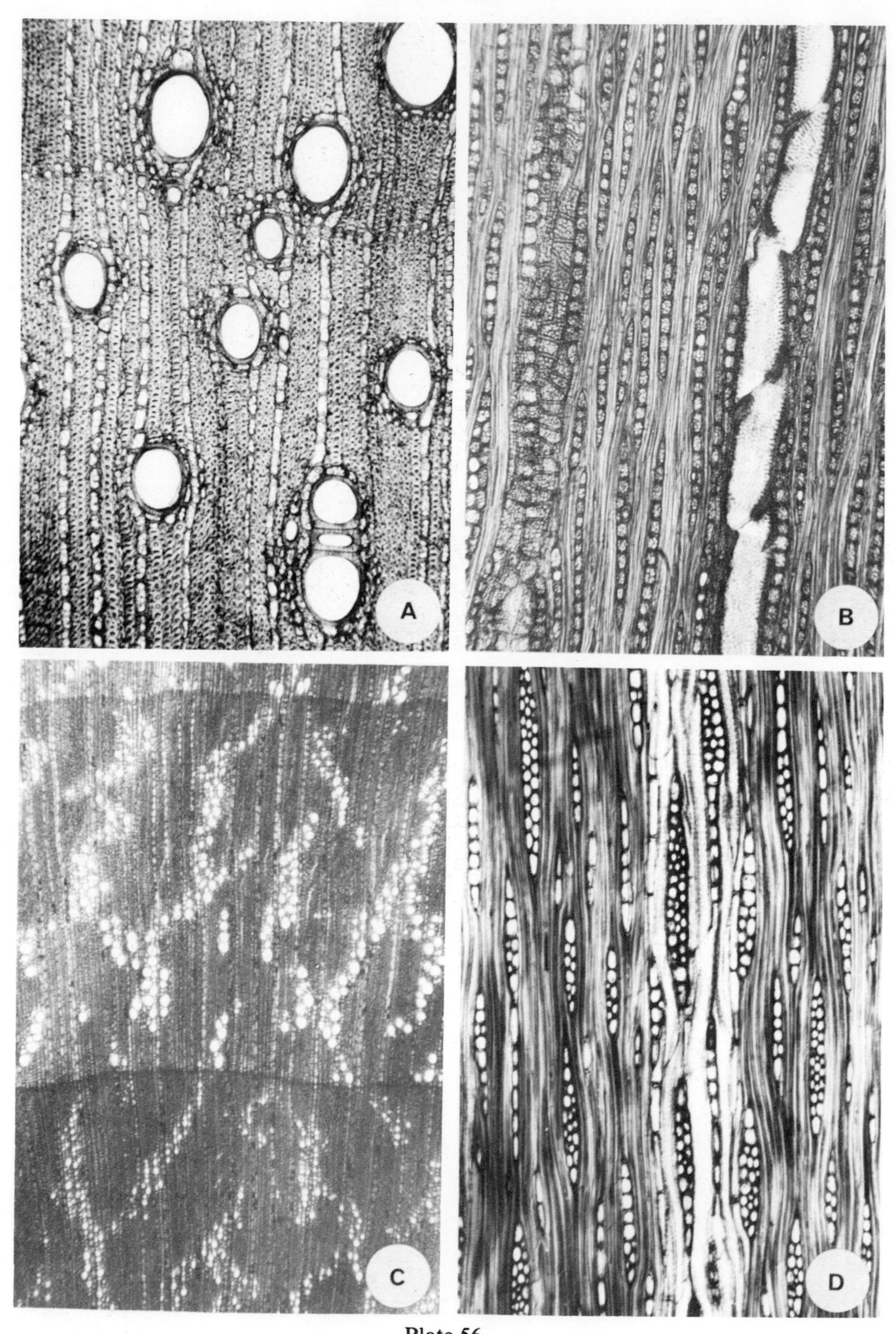

Plate 56.
A and B — *Paliurus spina-christi*, ×126. C and D — *Rhamnus alaternus*; C, ×42; D, ×126

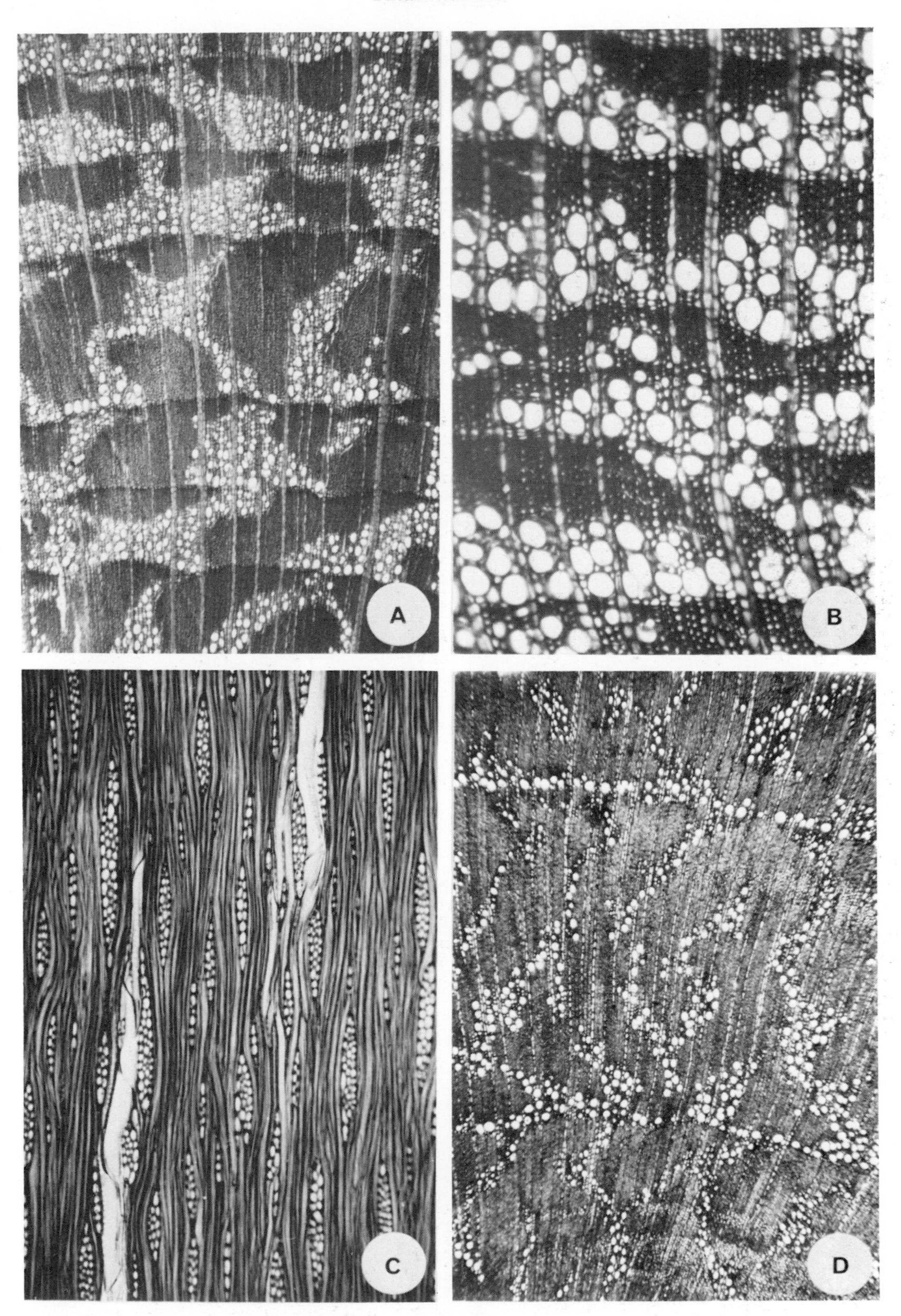

Plate 57.
Rhamnus. A — *R. dispermus*, ×42. B and C — *R. palaestinus*, ×126.
D — *R. punctatus*, ×42

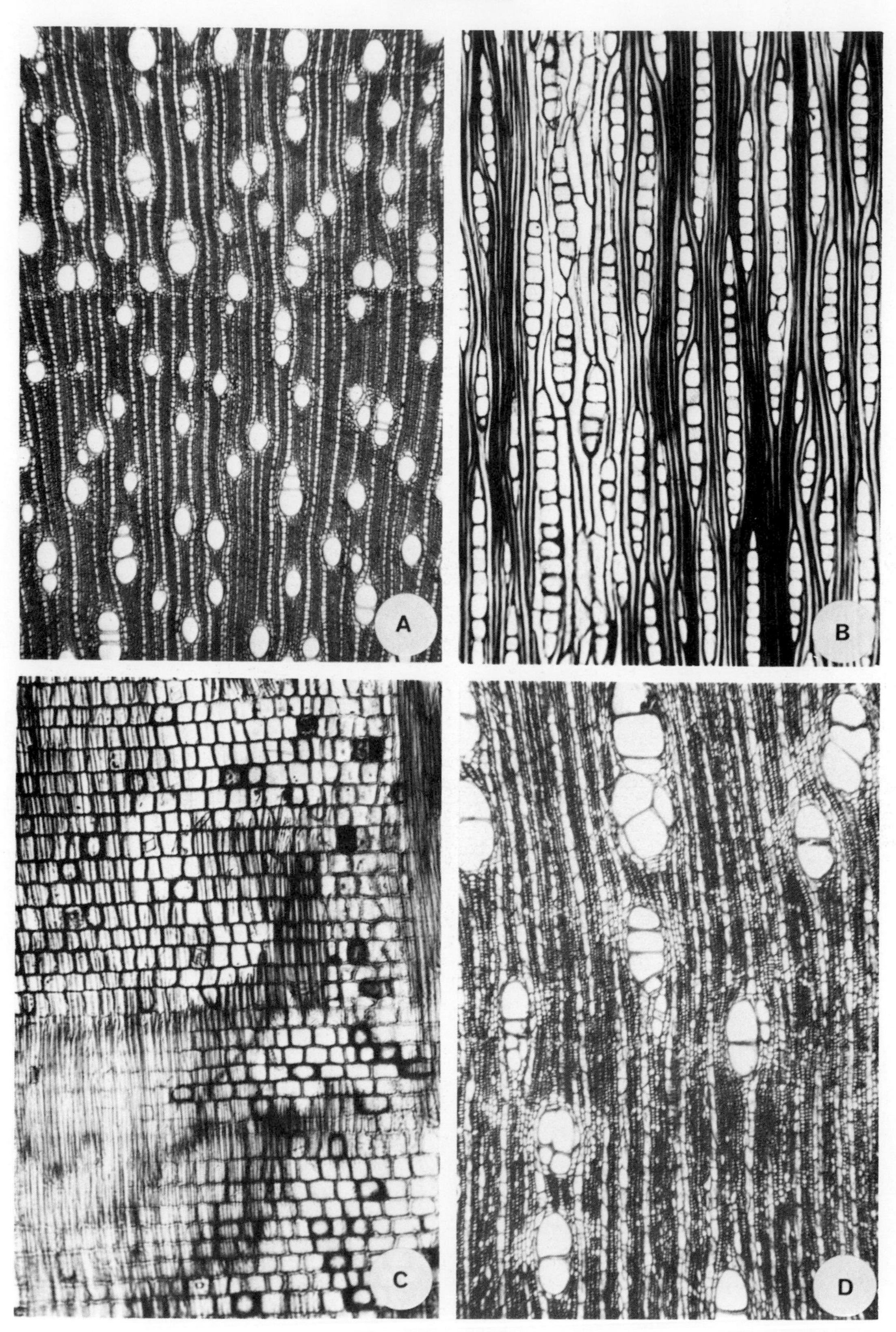

Plate 58.
Ziziphus. A–C — *Z. lotus*; A, ×42; B and C, ×126. D — *Z. spina-christi*, ×42

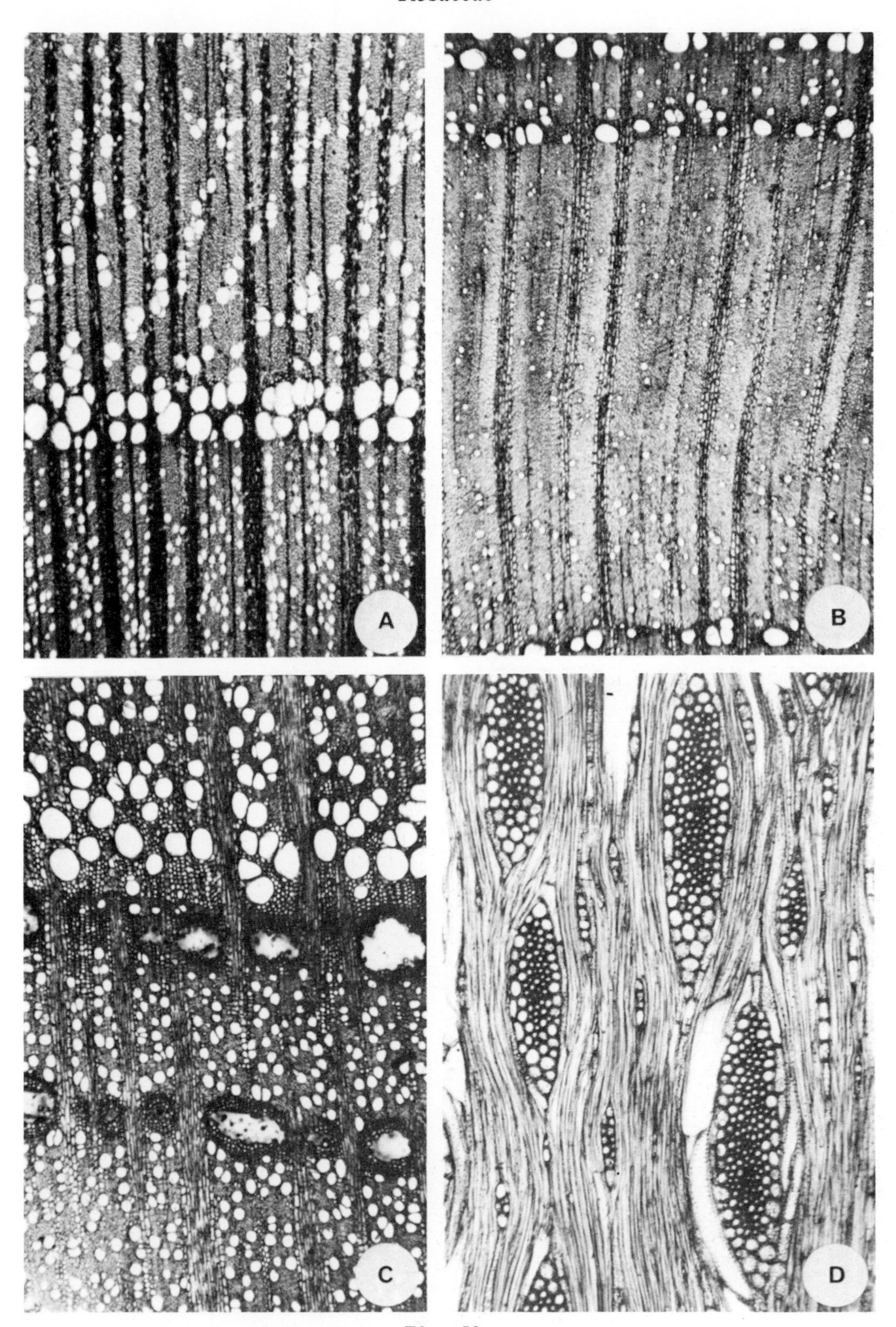

Plate 59.
A and B — *Amygdalus*, ×42; A — *A. communis*; B — *A. korschinskii.*
C and D — *Armeniaca vulgaris*; C, ×42; D, ×126

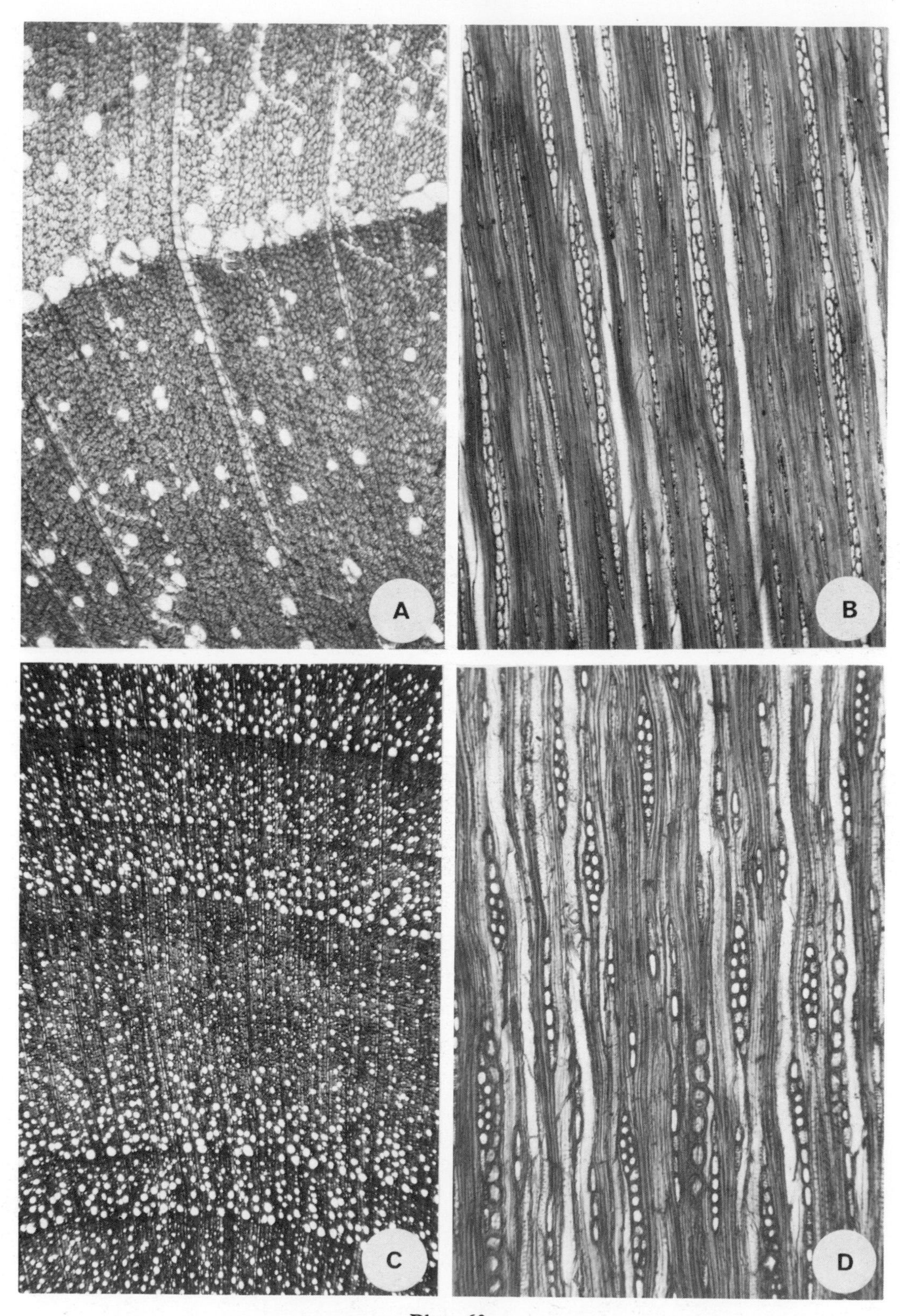

Plate 60.
A and B — *Cerasus prostrata*, × 126. C and D — *Cotoneaster nummularia*; C, × 42; D, × 126

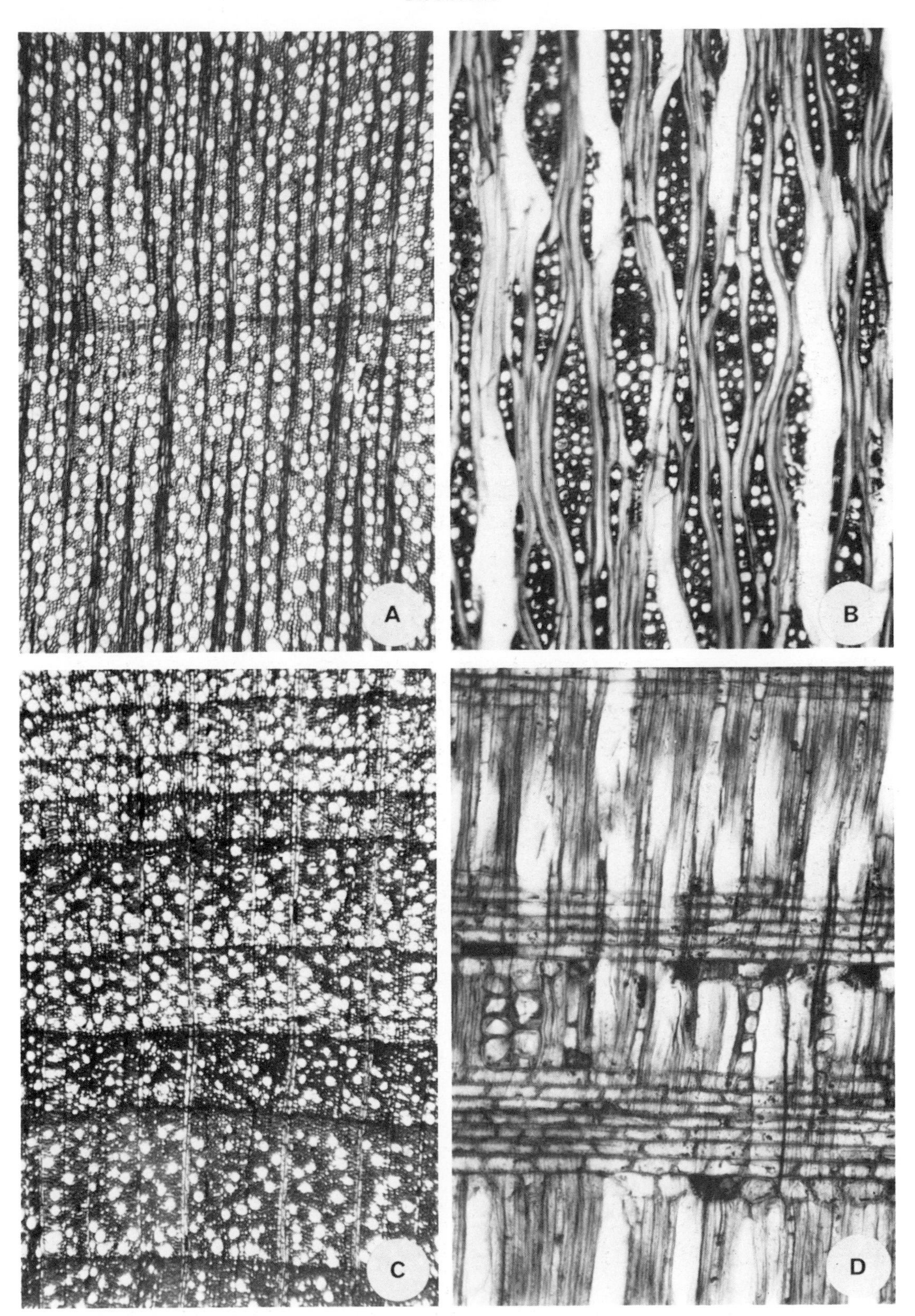

Plate 61.
Crataegus. A and B — *C. azarolus*; A, ×42; B, ×126. C and D — *C. sinaica*; C, ×42; D, ×126

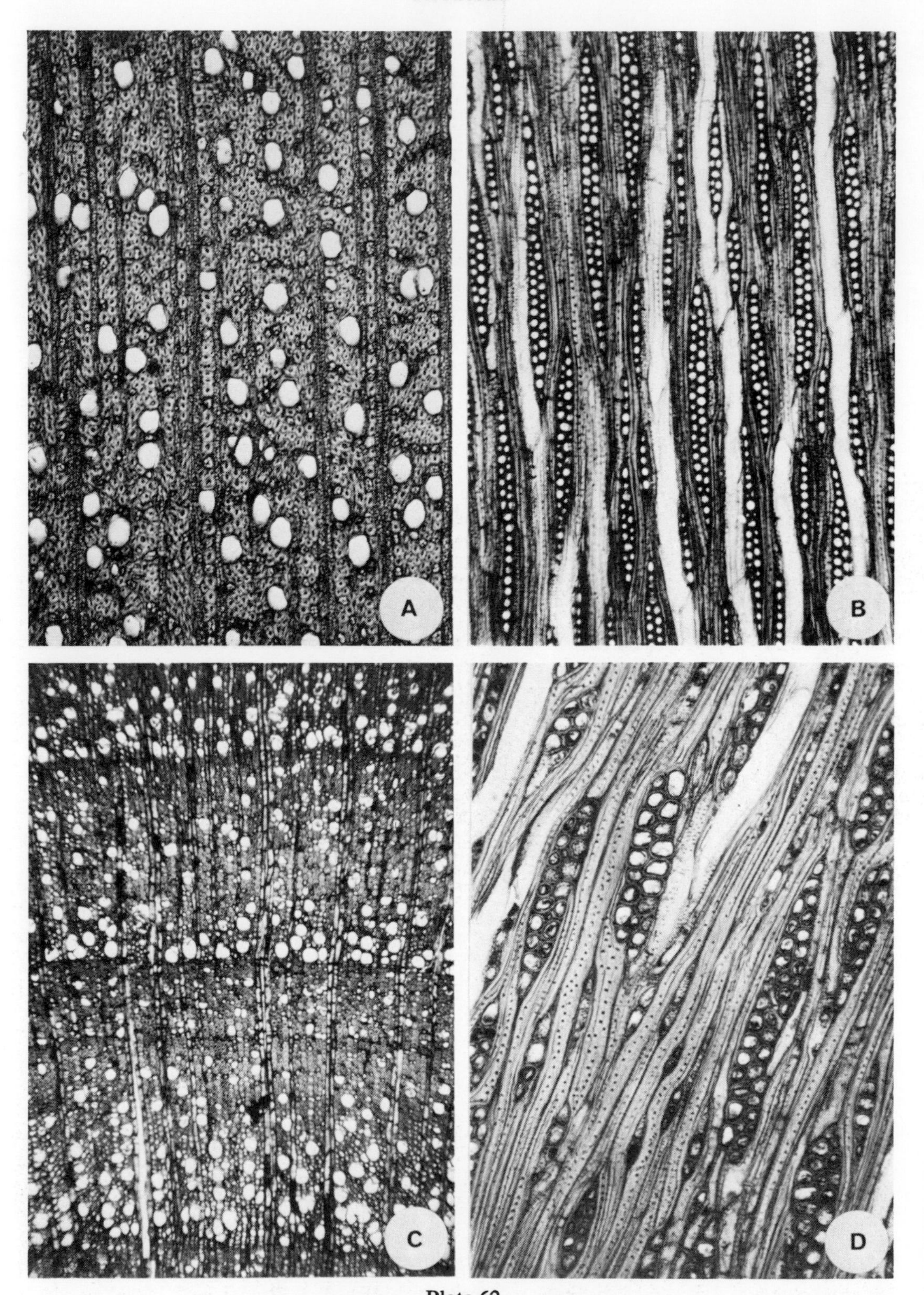

Plate 62.
A and B — *Cydonia oblonga*, ×125. C and D — *Eriolobus trilobatus*; C, ×42; D, ×126

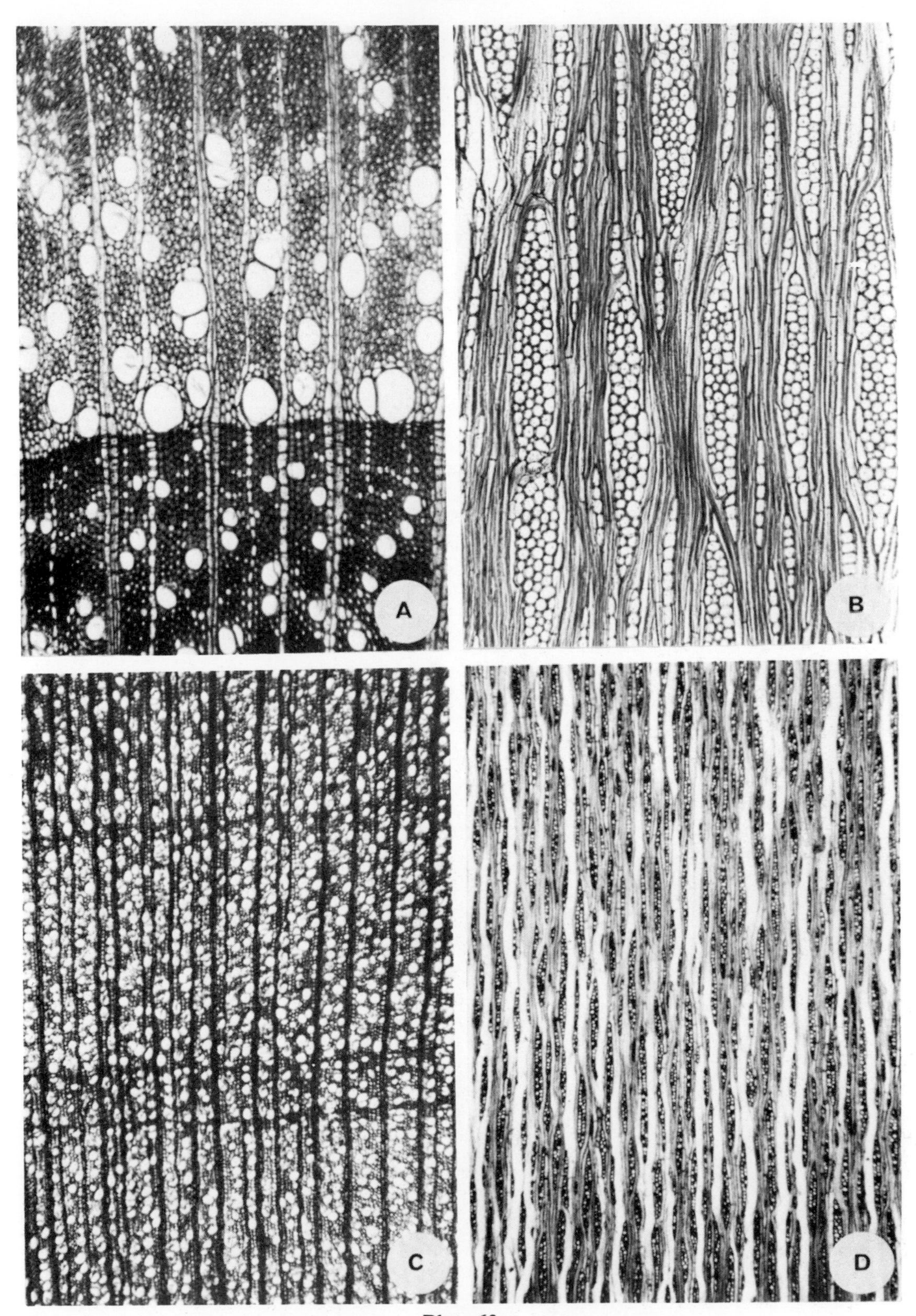

Plate 63.
A and B — *Prunus ursina*, ×126. C and D — *Pyrus syriaca*, ×42

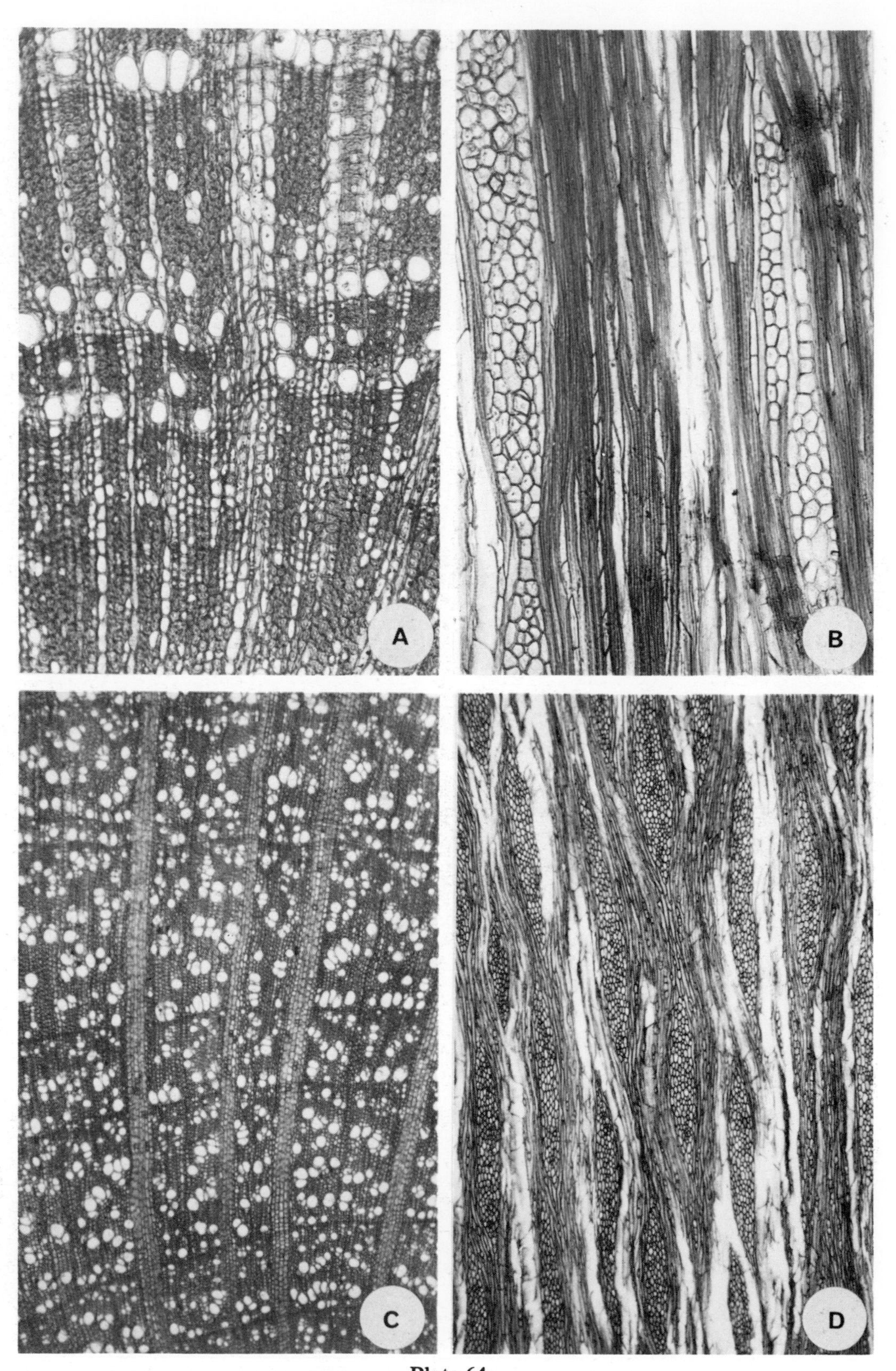

Plate 64.
A and B — *Rosa pulverulenta*, ×126. C and D — *Sarcopoterium spinosum*, ×42

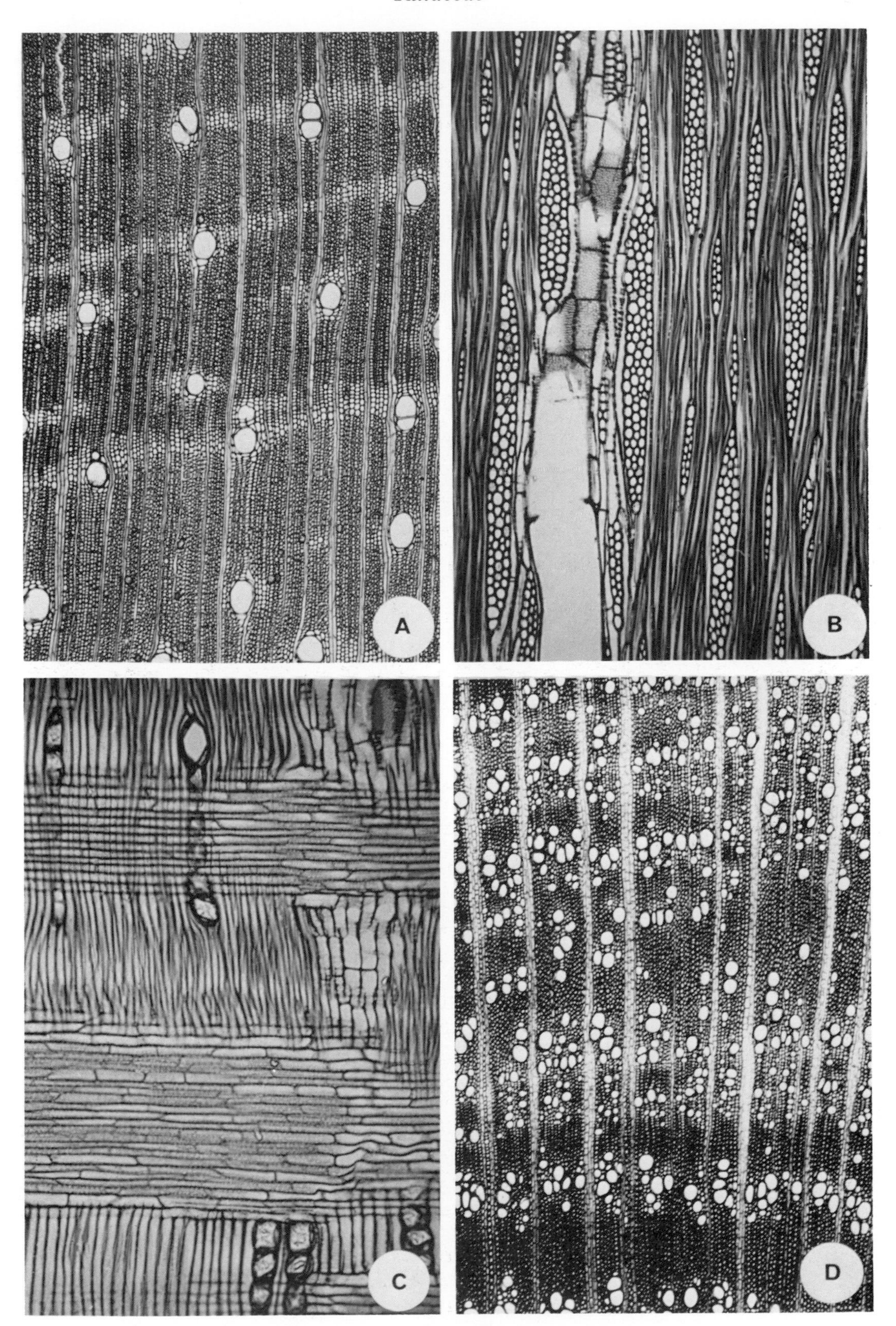

Plate 65.
A–C — *Citrus aurantium*; A, ×48; B and C, ×120. D — *Ruta chalepensis*, ×48

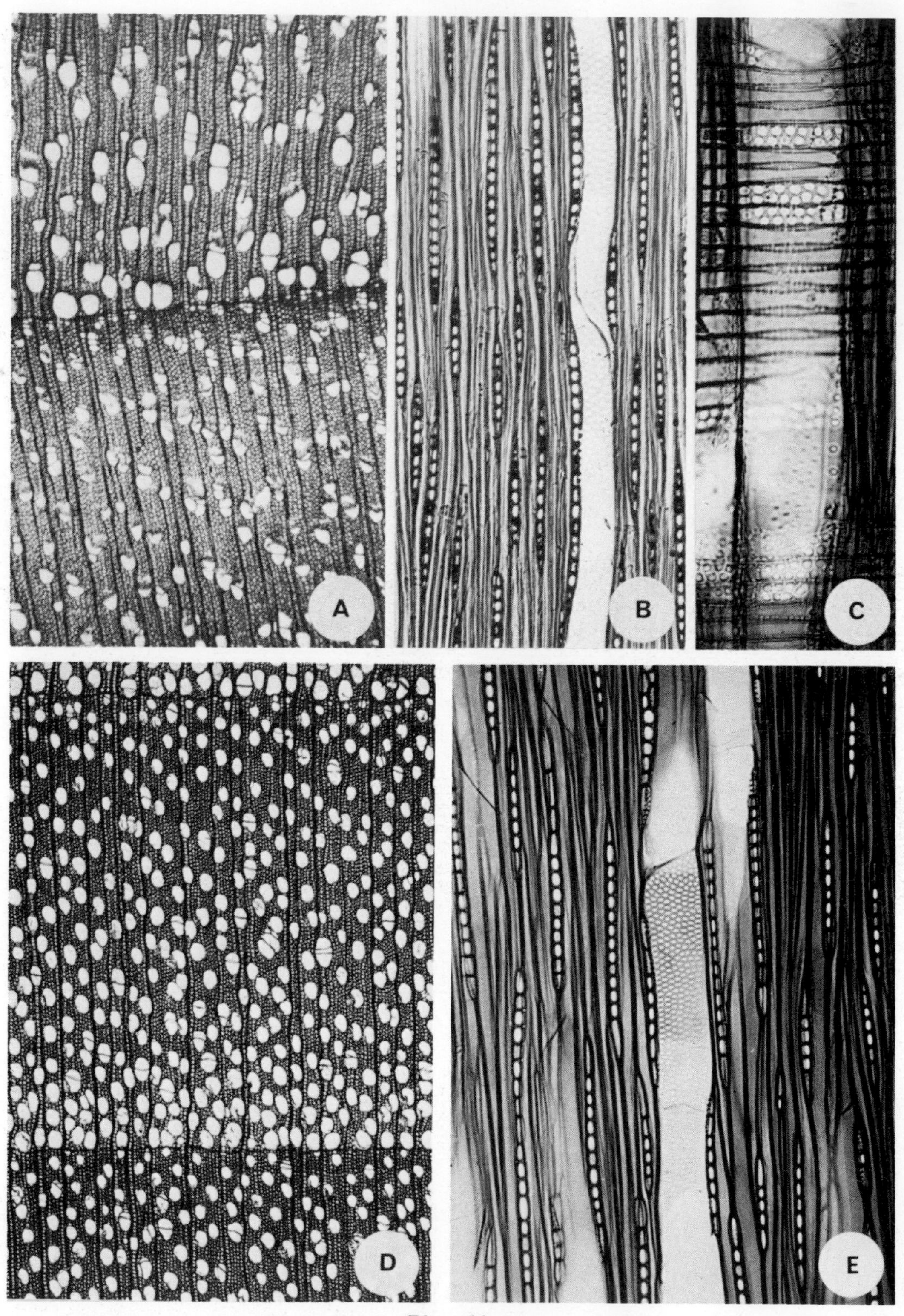

Plate 66.
A–C — *Populus euphratica*; A, ×42; B, ×126; C, ×300.
D and E — *Salix babylonica*; D, ×30; E, ×120

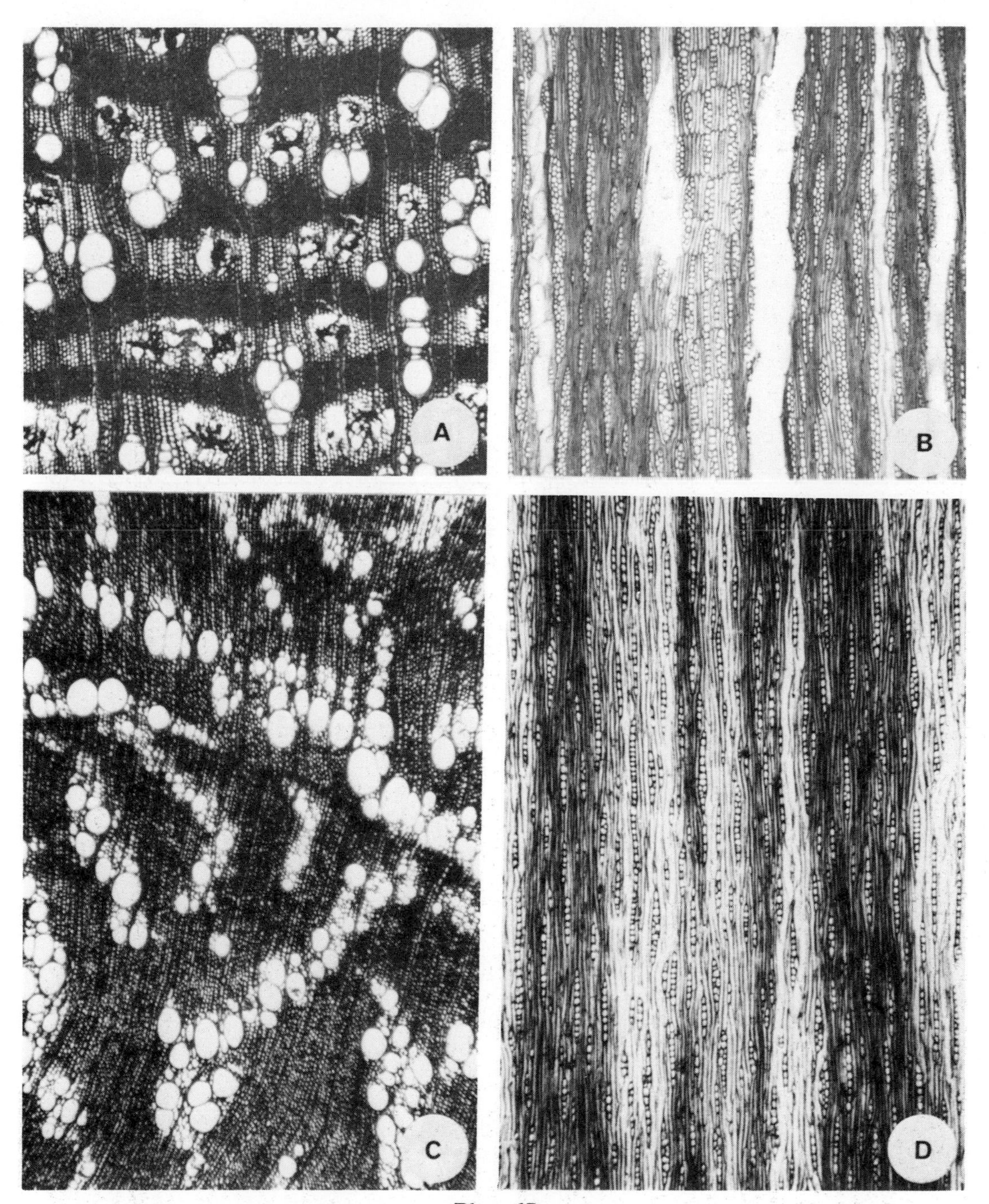

Plate 67.
A and B — *Salvadora persica*, ×42. C and D — *Lycium europaeum*, ×42

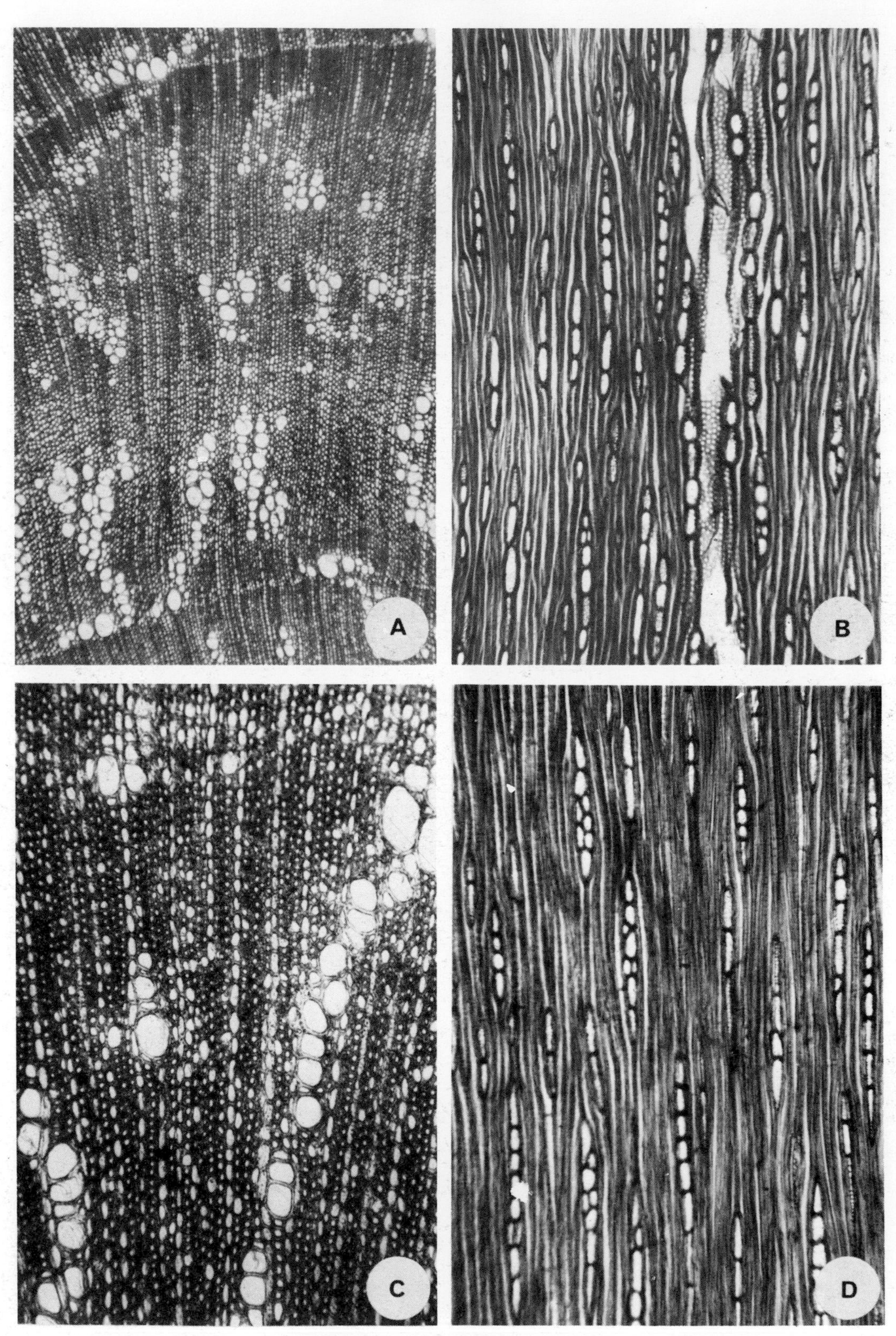

Plate 68.
Lycium. A and B — *L. schweinfurthii*; A, ×42; B, ×126. C and D — *L. shawii*, ×126

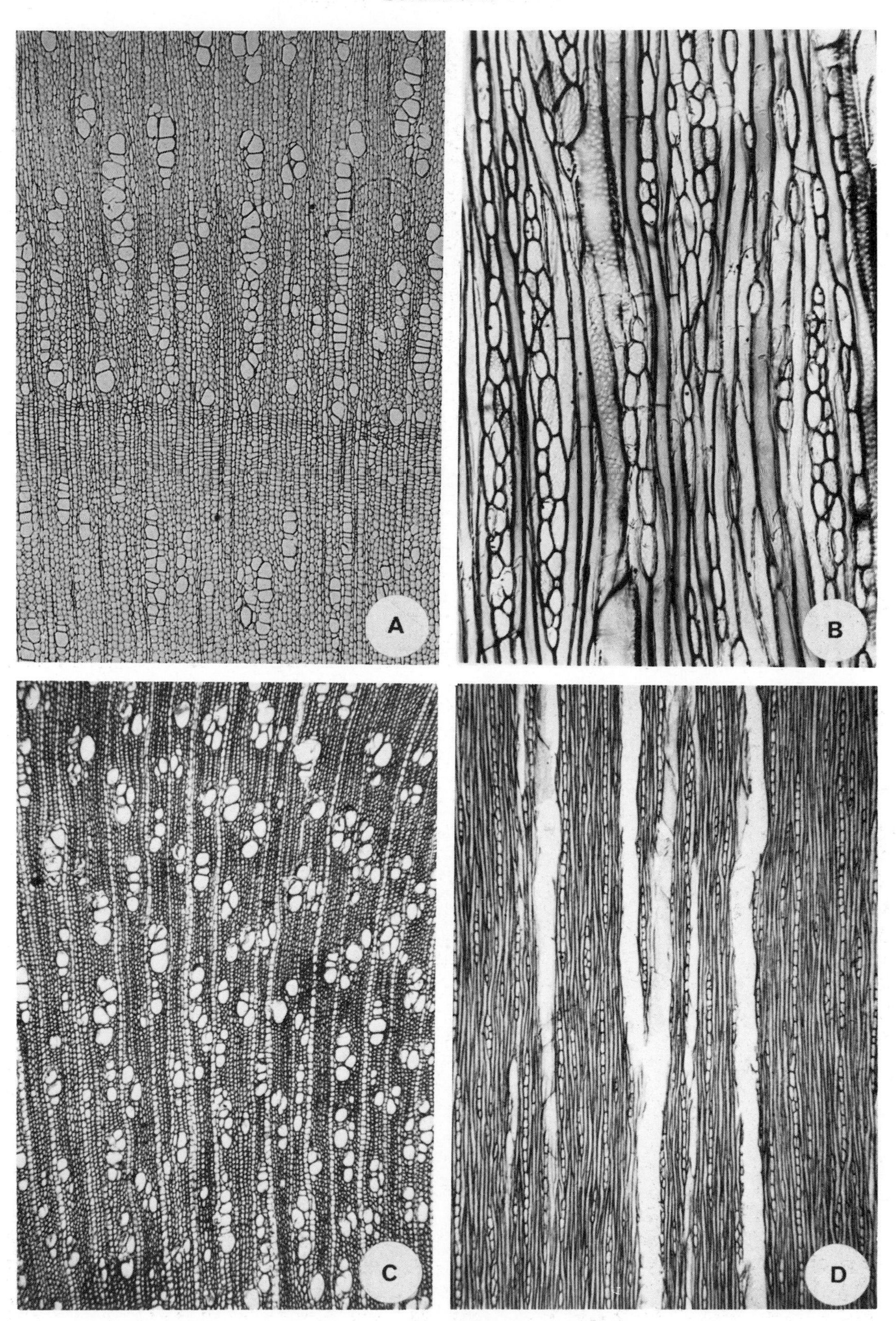

Plate 69.
A and B — *Nicotiana glauca*; A, ×48; B, ×126. C and D — *Withania somnifera*, ×42

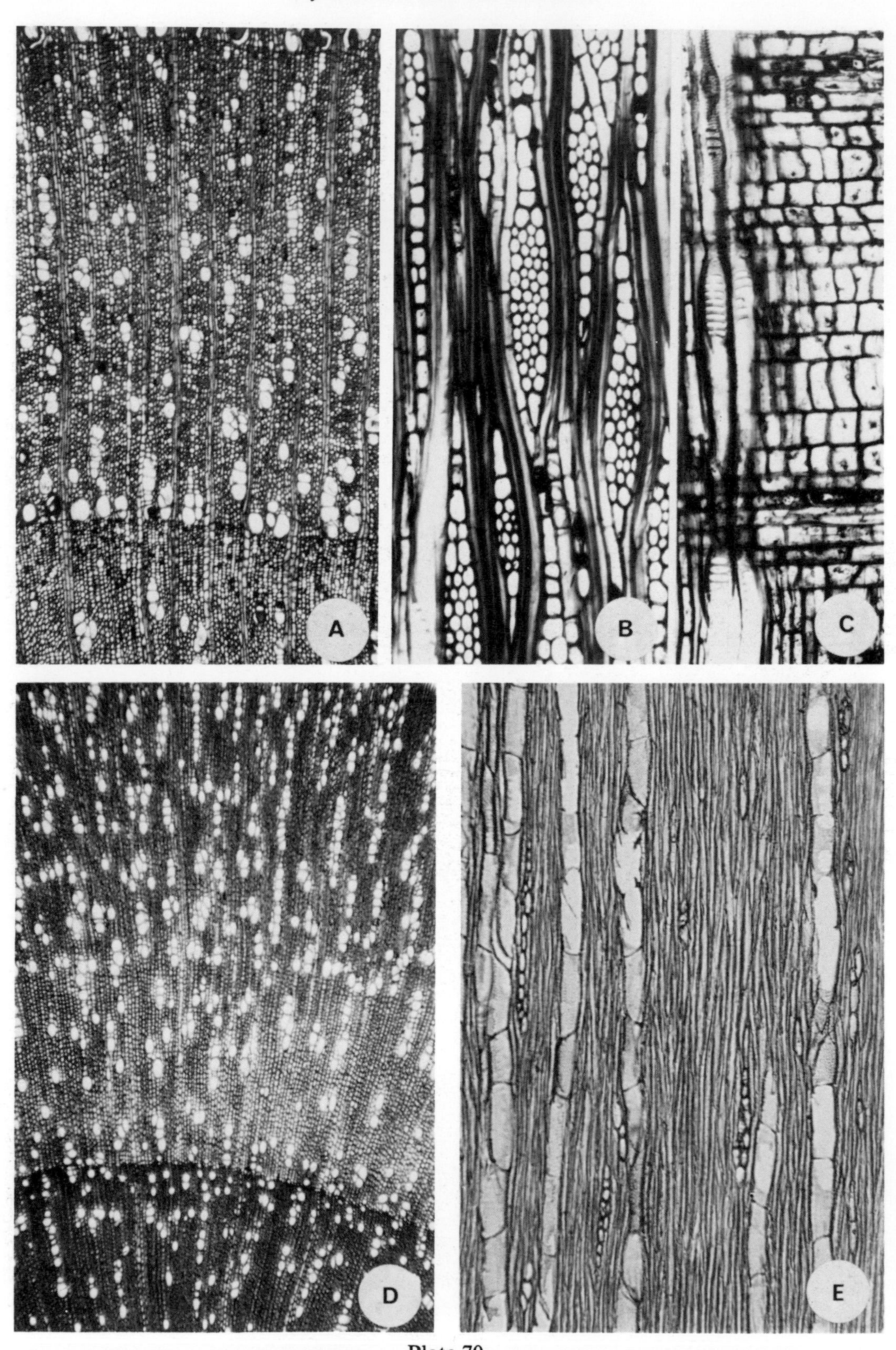

Plate 70.
A–C — *Styrax officinalis*; A, ×42; B and C, ×126. D and E — *Reaumuria hirtella*; D, ×42; E, ×126

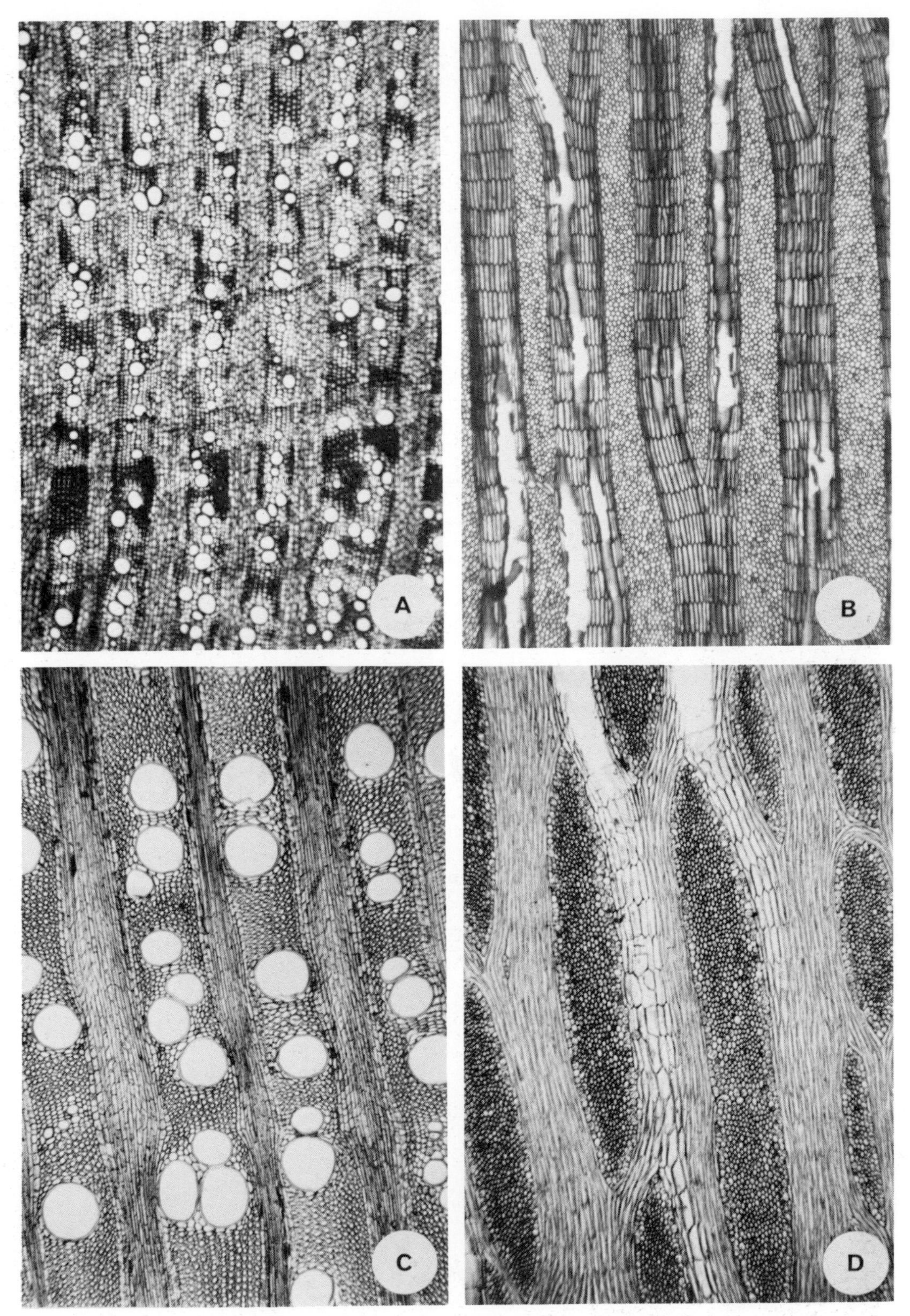

Plate 71.
Tamarix, ×42. A and B — *T. amplexicaulis*. C and D — *T. aphylla*

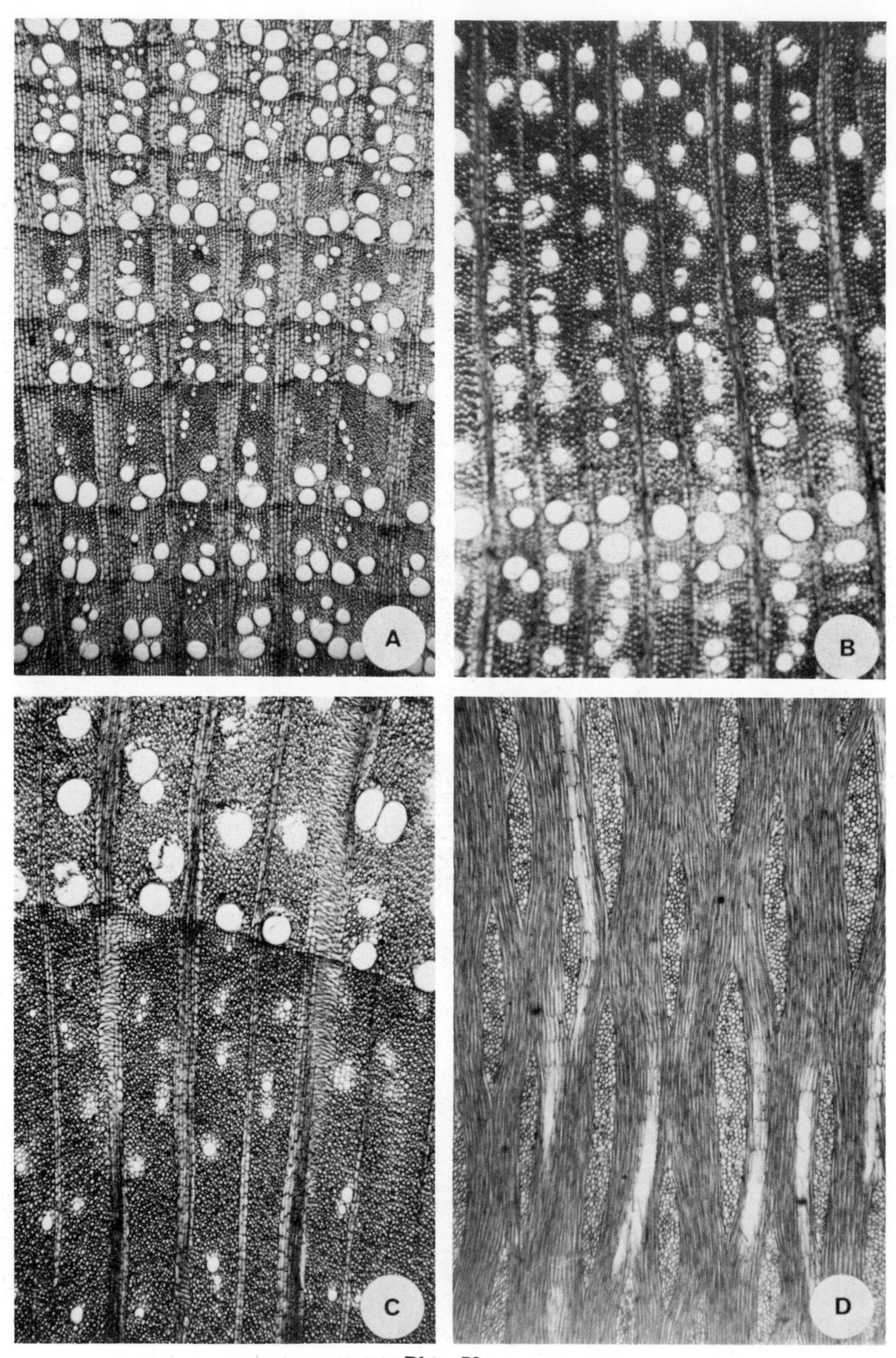

Plate 72.
Tamarix, ×42. A — *T. chinensis*. B — *T. gennessarensis*. C and D — *T. nilotica*

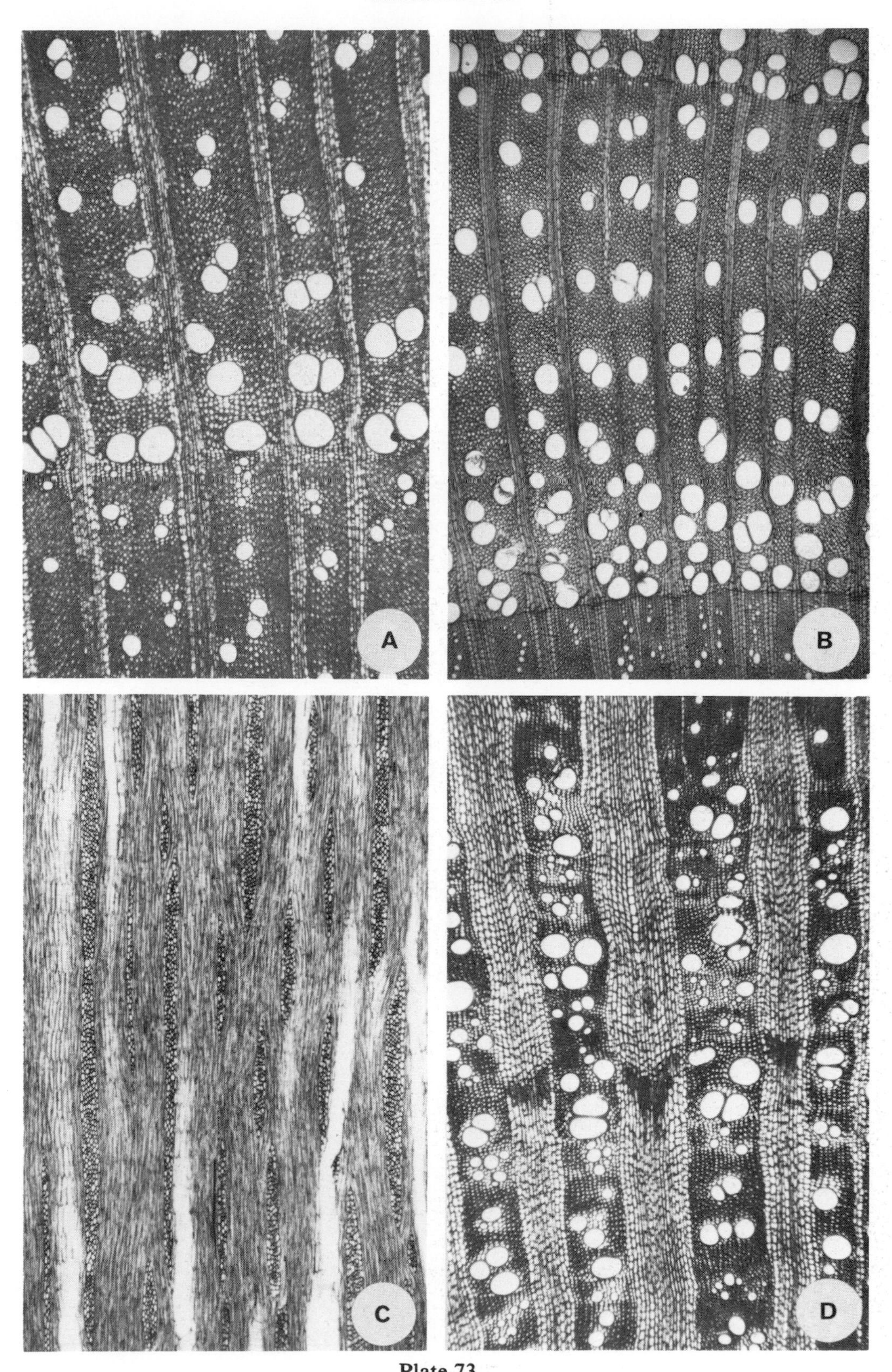

Plate 73.
Tamarix, ×42. A — *T. palaestina.* B and C — *T. parviflora.* D — *T. passerinoides*

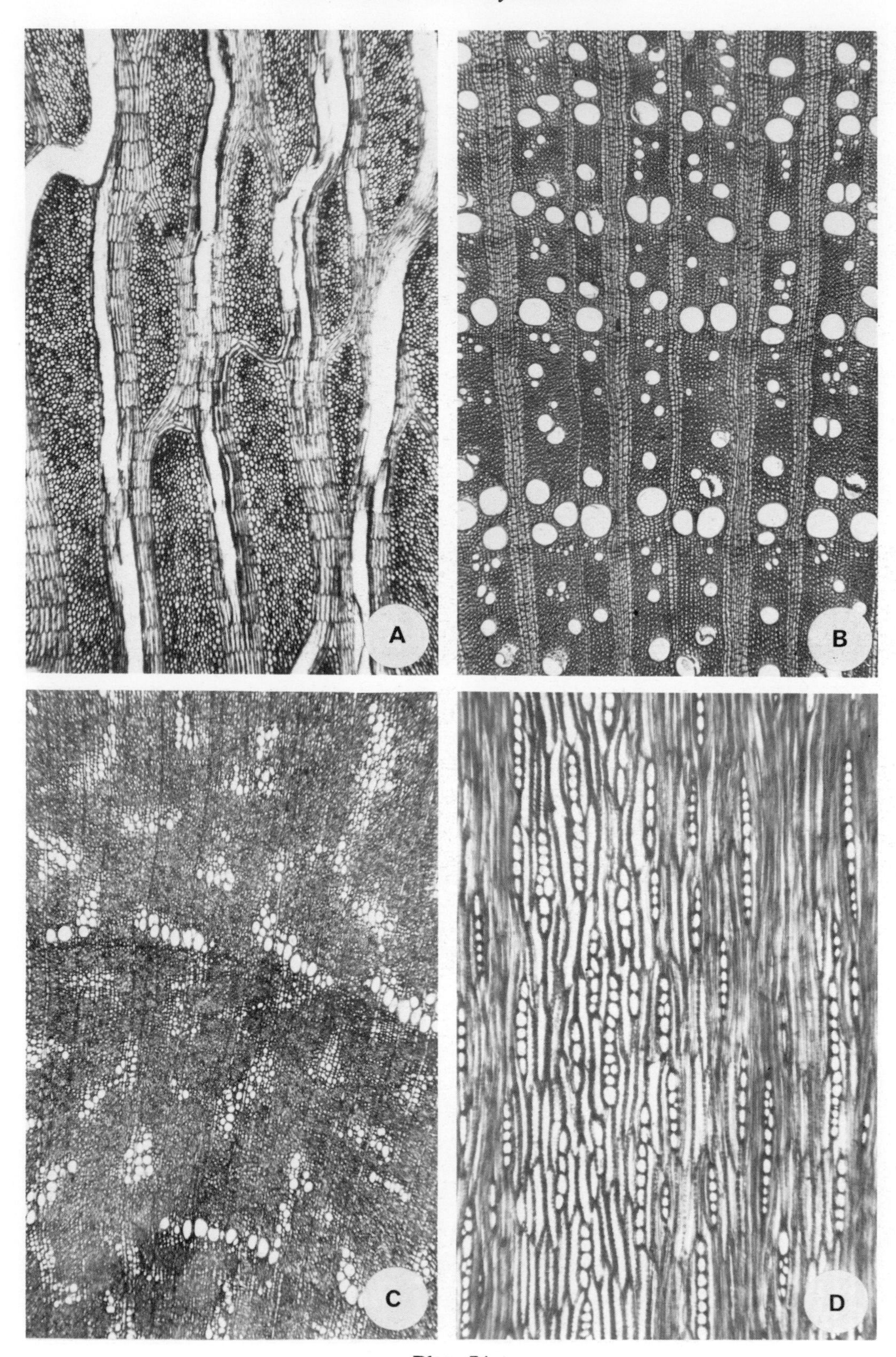

Plate 74.
A and B — *Tamarix*, ×42. A — *T. passerinoides*. B — *T. tetragyna*.
C and D — *Thymelaea hirsuta*; C, ×42; D, ×126

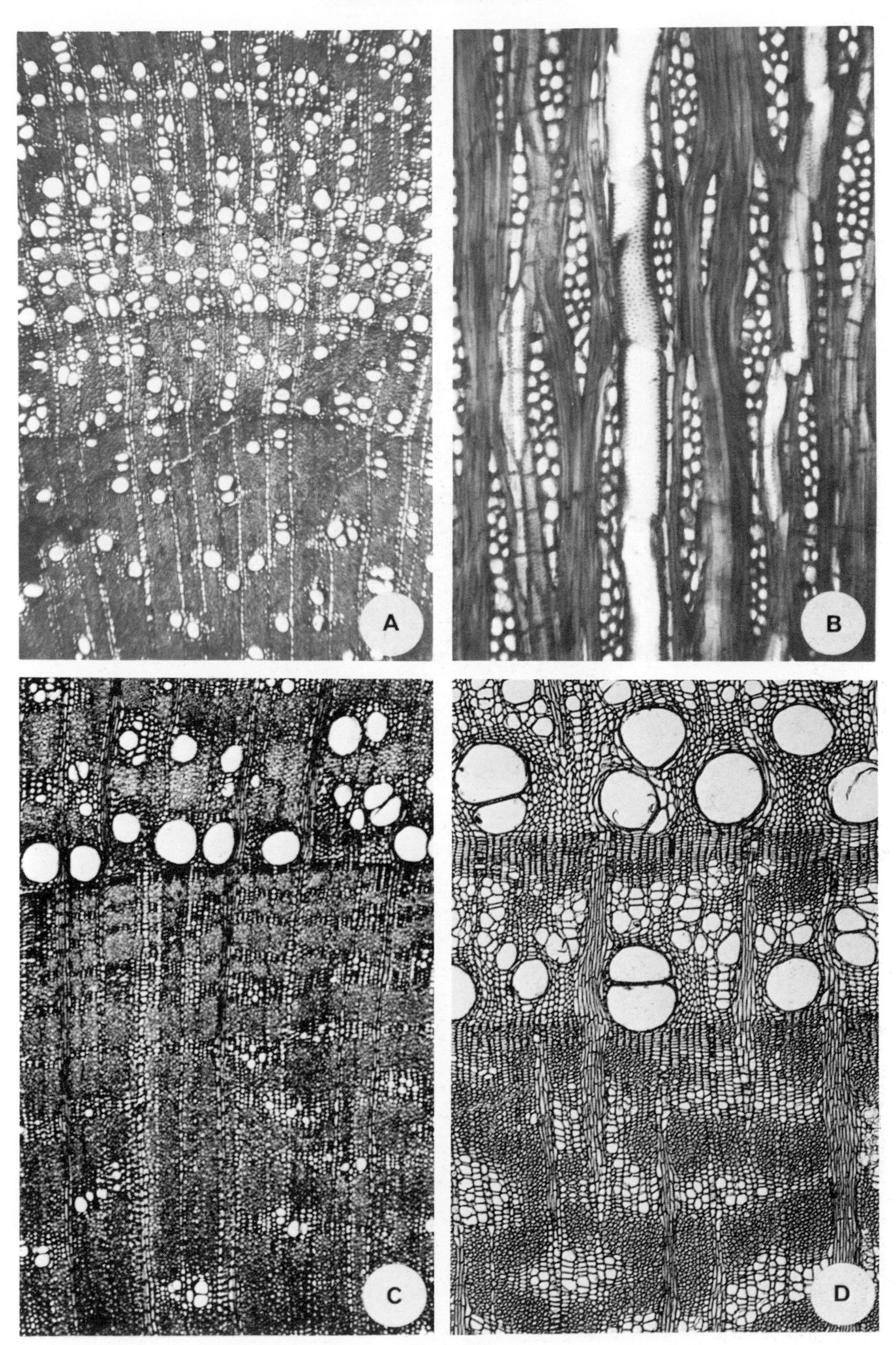

Plate 75.
A and B — *Grewia villosa*; A, ×42; B, ×126. C and D — *Celtis australis*, ×48

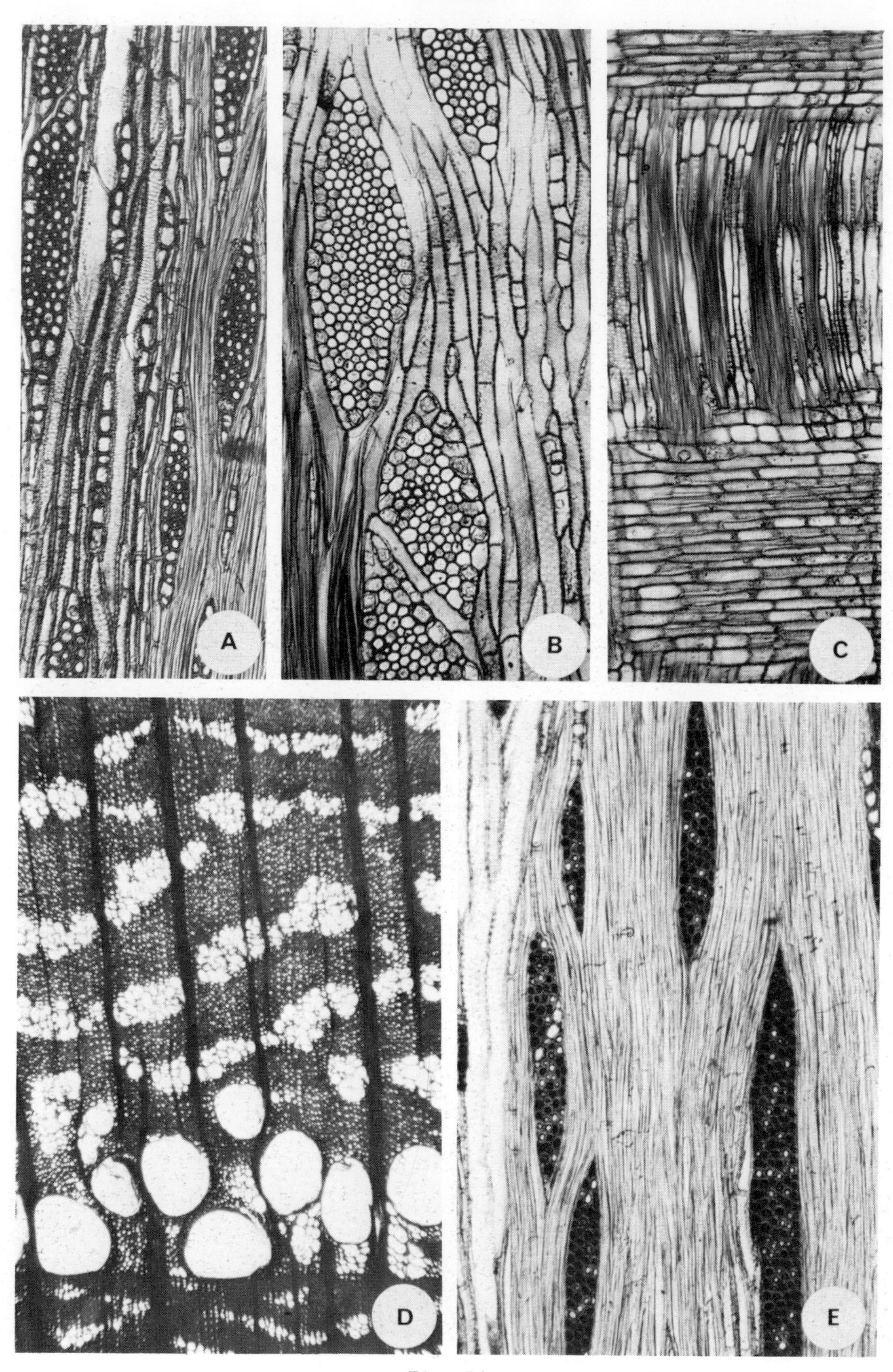

Plate 76.
A–C — *Celtis australis*, ×125. D and E — *Ulmus canescens*; D, ×42; E, ×126

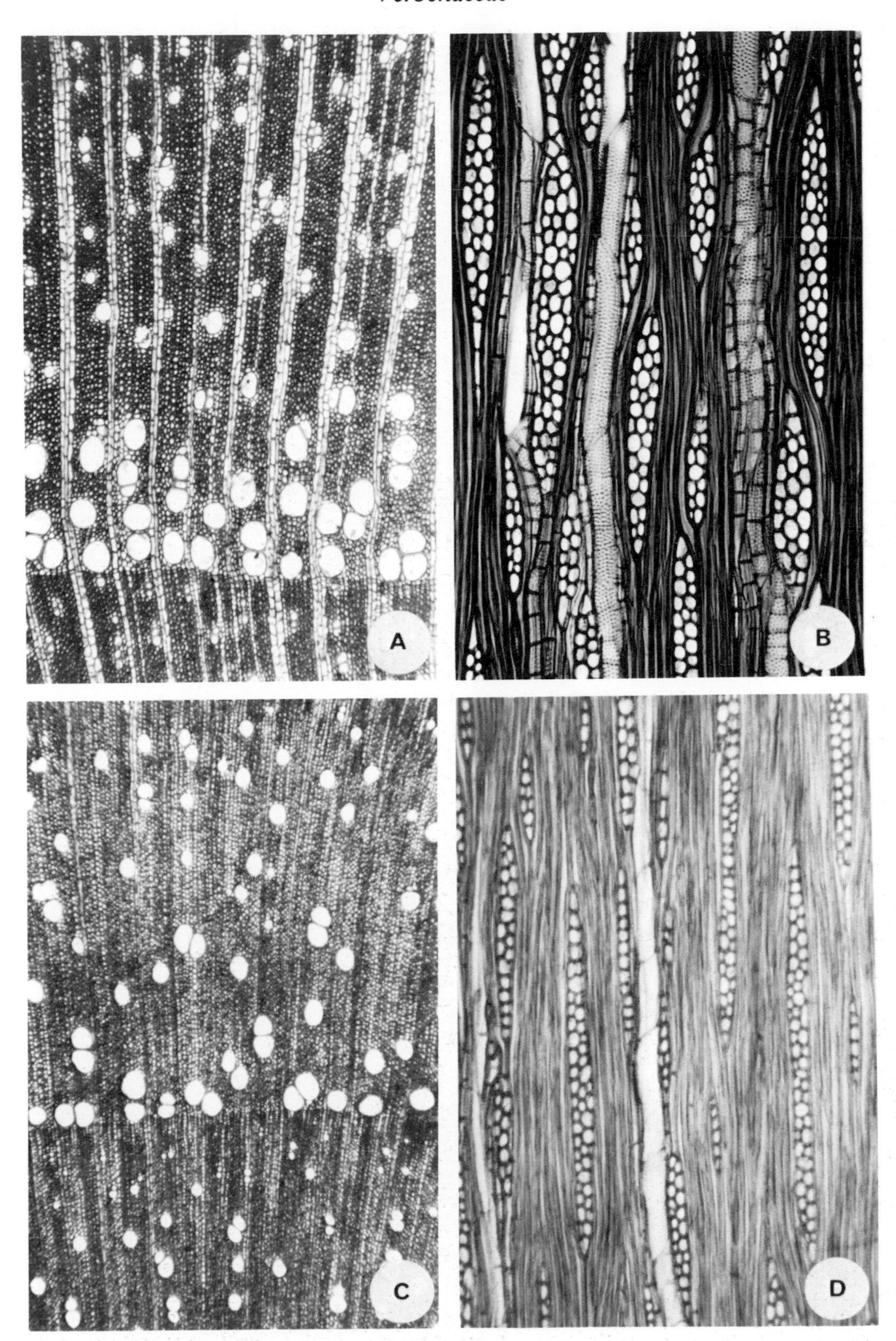

Plate 77.
Vitex. A and B — *V. agnus-castus*; A, ×42; B, ×120. C and D — *V. pseudo-negundo*; C, ×42; D, ×126

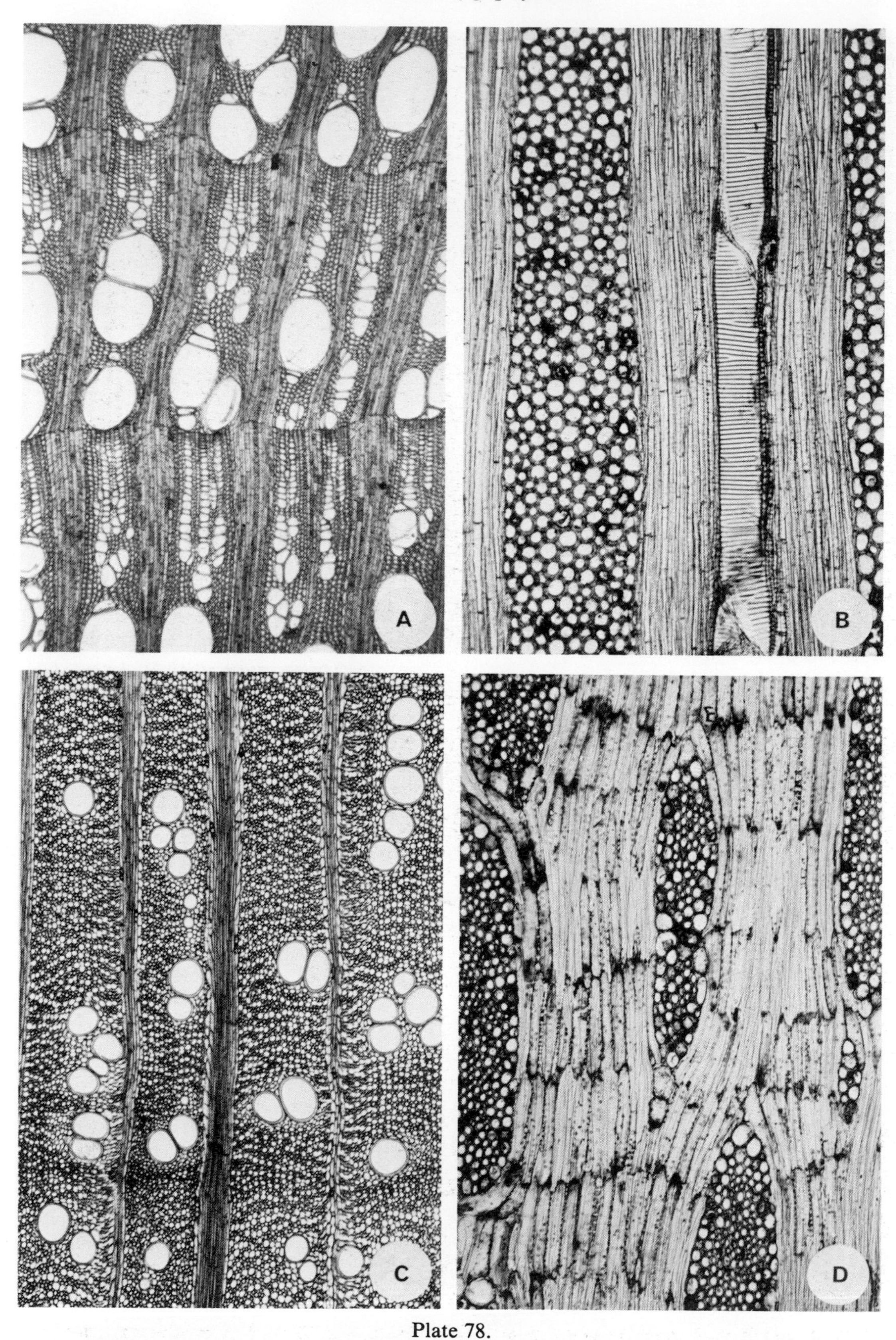

Plate 78.
A and B — *Vitis vinifera*; A, ×42; B, ×120. C and D — *Balanites aegyptiaca*; C, ×48; D, ×120

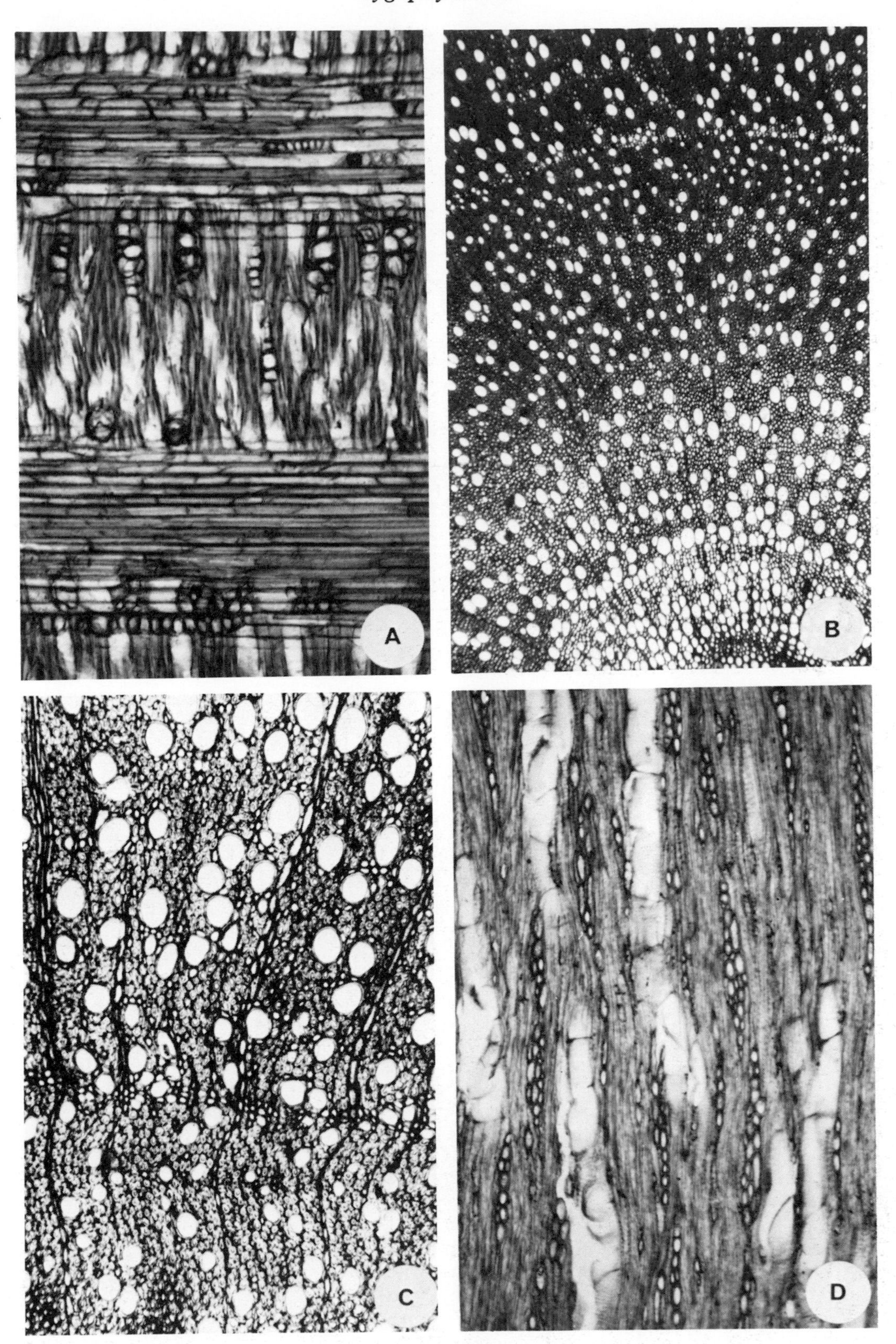

Plate 79.
A — *Balanites aegyptiaca*, ×126. B–D — *Fagonia mollis*; B, ×42; C and D, ×126

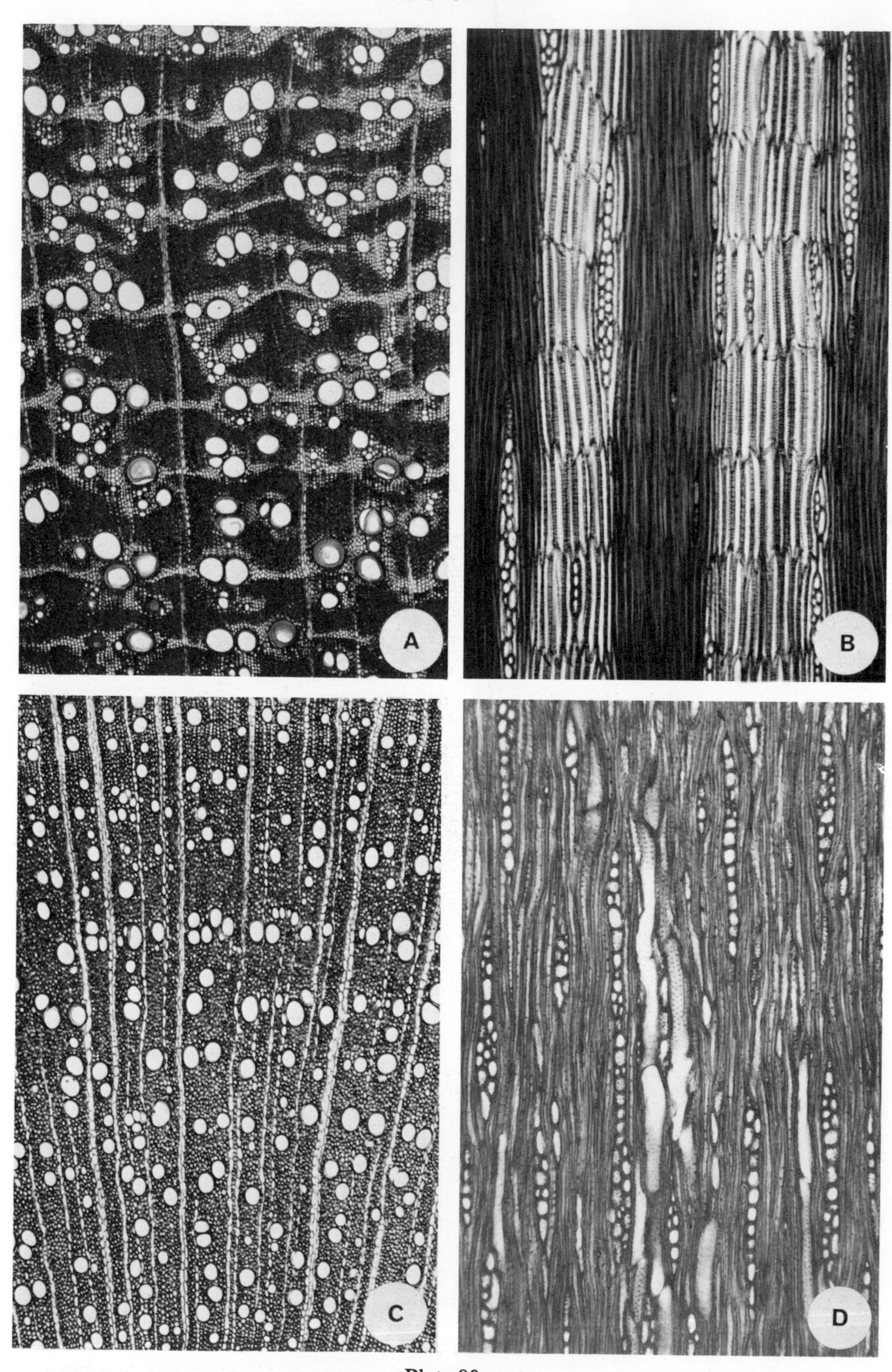

Plate 80.
A and B — *Nitraria retusa*; A, ×48; B, ×120. C and D — *Zygophyllum coccineum*; C, ×48; D, ×126

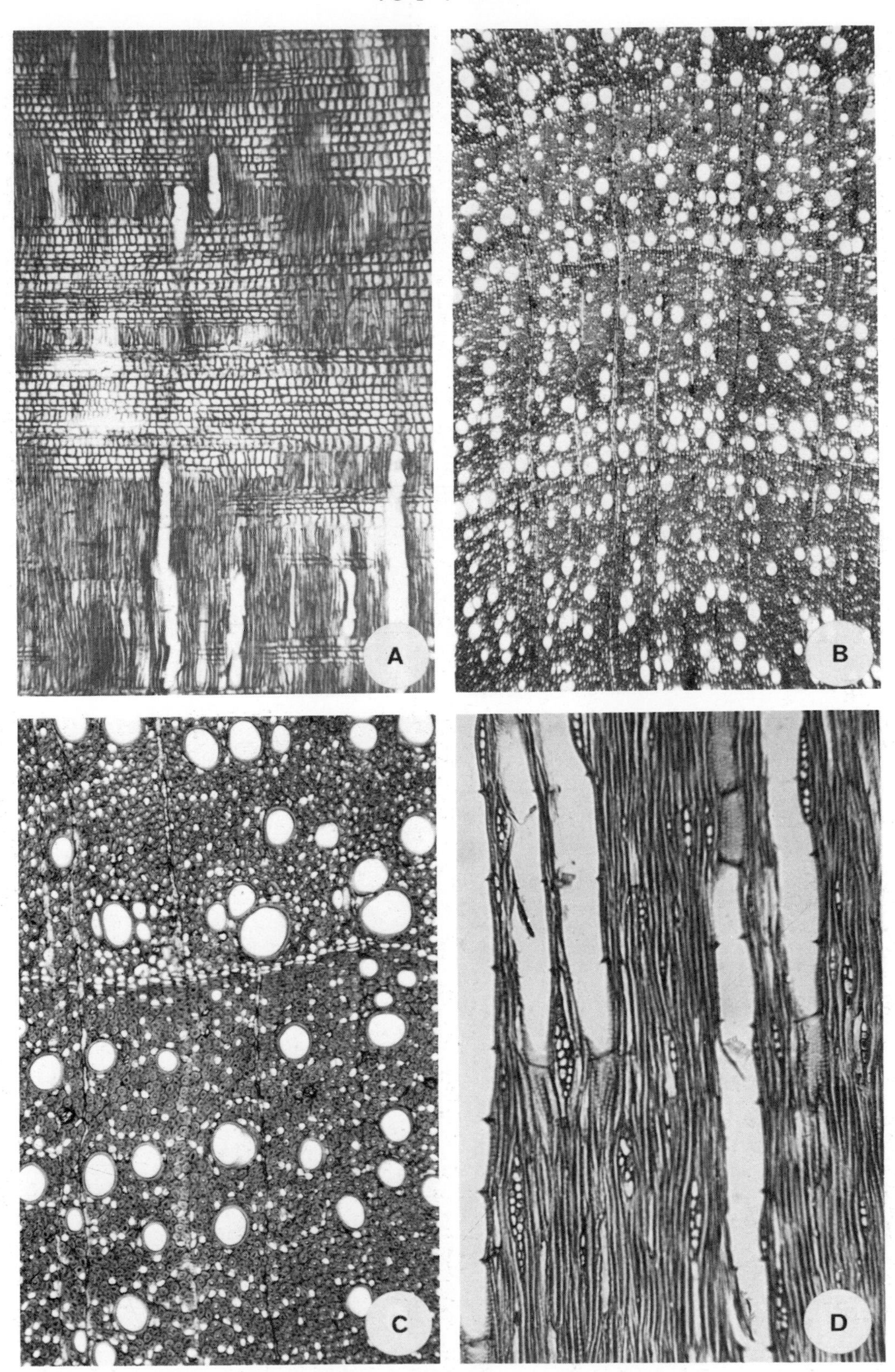

Plate 81.
Zygophyllum. A — *Z. coccineum*, ×42. B–D — *Z. dumosum*; B, ×42; C and D, ×126

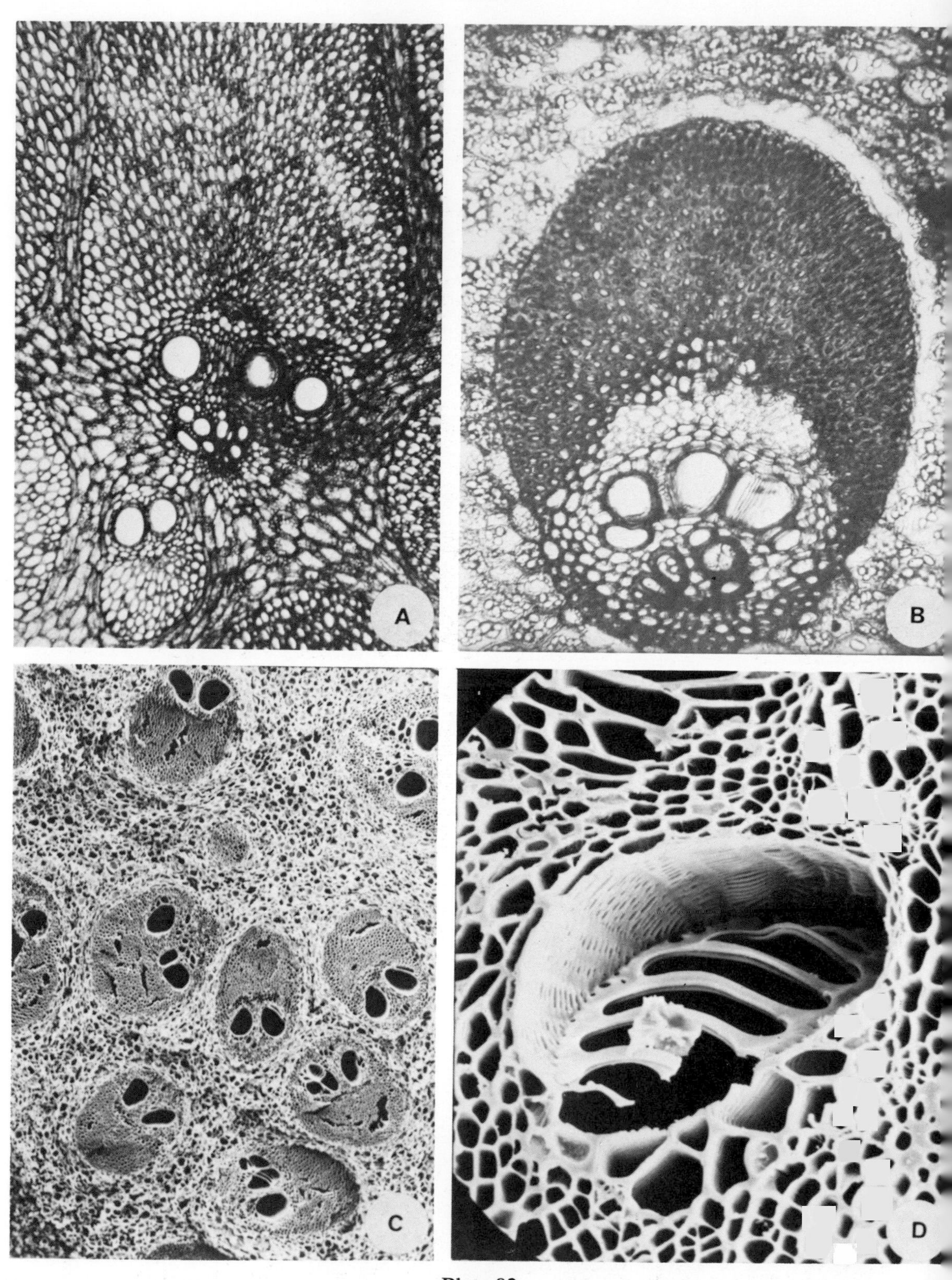

Plate 82.
A — *Hyphaene thebaica*, ×42. B–D — *Phoenix dactylifera*; B, ×140; C and D, scanning electron micrographs of charred stem pieces; C, ×20; D, ×200

כתבי האקדמיה הלאומית הישראלית למדעים
החטיבה למדעי הטבע

מבנה העצה של העצים והשיחים של ישראל וסביבתה וזיהוים

מאת

אברהם פאהן   אלה ורקר

האוניברסיטה העברית בירושלים

פיטר באס

העשבייה הממלכתית, ליידן

ירושלים תשמ״ו